Rubber Processing and Production Organization

Rubber Processing and Production Organization

Philip K. Freakley
Institute of Polymer Technology
Loughborough University of Technology
Loughborough, United Kingdom

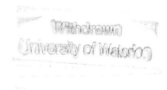

Plenum Press • New York and London

Library of Congress Cataloging in Publication Data

Freakley, Philip K.
    Rubber processing and production organization.

    Includes bibliographies and index.
    1. Rubber industry and trade. I. Title.
TS1890.F7   1985   678′.2   84-24835
ISBN 0-306-41745-6

©1985 Plenum Press, New York
A Division of Plenum Publishing Corporation
233 Spring Street, New York, N.Y. 10013

Printed in the United States of America

To Viv, for helping to unsplit infinitives

Foreword

The absence of a book dealing with rubber processing has been apparent for some time and it is surprising that a straightforward text has not been produced. However, this book goes far beyond the scope of a simple technical approach and deals with the full spectrum of activities which lead to successful and profitable product manufacture. The need to deliver a product to a customer at the right time, at the right cost, and at the right quality is a basic premise on which the book is based.

The increasingly stringent demands of customers for products that can be introduced directly into an assembly or production line without goods-inwards inspection, are placing increasing pressures on the manufacturer. As a result, it is becoming essential to achieve and sustain product quality and consistency, by the monitoring and control of manufacture, at a level which renders all products saleable.

The book has been written to satisfy the needs of practitioners in the rubber industry and is certainly not another descriptive text which is only read for interest when more important matters are not pressing. My close cooperation with Philip K. Freakley during the writing of the book has resulted in the incorporation of many of the viewpoints and methods which I have developed and refined during more than 38 years in the rubber industry.

Since I feel that my own company is at the forefront of the application of computers to rubber product manufacture, it has been possible to draw upon sound practical experience of their applications in both the management and technical areas of rubber company activity, and then to project these applications forward some five years. In many of its aspects, the subject matter of the book looks to the future and anticipates technical and organizational methods which rubber companies will need to assimilate and apply, in order to maintain and improve their performance. I strongly recommend this book for all company staff involved in rubber product manufacture.

R. W. Garfield

Works Director, Woodville Polymer Engineering Ltd.

Preface

The primary aim of this book is to provide technical, engineering, and management staff in rubber product manufacturing companies with a detailed and practical guide to manufacturing systems.

In recent years the need to achieve greater productivity and precision in manufacture, in order to satisfy the increasingly stringent demands of customers, has stimulated a much greater interest in manufacturing methods than has hitherto been apparent. The changes thus initiated have now been substantially accelerated by the advent of the microprocessor, with all its attendant implications for automation, process monitoring, and production organization. The style of the book owes much to its emphasis on computer methods and the adoption of the systems approach to manufacturing, both of which have been extensively developed in the area of general production engineering and management. It is also unique in using the commercial viewpoint to put the technical details of processing operation performance into an economic perspective.

The basis of the systems approach is the recognition that the development or improvement of a manufacturing facility can be divided into two major stages, which are known as *analysis* and *synthesis*. In analysis the basic elements of a manufacturing system are identified and dealt with separately, with the objective of understanding and subsequently improving them. Synthesis is then concerned with the reassembly of the improved elements into a system which fully exploits all the advantages to be gained from them. However, because of the interactions and dependencies which inevitably occur among elements, improved system performance cannot be guaranteed without an understanding of the principles and practice of synthesis.

Many technical activities (and texts) concentrate upon analysis to the almost total exclusion of synthesis. This may account for the difficulties which are often experienced in demonstrating that improved profitability accrues from technical progress in manufacturing methods. In this book Chapters 2 to 6 are primarily concerned with analysis, dealing with material behaviour and the main processes of the rubber industry. Chapters 7 to 11 then address the complex task of assembling or synthesizing the individual processes into efficient manufacturing systems, capable of achieving high productivity and profitability.

Because of its objective of providing a practical guide to product manufacture, the book concentrates upon the efficient utilization of currently available equipment. Design of machinery is only covered insofar as it is necessary to support the selection of equipment for purchase; theoretical methods are only introduced where they can be directly applied to the improvement of process performance. While a detailed understanding of microprocessor systems and mathematical methods may be necessary to implement some of the techniques dealt with in the book, the concepts involved and their applications should be readily apparent. Managers will be able to gain a clear overview of the resources needed to implement the suggested techniques and the benefits to be derived from them.

Philip Freakley

Institute of Polymer Technology
Loughborough University of Technology, U.K.

Acknowledgments

A number of people have given generously of both their time and expertise to ensure that the treatment of the very broad range of topics covered in this book is both accurate and consistent in contributing to the subject of "rubber product manufacturing systems."

First and foremost, Roy Garfield, Works Director of Woodville Polymer Engineering Ltd., was unable to carry through his original intention of coauthorship, due to unavoidable pressures of work, but has maintained an encouraging interest and enthusiasm for the book. He undertook the substantial task of reading and discussing the manuscript with the author as it was produced, with the result that an eminently practical and industrially relevant approach has been maintained throughout. Chapters 10 and 11 were read by Dr. Malcolm R. Hill of the Management Studies Department of Loughborough University of Technology and the former chapter by his colleague, Mr. Nigel Coulthurst. Many useful comments and suggestions flowed from their detailed appraisal of these two chapters, which have contributed substantially to their coherence.

A number of processing machinery manufacturers have provided detailed illustrations and have assisted the author by discussion of the design, operation, and control of their equipment. The majority of these contributions are acknowledged in the text.

Last, but certainly not least, the author would like to thank Mrs. Joyce Deaville of Woodville Polymer Engineering Ltd., who typed the manuscript and coped with some very substantial additions and modifications, and Mrs. Barbara Green of the Institute of Polymer Technology, who undertook the ⸱cting task of labeling the illustrations.

Contents

1. RUBBER PRODUCT MANUFACTURING SYSTEMS 1

1.1. Introduction 1
1.2. The Systems Concept 3
1.3. The Selection and Operation of Tests for Unvulcanized
     Rubber 5
1.4. The Prediction, Monitoring, and Control of Process
     Performance 7
1.5. Production Organization 8
     References 13

2. MATERIALS BEHAVIOR AND TESTING 15

2.1. Introduction 15
2.2. Flow Properties of Raw Elastomers and Rubber Mixes . 16
2.3. Measurement of Flow Properties 21
2.4. Thermal and Heat-Transfer Properties 29
2.5. Vulcanization Characteristics 32
     References 40

3. PRINCIPLES OF MIXING AND INTERNAL MIXERS 43

3.1. Introduction 43
3.2. The Mechanisms of Mixing 44
3.3. Elements of Internal Mixer Design 48
3.4. Practical Mixing Variables 56
3.5. Flow Instabilities 64
3.6. Laboratory Simulation of Full-Scale Mixing 65
     References 67

4. SCREW EXTRUSION AND CONTINUOUS MIXING 69

4.1. Introduction 69
4.2. Elements of Extruder Construction 70
4.3. Hot-Feed Extruders 74
4.4. Cold-Feed Extruders 75
4.5. Design of Extruder Heads and Dies 79
4.6. Determination and Control of Extruder Operating
     Characteristics 94
4.7. Continuous Mixing 100
     References 108

5. CALENDERING AND MILLING 111

5.1. Introduction 111
5.2. The Operating Characteristics of Two-Roll Mills 111
5.3. Mill and Calender Roll Temperature Control 118
5.4. Calender Configurations and Operations 119
5.5. Roll Deflection and Methods of Correction 121
5.6. Feeding, Sheet Cooling, and Batch-Off Equipment . . . 124
5.7. Determination and Control of Calender Operation
     Characteristics 126
     References 129

6. HEAT TRANSFER AND VULCANIZATION METHODS 131

6.1. Introduction 131
6.2. Heat Transfer 131
6.3. Prediction of State of Cure 140
6.4. Molding 146
6.5. Batch Vulcanization 172
6.6. Continuous Vulcanization 174
     References 181

7. PROCESS CONTROL AND QUALITY CONTROL 183

7.1. The Interaction of Process Control and Quality Control . 183
7.2. Specifications 184
7.3. Process-Capability Studies 192
7.4. Process Monitoring 217

7.5.   Process Control 241
7.6.   Quality Control 254
       References 264

8. PLANT LAYOUT AND OPERATIONS METHODS 267

8.1.   General Considerations 267
8.2.   Transport and Storage in Manufacture 271
8.3.   Handling Methods and Operations at Work Stations . . 280
8.4.   Planning and Allocating Space 288
8.5.   Layout Synthesis and Evaluation 296
8.6.   Installing and Commissioning a Layout 309
       References 313

9. COMPANY PHILOSOPHY, ORGANIZATION, AND STRATEGY   315

9.1.   Philosophy 315
9.2.   Company Organization 316
9.3.   Market Research and Company Development 336
       References 351

10. THE ECONOMICS OF MANUFACTURING OPERATIONS . . . 353

10.1. The Flow of Cash Through a Company 353
10.2. Cost Identification and Analysis Methods 356
10.3. Standard Costs 369
10.4. Business Plans and Budgets 377
10.5. Budgetary Control 391
       References 395

11. PRODUCTION MANAGEMENT 397

11.1. Production Planning 397
11.2. Purchasing and Inventory Control 412
11.3. Implementing the Production Plan 431
       References 442

INDEX 445

Rubber Product Manufacturing Systems

1.1. INTRODUCTION

The manufacturing technology of the rubber industry is based on a number of key processes and operations, some of which have undergone a continuous evolution since the founding of the first rubber companies. These processes and operations form the building blocks from which a complex and diverse range of manufacturing systems are assembled. For conventional vulcanized rubber products, on which this book concentrates, a sequence of processes and operations is necessary to complete manufacture. It is possible to draw complex generalized flowcharts for the manufacture of typical products, but in the majority of cases these flowcharts can be reduced to three fundamental stages:

The techniques used for each of these three stages will exert a substantial influence on productivity and on the quality of the finished products, both singly and in combination with each other. For example, it is well known that the quality and uniformity of mixing exerts a profound influence on the performance of downstream processes. Consequently, it is necessary to deal with rubber product manufacturing systems at three levels:

1. Processing behavior of raw elastomers and rubber compounds.
2. Unit processes.
3. Manufacturing systems.

Determination of processing behavior is essential for the effective setting up and operation of unit processes, which include familiar operations such as mixing, extrusion, calendering, and molding. The unit processes then have to be assembled into a viable manufacturing system, which requires both organizational and technical skills.

Traditionally, the technical expertize of the rubber industry has been biased toward physical chemistry and materials technology, with processing and manufacturing methods generally receiving far less attention than rubber compound development. Indeed, technologists have been very successful in modifying compounds to overcome some of the worst problems of aged, poorly controlled, or inappropriate processing equipment. However, it must be acknowledged that in most cases where such modifications have been necessary the resulting productivity is low and product quality is variable.

In recent years the introduction of rubber compounds with widely differing processing characteristics, coupled with the need to achieve greater precision in manufacture to satisfy the increasingly stringent demands of customers, has stimulated a much greater interest in manufacturing methods as a whole. The changes thus initiated have now been substantially accelerated by the advent of the microprocessor, with all its attendant implications for automation, process monitoring, and production organization.

The introduction of computer methods into company operations is already well advanced, with most accounting and stock-control systems being computer-based and manufacturing data processing becoming commonplace. However, these applications are essentially peripheral to the manufacturing system, involving the replacement of routine clerical work and the improvement of information handling. More recently, computer technology has been introduced at the process level, with an increasing number of machines being equipped with microprocessor controls. These generally confer greater precision of operation, in comparison with conventional "hard-wired" machines, but their greatest attributes arise from the versatility and power of programmable systems. In addition to providing an extremely broad choice of operating sequences and conditions, a microprocessor system can be used to monitor both the quality and quantity of production. Linking the process-level microprocessor systems to the management-level computers is then a logical step to speed up the reporting of production performance.

Computer methods also have a substantial role to play in the support of manufacture. The increasing range of options for process operating sequences and conditions, with the accompanying potential for high productivity and quality, require improvements in the methods used for the selection of those options which give optimum performance for specific products.

Efficient and practical methods of determining process capability are now available which rely entirely on computer analysis. Similarly, die and mold design methods are substantially improved by adopting a computer-aided approach.

Integrating the introduction of computer methods into a company at a number of different levels, with all the attendant considerations of their influence on operating and organizational methods, is a complex undertaking. Fortunately, the systems approach, which is well-established in the production engineering and management fields, provides the concepts and techniques necessary to facilitate this integration.

1.2. THE SYSTEMS CONCEPT

The systems approach to manufacturing is based on making a clear distinction between analysis and synthesis.[1] Analysis is concerned with the identification, detailed study, and improvement of the fundamental elements of manufacturing systems. A system can be identified as being any activity that has well-defined inputs and outputs, which indicates that the scale of analysis can be established at many different levels. For example, an extrusion line can be treated as a complete system which has well-defined inputs and outputs and a number of interdependent elements, as shown in Figure 1.1. Moving to a larger system, the extruder line can be treated as a single element in a sequence of operations necessary for the manufacture of a product.

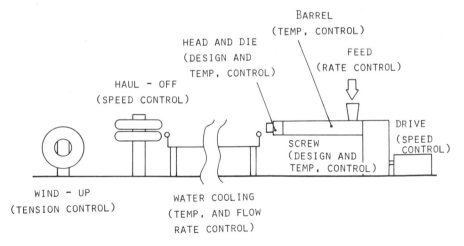

FIGURE 1.1. Elements of an extruder line which interact to determine its overall performance.

Synthesis is the assembly of a viable manufacturing system from the elements dealt with during analysis. It can take two main forms. The first form entails using existing elements in a new configuration, to obtain new or improved system capabilities. The second form involves the integration of new or improved elements into a manufacturing system in a manner which ensures their full utilization and exploitation. Consequently, system synthesis requires an understanding of the interaction and dependencies between the elements being assembled together, otherwise the expected system performance may not be achieved.

The distinction between analysis and synthesis aids both the planning and management of manufacturing operations. In this book Chapters 2–6 are concerned with analysis, whereas Chapters 7–11 deal with synthesis. Synthesis develops when adequate differentiation is drawn between[1]:

1. *Planning* the production system with respect to products, processes, and facilities.
2. *Implementing* the production system design.
3. *Monitoring and controlling* operations at various levels of systems involvement with computer methods, automation, and the concept of management by exception.

A viable decision framework or procedure is a great advance in achieving synthesis. Separate analytic results that indicate a number of best strategies for elements or subsystems will seldom combine together to yield an overall best strategy. A decision situation is composed of five basic elements[1]:

1. *Strategies or plans* constructed of controllable variables.
2. *States of nature* composed of noncontrollable variables.
3. *Outcomes* which are observations of results which occur when a particular strategy is employed and a particular state of nature exists.
4. *Forecasts* of the likelihood of each state of nature occurring.
5. *The decision criterion* which dictates the way in which the preceding information will be used to select a single plan to follow from a number of alternatives.

To achieve synthesis it is necessary to identify frameworks that will permit the simultaneous consideration of all the important production factors within the region being investigated. For example, if a process is to be selected to manufacture both product A and product B, it is not often possible to determine the best process for each independently and then combine the two. There will probably be a different best process for products A and B when they are considered together. The identification of decision frameworks which enable multiple factors to be considered simultaneously is an integral part of the latter half of this book.

Decision making is the process by which the selection of a plan from a number of alternatives is made. It is *not* the process by which plans are formulated. Planning is an innovative activity which must be guided toward the production of viable alternatives for the selection decision.

Having established that the systems approach is useful for dealing with the manufacturing technology of the rubber industry, some of the main themes of this book can now be introduced. In the following three sections the contributions of materials testing, process performance, and production organization to successful and efficient manufacture are discussed; and the concepts which underlie later chapters are examined.

1.3. THE SELECTION AND OPERATION OF TESTS FOR UNVULCANIZED RUBBER

An important step toward the prediction, optimization, and control of process performance is the use of testing methods which produce good indicators of the processing behavior of the materials being tested. There are two distinct areas in which test results contribute to these objectives:

1. Process development and problem solving.
2. Routine "quality-control" testing.

Each of these generates its own requirements with respect to testing procedures, conditions, and the treatment of results; although it is often possible to use similar test equipment in each area. For process development, test results for each of the following three physical-property groups are generally desirable:

1. Flow behavior.
2. Vulcanization characteristics.
3. Heat-transfer properties.

Tests for problem solving and routine quality control concentrate upon the measurement of vulcanization characteristics and flow behavior.

The trend in processing methods toward complex operation, high rubber temperatures, and high output rates has generated requirements for testing conditions and procedures which cannot be achieved by many of the traditional rubber industry testing instruments. In addition to being designed for service in conjunction with processes operating at far less demanding conditions than those in current use, the limitations of instruments, such as the Mooney viscometer, arise from their roles as routine testing tools. Consequently, the requirements for rapid tests and simple results, usually expressed in the form of a single number, have outweighed other considerations. However, it is important to remember that these limitations are largely due to the manual operation and recording methods used, which need to be within the scope of semiskilled operators. Provided that the basic concept of a test is appropriate for the manufacturing operations it is intended to support, most of the limitations of manual methods can be overcome by the use of microprocessor technology.

There is a sharp divergence between the testing procedures and treatment of results from routine testing and those for process development and problem solving, although both benefit from the test-to-test repeatability which accompanies microprocessor control. Routine testing procedures and conditions are required to produce results which correlate well with the performance of the materials being tested in the processing operations which they are intended to support. The best correlation is usually obtained using conditions which approximately simulate those encountered in the process; it militates against using standard conditions in all cases. For routine testing the following facilities should be available, either from the testing-instrument microprocessor or, as is more likely, from the laboratory computer to which it is linked:

1. Recording and analysis of data.
2. Comparison of results with programmed target values or tolerance bands.
3. Notification of out-of-tolerance results.

In contrast to the specific nature of routine tests, problem solving is an exploratory activity, in which a range of testing procedures and conditions may need to be tried in order to identify the cause of a problem. Process development tests also need to span a broad range of conditions, to determine the way in which material properties change over the operating range of the process being investigated.

Test results which are to be used in a scheme of process development generally need to be expressed in fundamental units of measurement, in comparison with the arbitrary units associated with the Mooney viscometer and other similar "quality-control" instruments. This enables the results to be used in quantitative analytical and predictive procedures, as well as enabling results obtained from different instruments to be compared.

1.4. THE PREDICTION, MONITORING, AND CONTROL OF PROCESS PERFORMANCE

The precision with which a rubber product manufacturer can predict, monitor, and control process performance will have a profound influence on the productivity of the whole manufacturing system. This starts with the assignment of new products to the manufacturing equipment best suited for them and continues into the optimization of the selected operations and the establishment of procedures and specifications which ensure that both quality and productivity are maintained. When process capability is well-defined, the dangers of increasing unnecessarily manufacturing costs by using a machine overly sophisticated for the task in hand, or of assigning a product to a machine inherently incapable of the required precision, are much reduced.

The techniques available for the prediction, monitoring, and control of performance are equally applicable to most of the unit processes of the rubber industry, which include mixing, extrusion, calendering, and molding. This enables a unified approach to be adopted, in which each process can be treated as a system with well-defined inputs and outputs. The output of a rubber processing operation is invariably dependent on complex interactions between the machine and the rubber compound being processed; which fact leads on from the previous section on testing methods, where a prime objective was the characterization of the input. Consequently, the main factors influencing the quality and quantity of the output from a process are:

1. The properties of the input.
2. The machine geometry.
3. The operating conditions and procedures.
4. The influence of uncontrollable variables (states of nature).
5. The form of the output.

Item 1 is concerned with the characteristic properties of the input which differentiate it from other types of input. Hence, for a given machine each input-type/product combination will have an operating window, which is defined by *all the combinations of operating conditions and procedures which enable a product of acceptable quality to be produced.* Rubber processes often have large operating windows, enabling conditions which give an acceptable product to be quickly established. However, the productivity of a process will vary over a substantial range within the operating window; and most of the viable combinations of conditions which combine acceptable quality with high productivity generally lie within tightly constrained regions of the operating window, where product quality is very sensitive to changes in uncontrollable variables. Locating these regions of high productivity requires a systematic approach to process trials, analysis of performance, and optimization; and operating in them demands good process control. They are also strongly influenced by factors such as mold and die design.

The effect of changes in uncontrollable variables, which can derive from the input, the processing equipment, or the environment in which it is situated, is to change the size and shape of the operating window. Since uncontrolled changes in a process generally result in a deterioration rather than an improvement, this will increase the probability of defective work being produced. For a process being run near the limits of the operating window, the probability of an uncontrolled change causing defective work will be high.

Some process variability is inevitable and a compromise between precision of control, product quality, and process productivity can only be reached by reference to the economics of manufacture. Improvements in process control can have two effects. It will stabilize the boundary of the operating window, enabling the process to be operated near to it without incurring an unacceptably high probability of defective work. Alternatively, it will extend the boundary of the processing window into regions of higher productivity.

The ability of a control system to cope with all the abnormalities which arise to cause defective output is limited, requiring occasional intervention by technical and managerial staff for remedial action. To ensure that problems are detected and reported before the mounting cost of defective products becomes excessive, monitoring systems are required which can be used in conjunction with the control systems. These monitoring systems are appropriate for operations in which the product quality is largely dictated by machine performance, rather than by manual skills; this type of operation includes all the main unit processes. Monitoring involves the measurement, at closely spaced intervals, of some aspect of process performance which provides a good indicator of product quality. It is essentially a computer-based operation, in which the measured quantity is compared with a target value; and the staff responsible for the process are notified immediately when an abnormality is detected.

Systems used for monitoring the quality of the output are also employed to monitor the quantity of output, for production planning and management purposes.

1.5. PRODUCTION ORGANIZATION

The operation of a manufacturing system at a level of performance which achieves the objective specified for it can only be accomplished with effective production organization and management, which identifies them as crucial elements to be included in systems synthesis. An understanding of the interactions and dependencies which occur between the different activities contributing to successful manufacture is essential. Although generalized production management techniques can be used in a wide variety of manufacturing systems, their effective implementation depends on a detailed understanding of the product requirements and manufacturing technology of specific systems. Conversely, technical expertize alone cannot result in a viable manufacturing system. For these reasons both have to be included in any discussion of overall manufacturing system performance.

A first step toward effective management is the identification of the different decision levels needed to implement company objectives. The main differentiation is between strategic and tactical decisions. In simple terms, the former are concerned with formulating company policy while the latter are concerned with its implementation. The selection of markets in which a company will operate and the choosing of manufacturing and other systems necessary to achieve the company's objectives in its markets derives from strategic decision making by senior management. In contrast, the day-to-day operation of manufacturing systems involves a very large number of tactical decisions made by middle and junior managers.

The provision of alternative schemes and courses of action for selection at the strategic decision level and the implementation of the one chosen are project-type activities. These have well-defined beginning and end points, although it is often difficult to assess performance at intermediate stages. Again, this is in contrast to the day-to-day operation of manufacturing systems, which has no readily definable beginning or end points, although techniques for the frequent or continuous measurement of performance are well-developed. Figure 1.2 shows a simple manufacturing system in the form of a control diagram. Objectives or production targets are determined by the delivery dates and quantities set against the incoming orders or sales returns. All the decisions which have to be made in order to achieve these objectives using the available resources, including those needed to overcome problems such as malfunctioning equipment, material shortages, and operator absenteeism, are tactical in nature.

If the performance of a manufacturing system does not achieve a company's objectives, despite observably sound tactical decision making by production management, it is usually due to the system being inappropriate or inadequate for the tasks assigned to it and strategic decisions are then required for remedial action. This example is intended to show both the distinction and interaction between strategic and tactical decision making. Senior managers should avoid tactical decisions, which will result in them becoming "bogged down in detail" and unable to give adequate attention to

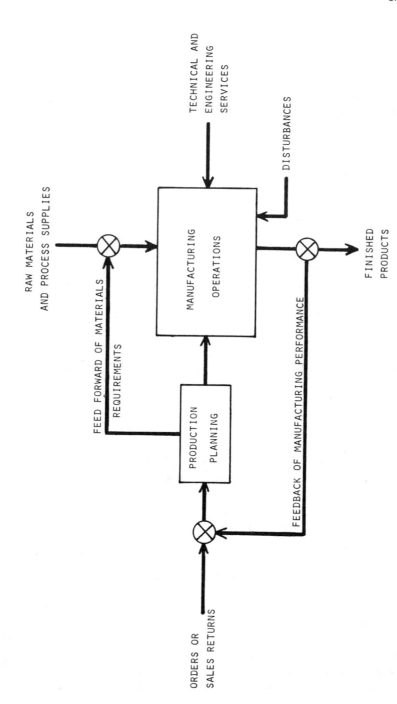

FIGURE 1.2. Production management control diagram.

their main task of making strategic decisions, on which the survival of a company depends. Similarly, middle and junior managers should not engage in strategic decision making, since this is likely to result in far-reaching changes being implemented to solve local problems, with a consequent distortion of company activities away from objectives on which survival depends.

Returning to Figure 1.2, the quantity of finished products is compared with the target derived from the orders or sales returns and, if the required quantity can be completed on or before the due date, no control action is required. This is a very simple application of the principle of *management by exception*, for which the manufacturing system must have self-regulating features and only require control action when abnormal conditions arise. With the introduction of computer information gathering, processing, and communication, this principle must also be applied to the feedback of information to managers. All the restrictions on the quantity of information imposed by the practical limitations of manual clerical work have now been lifted; and only *the information which is needed to support action* at each level of management should be transmitted. Otherwise, the time spent by managers in dealing with a flood of superfluous information can totally negate the undoubted benefits of computer methods.

The quality, nature, format, and method of presentation of manufacturing performance information will all exert a strong influence on the effectiveness of management. The tasks which most managers undertake can be divided into two groups for the purpose of establishing the requirements of an information system: control and planning or "doing." Control is a continuing activity, being concerned with the day-to-day running of the manufacturing system, whereas planning and "doing" are project-type activities. The term "doing" is used to denote a task carried out personally by a manager, rather than being delegated in the usual way.

The analogy between the feedback of information for process control and for production management, illustrated by the control diagram in Figure 1.3, can be used to establish the basic requirements of information flows to each level of management. In moving from junior to senior management it can be seen that the time scales of activities become substantially longer. It is the junior manager's responsibility to control the short-term fluctuations in manufacturing performance, while the senior manager is more concerned with the direction of long-term trends.

The information flows to junior managers must be sufficiently detailed to give an accurate measure of short-term fluctuations in manufacturing performance and be available quickly enough for courses of action based on this information to be effective. This introduces the concept of response time from process-control theory. For control to be effective the response time for

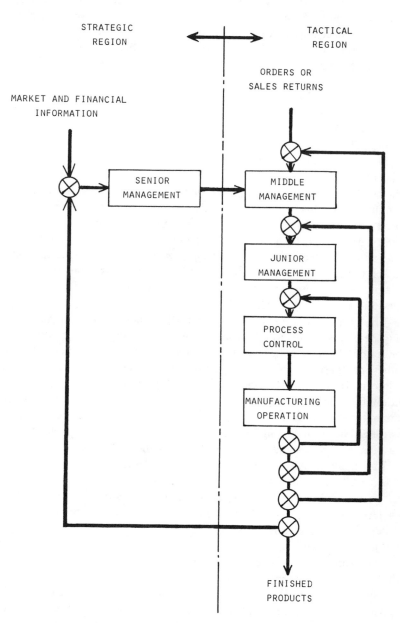

FIGURE 1.3. Levels of control for manufacturing.

feeding back information and taking action must be appropriate to the typical time scales over which fluctuations in performance occur. If the response time is long, the circumstances causing a deterioration in performance may have changed before corrective action can be taken. At best, this will grossly reduce the value of the information flow. At worst, persevering with corrective action based on out-of-date information will itself cause a deterioration. In control engineering this is known as instability. It can result in a rapid and progressive deterioration as a succession of inappropriate courses of "corrective" action are implemented.

In designing information flows for senior managers, the main problem is one of distinguishing between short-term fluctuations in performance and genuine long-term trends. If this is not achieved, the senior manager will be induced to respond to short-term fluctuations, which are properly the responsibility of junior and middle managers. It will also mean that the quantity of information converging on the senior manager will be far too great to be assimilated and used effectively. This calls for a lengthening of the response time to a value appropriate to the rate of change of the long-term indicators of company performance and can be achieved by using moving averages of individual information flows. As the period time over which performance is averaged is increased, the influence of short-term fluctuations becomes smaller.

As with all aspects of manufacturing, the cost of providing detailed information flows must be compared with the benefits to be derived from them. This is particularly pertinent at the junior management level, where sophisticated methods are often needed to provide a rapid feedback. These must be justified by a general increase in productivity and the reduction of the cost of defective work produced during short-term deteriorations in manufacturing performance.

REFERENCES

1. Starr, M. K., *Production Management: Systems and Synthesis*, 2nd ed., Prentice-Hall, Englewood Cliffs, N.J. (1972).

2

Materials Behavior and Testing

The "processability" of a rubber mix depends on three main aspects of materials behavior:

1. The flow properties.
2. The thermal or heat-transfer properties.
3. The vulcanization characteristics.

The extent to which the performance of rubber processing operations may be predicted, optimized, and controlled is dependent, in large measure, on the characterization and understanding of these properties. Before dealing in detail with individual properties and their measurement, it is important to review their influence on process behavior.

Rubber is a viscoelastic material and even in the unvulcanized state is capable of displaying behavior ranging from predominantly viscous to predominantly elastic, depending on temperature and rate of deformation. In processing, the main concern lies with flow and shaping operations, requiring that material temperatures and rates of deformation be adjusted so that the behavior is predominantly viscous. However, the time for which a conventional rubber mix may be held at an elevated temperature is limited by the onset of cross-linking, which effectively prohibits further flow and shaping operations.

Rubber processing can now be seen to be a compromise between conflicting requirements; raising the processing temperature generally results in the possibility of a higher output rate but brings with it the danger of the onset of cross-linking or scorch. Also, a temperature rise is usually an inevitable consequence of an increased output rate, arising from the conversion of mechanical energy to heat energy in the rubber (shear heating or viscous dissipation).

The temperature rise due to viscous dissipation is, as the name suggests, dependent on the viscosity of the rubber. More energy is needed to maintain an equivalent flow rate with a high viscosity material than with one of low viscosity. However, increasing the temperature of a rubber reduces its viscosity, which indicates that there will be a trend toward a temperature – viscosity equilibrium in a flow process. Heat transfer to and from a process can therefore exert a considerable influence on its operating characteristics; and will depend on both the temperature – viscosity relationship of the rubber and the mode of heat transfer. The latter is determined by the flow patterns generated in the rubber. For simple laminar flow, or in the absence of flow, the primary mode of heat transfer is conduction; whereas in a process generating complex flow fields, physical movement of material to the metal surfaces of a machine at which heat transfer occurs results in far more effective heating or cooling. This mechanism is known as forced convection and is one of the primary reasons for the use of mixing screws in extruders.

Conductive heating results in pronounced temperature gradients due to the slow rate of heat transfer through rubber. This is particularly important for vulcanization, where the main aim is to heat the rubber as quickly and uniformly as possible to minimize the cure time. The selection of a suitable vulcanization process and optimization of the operating conditions first requires the measurement of the vulcanization characteristics and the heat-transfer properties of the rubber mix. Predictions of cure time and uniformity of cure can then be made using these measurements.

2.2. FLOW PROPERTIES OF RAW ELASTOMERS AND RUBBER MIXES

2.2.1. Viscous Flow

For the case of laminar flow between parallel plates, one of which has a velocity V relative to the other plate (Figure 2.1), shear stress τ is defined as the force required to maintain the relative plate velocity V divided by the

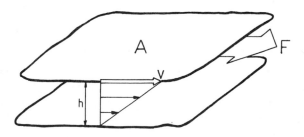

FIGURE 2.1. Shear flow of fluid contained between two parallel plates, resulting from their relative motions.

area over which the force acts; τ is in pascals (Pa). The rate of flow, expressed as the rate of change of shear strain, more commonly called shear rate $\dot{\gamma}$, is defined by the relative velocity V divided by the distance between the plates. Since shear strain is a dimensionless quantity, shear rate has units of 1/seconds (s^{-1}).

The relation between shear stress and shear rate will depend on the nature of the fluid between the plates. Newton postulated that shear rate is directly proportional to shear stress, the proportionality constant being the viscosity of the fluid η_N:

$$\tau = \eta_N \dot{\gamma} \tag{2.1}$$

However, the assumption of constant viscosity is not correct for either raw elastomers or rubber mixes at practical processing shear rates, although the concept of viscosity is extremely useful. An apparent viscosity is usually quoted, being the viscosity measured at a specified shear rate.

The change in viscosity with shear rate is most often described by a "power-law" relationship, which gives a reasonable approximation to the observed steady-state flow behavior under normal processing conditions:

$$\tau = \eta_0 \dot{\gamma}^n \tag{2.2}$$

Here n is the non-Newtonian or power-law index, which defines the deviation from the simple proportionality of a Newtonian liquid; when $n = 1$ the apparent viscosity at $1\,s^{-1}\eta_0$ is equal to the Newtonian viscosity η_N. For rubbers n is generally in the range 0.15–0.4, defining rubber as pseudoplastic material since $n < 1$ and showing that the apparent viscosity η_a decreases as shear rate increases. Combining Eqs. (2.1) and (2.2) gives

$$\eta_a = \eta_0 \dot{\gamma}^{n-1} \tag{2.3}$$

or

$$\eta_a = \eta_0^{1/n} \tau^{(n-1)/n} \tag{2.4}$$

If a raw elastomer or rubber mix conforms reasonably well to power-law behavior, a plot of $\log \tau$ *vs.* $\log \dot{\gamma}$ will give a straight line of slope n and an intercept of η_0 on the $\log \tau$ axis. Over more than one or two decades of shear rate the experimental readings will tend to show significant deviations from a straight line; and the constants η_0 and n must be determined from the portion of the curve relevant to a specific processing operation.

The temperature dependence of apparent viscosity of a polymer is usually described by

$$\eta_a = A e^{-bT} \tag{2.5}$$

where A and b are constants and T is absolute temperature. A more useful form of Eq. (2.5) relates apparent viscosity at one temperature to that at a different temperature:

$$\eta_{a1} = \eta_{a2} \exp[-b(T_1 - T_2)] \qquad (2.6)$$

Combining Eqs. (2.3) and (2.6) gives an expression which describes both the shear rate and temperature dependence of viscosity:

$$\eta_{aT} = \eta_0 \exp[-b(T - T_0)] \, \dot{\gamma}^{n-1} \qquad (2.7)$$

From Eq. (2.7) the apparent viscosity at any shear rate $\dot{\gamma}$ and any temperature T may be determined, provided that the reference viscosity η_0 (at $1s^{-1}$) derived from the log τ *vs.* log $\dot{\gamma}$ curve at temperature T_0 is known and the constants b and n are known.

The index b, describing the temperature dependence of viscosity, may be determined from a plot of $\ln \eta_a$ *vs.* T, from results obtained at a constant shear rate. However, practical results for rubbers show that a change of slope may occur in the region of 100°C and that two values of b are often needed to define the temperature dependence of viscosity.

Thus far the relationships described have been limited to the shear mode of flow. Recent work has shown that extensional flow behavior has a significant influence on the characteristics of some processing operations. Similar relationships to those for shear flow can be used to describe extensional flow.

Assuming Newtonian flow behavior, the extensional strain rate or elongation rate $\dot{\varepsilon}$ and the tensile stress σ can be related by an elongational viscosity λ_N:

$$\sigma = \lambda_N \dot{\varepsilon} \qquad (2.8)$$

However, as for shear flow, a power-law relationship gives a more representative description of viscous flow behavior:

$$\sigma = \lambda_0 \dot{\varepsilon}^P \qquad (2.9)$$

To obtain a constant elongation rate in a simple tensile test, where a strand of unvulcanized rubber is stretched, the grip separation rate must increase exponentially, as described by

$$\dot{\varepsilon} = \frac{1}{t} \ln \left(\frac{l(t)}{l_0} \right) \qquad (2.10)$$

where l_0 is the initial specimen length and $l(t)$ is the length at time t. Alternatively, the initial length l_0 can be maintained constant by using the pulley wheel configuration of Figure 2.2(b) giving

$$\dot{\varepsilon} = \frac{r}{l_0}\frac{d\theta}{dt} = \frac{\omega r}{l_0} \qquad (2.11)$$

However, tensile tests of this type are transient in nature and yield a result which reflects the viscoelastic behavior of the rubber, which must then be resolved into its viscous and elastic components.

Comparisons of shear flow and extensional flow behavior for butyl rubber[1] show that

$$\lambda_N = 3\eta_N \qquad (2.12)$$

which is described as Troutonian[2] behavior (after Trouton); but this simple relationship breaks down at practical strain rates for most polymers and independent determination of shear and extensional flow properties is necessary.

Extensional tests can also be used to determine the failure behavior of raw elastomers and unvulcanized rubber compounds, which is generally referred to as "green strength." Processing failures due to inadequate green strength include crumbling on a two-roll mill or in an internal mixer, and

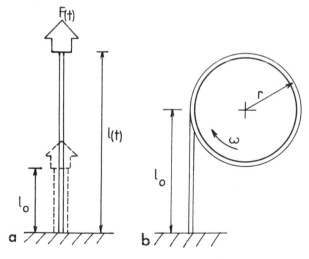

FIGURE 2.2. Configurations of extensional rheometers. (a) Extension by grip separation. (b) Extension by pulley "windup."

melt fracture of extrudates. However, green strength is dependent on both temperature and rate of deformation pointing to the sensible use of these variables in practical processing to avoid failure.

2.2.2. Viscoelasticity

The viscoelastic behavior of a raw elastomer or a rubber mix is difficult to measure and to express in terms which are readily related to processing operations. It is necessary to postulate a model, composed of discrete viscous and elastic elements, to represent the material being studied and then to assign values to the elements from experimental measurements. The major problem with this approach, apart from experimental difficulties, is that complex models are required to give a good analogue of the viscoelastic responses of rubber materials; and these complex models are extremely difficult to apply quantitatively to processing operations.

The Maxwell model shown in Figure 2.3(a) is one of the simplest viscoelastic models, being a Newtonian dashpot in series with a Hookian spring. The strain rate of the dashpot (here shear units are used but the model is equally applicable to tensile deformations) is defined, from Eq. (2.1), as τ/η_N. The strain in the Hookian spring is given by τ/G, where G is the shear modulus; τ/G must now be differentiated to obtain the time derivative, since the stress in the spring is equal to and depends on that in the rate-dependent dashpot. Hence

$$\frac{d\gamma}{dt} = \frac{1}{G}\frac{d\tau}{dt} \tag{2.13}$$

Adding this to the viscous component gives

$$\dot{\gamma} = \frac{1}{G}\frac{d\tau}{dt} + \frac{\tau}{\eta_N} \tag{2.14}$$

which determines the mechanical response of the model to a constant shear stress or strain rate. If $\dot{\gamma}$ is set to zero, Eq. (2.14) can be integrated to give

$$\tau = \tau_0 e^{-Gt/\eta_N} = \tau_0 e^{-t/\Gamma} \tag{2.15}$$

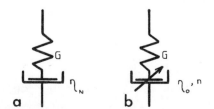

a   b

FIGURE 2.3. Models for viscoelastic behavior of rubbers. (a) Basic Maxwell model. (b) Maxwell model with power-law dashpot.

which indicates that if a strain γ is achieved and held, there will be an exponential decay of stress from a peak value τ_0, the rate of which is determined by the relaxation time Γ. This quantity has some practical utility since the extrusion[3] and milling[4] behavior of raw elastomers and rubber mixes has been shown to depend, in some measure, on the relaxation time.

While experimental evidence indicates that it is reasonable to assume linear elasticity in viscoelastic models for rubber materials,[5] the viscous component is usually strongly nonNewtonian, requiring a dashpot which has a power-law flow characteristic, as shown in Fig. 2.3(b). For comparison, the differential equation of the power-law Maxwell-type model is

$$\dot{\gamma} = \frac{1}{G} \frac{d\tau}{dt} + \left(\frac{\tau}{\eta_0}\right)^{1/n} \tag{2.16}$$

and the stress relaxation of this model in time t is

$$\tau^q = \tau_0^q - \frac{G^q}{q} \left(\frac{G}{\eta_0}\right)^{1/n} t \tag{2.17}$$

where $q = (n-1)/n$. While a unique relaxation constant cannot be resolved from this model, its closer correspondence to the behavior of the real materials gives a better prediction of processing behavior than is possible with the basic Maxwell model.

Expressions similar to Eqs. (2.16) and (2.17) can be used to characterize viscoelastic behavior due to extensional deformation, using the Young's modulus E in place of the shear modulus G and extensional power-law constant p and reference viscosity λ_0 in place of n and η_0. Tensile stress σ will also replace τ. In practical processes extensional flow usually occurs either as a result of convergence in a flow path, where shear flow is also present, or during the haul-off from a calender or extruder. In the former case the extensional component of flow controls the onset of melt fracture, which places definite limits on the speed of some processing operations. The consequences of extensional flows for die design are examined in Section 4.5.3.

2.3. MEASUREMENT OF FLOW PROPERTIES

2.3.1. Rotational Rheometers and Viscometers

In these instruments the torque necessary to maintain a constant rate of rotational and/or twisting flow, induced by the angular motion of a rotor, is measured. The Mooney viscometer, which is used throughout the rubber industry, embodies both these modes of shear flow and, since the instrument is familiar to most technologists, it is reasonable to use it as an example from which the principles of rotational rheometers can be developed. However, before proceeding it is worth defining the difference between a viscometer and a rheometer. In this context a viscometer is a single-speed machine, giving a single-point measurement which can only be resolved into the fundamental units of shear stress and shear rate if basic assumptions are made about the flow behavior of the rubber. A rheometer is a variable-speed machine, permitting a relationship between shear stress and shear rate to be established and having a geometry which also enables results to be expressed in fundamental units without any prior assumptions about the flow behavior of the material under test.

Rotational instruments are particularly suitable for testing rubber mixes because of the control which can be exercised over the total shear strain input to the test sample. This is necessary to overcome the problems of thixotropic effects masking the steady-state flow properties. Thixotropy, which results from the breakdown of weak rubber-filler linkages, causes a high initial torque value, which then decays in an approximately exponential manner to a steady-state value. The delta-Mooney[6] measurement assesses the initial thixotropic contribution to torque, while the Mooney $ML(1 + 4)$[6] test, involving 1-min preheat time and 4-min running time prior to a measurement being made, is designed to give the steady-state torque value. Both of these times are rather short due to the speed of test requirements for quality control.

The Mooney instrument also displays the main disadvantage of rotational viscometers: to avoid a significant temperature rise in the test sample, due to shear heating, or more precisely, viscous dissipation, the shear rate must be low. With the Mooney large rotor the shear rate varies from zero at the center to $1.7 s^{-1}$ at the periphery, as a result of the rotor geometry, which is shown in Figure 2.4(a). The shear rate at any radius r is given by

$$\dot{\gamma}(r) = \omega r/t \qquad (2.18)$$

where ω is the angular velocity of the rotor in radians/second (rad/s), while the annular gap between the rotor and the chamber gives concentric cylinder or Couette flow, which again cannot be resolved into fundamental units without postulating a relationship between shear stress and shear rate.

The very low shear rates and single-point measurement of the Mooney viscometer identify it as a quality-control instrument rather than a tool for development and problem solving. Even then, the correlation between

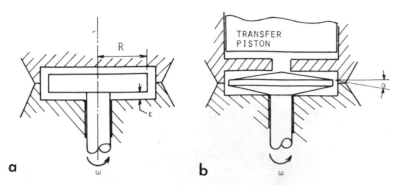

FIGURE 2.4. Test cavity configurations of rotary viscometers. (a) Mooney
viscometer. (b) Biconical rotor "TMS" rheometer.

Mooney viscosity, which is a torque measurement, and processing behavior
should be established before proceeding to utilize it as a measure of material
consistency.

For a uniform shear rate between $r = 0$ and $r = R$, it is necessary for
$r/t = \text{const}$. This is achieved by using the biconical rotor of Fig. 2.4(b), for
which $r/t \simeq \alpha$, where α is the cone angle in radians, provided that $\alpha \leqslant 6°$.
The shear rate is then

$$\dot{\gamma} = \omega/\alpha \qquad (2.19)$$

The total torque on the rotor T is the sum of the torque due to the double
surface of the cone $2T_c$ and the Couette flow at the edges T_e. The torque on
the rotor due to flow over and under the conical surfaces can be evaluated by
considering a typical circular element of width dr at radius r and then
integrating between limits of $r = 0$ and $r = R$ to give

$$2T_c = \tfrac{4}{3}\pi R^3 \tau \qquad (2.20)$$

Assuming Newtonian behavior, the torque due to the region of Couette flow
is

$$T_e = 2\pi R^2 Y\tau \qquad (2.21)$$

which is equivalent to the analysis for shear flow between parallel plates
moving at a differential velocity V. Now

$$T = 2T_c + T_e = \tfrac{4}{3}\pi\tau R^2 (R + \tfrac{3}{2}Y) \qquad (2.22)$$

The errors resulting from the assumptions made here can be minimized by reducing the depth Y of the rotor edge as far as is possible, consistent with the requirements of mechanical rigidity.

The temperature rise associated with viscous dissipation must be taken into account at shear rates which are of interest for the evaluation of processing behavior. It has been found[7] that the deviation from the set temperature is sufficiently small for most purposes if the measurement is taken immediately after the speed and torque stabilize. Alternatively, a temperature sensor designed to give a fast response to changes in temperature may be introduced into the test cavity.

The small clearances between rotor and die cavity surfaces, resulting from the biconical rotor design, must be capable of being accurately set prior to each test run, to give a reproducible shear rate. A compression molding technique, such as that used in the Mooney viscometer to shape the sample, is inadequate due to the much larger influence on the shear rate of the variable thickness of "flash" or "mold spew" between the clamping surfaces of the die halves. This problem is overcome in the SPRI Ltd. TMS instrument of Figure 2.4(b) by the use of a transfer molding method. This allows the die cavity to be closed prior to filling, which eliminates the flash problem and also enables the pressure in the die cavity to be controlled accurately. A further benefit of the transfer molding technique is that the rubber surfaces which come into contact with the cavity and rotor surfaces are newly generated. This is extremely important for wall-slip measurements.

The relation between torque and angular velocity can be further influenced by the occurrence of slip between the rubber and the rotor. Consequently, the angular velocity ω resulting from a given torque is the sum of the viscous shear component ω_v and the angular slip component ω_s. Hence

$$\omega_v = \omega - \omega_s \qquad (2.23)$$

and the shear rate is

$$\dot{\gamma} = \frac{\omega_v}{\alpha} = \frac{\omega - \omega_s}{\alpha} \qquad (2.24)$$

Results from nominally similar tests with grooved and smooth-surfaced rotors give different values of angular velocity at a given torque level, for most rubber mixes. These differences are attributed to wall slip, since observations[7] indicate that a heavily grooved rotor will inhibit slippage, enabling an angular slip velocity ω_s to be determined. Transposing the angular slip velocity to a linear velocity presents problems since the latter quantity is dependent on radius. However, it can be argued that the peripheral region of the rotor will make the greatest contribution to slippage, giving the approximate relation

$$V_s = \omega_s R \tag{2.25}$$

The slip velocity is then found to be related to shear stress by the empirical equation

$$V_s = k\tau^m \tag{2.26}$$

which indicates that slip is due to a boundary layer providing lubrication between the rubber and the metal surface. Experimental results tend to confirm this view, showing increased slip for mixes incorporating oils, waxes, and fatty acids. Moore and Turner[8] report that angular velocities differing by a factor of 180 have been observed between grooved and polished rotors operating at a similar shear stress. Examination of the values[9] of the constants k and m in Eq. (2.26), derived from a range of tests performed on a number of mix types, shows m to remain constant irrespective of rotor surface finish and operating conditions. Hence it is appropriate to define it as a material constant. The value of k, however, varies with changes in rotor surface finish, hydrostatic pressure, and temperature; k increases with surface polish and temperature but decreases slightly with an increase in pressure.

The viscoelastic behavior of the test sample may be assessed by either a stress relaxation or a viscoelastic recovery experiment. Stress relaxation characteristics can be measured by stopping the rotor after a viscous shear flow test procedure and recording the decay in torque with respect to time. If the Maxwell model described by Eq. (2.15) is assumed to describe the material behavior, the relaxation time Γ is the time for the torque to decay to 63.2% of its initial value. Alternatively, using the more realistic model described by Eq. (2.17), a shear modulus can be calculated.

2.3.2. Capillary Rheometers

The capillary rheometer is a standard instrument for investigating the flow properties of plastics and is now being increasingly used to measure the flow properties of raw elastomers and rubber mixes. The material under test is induced to flow through a die (capillary) of circular cross section at a constant rate under conditions of telescopic flow. The importance of this mode of laminar flow derives from the capability of the capillary rheometer to achieve the high shear rates typical of modern processing operations and from the similarities between flow in a capillary and flow in extruder dies. However, in deriving the expressions for shear stress and shear rate, which

usually refer to their values at the capillary wall, the following assumptions
are initially made:

1. There is no slip at the capillary wall.
2. The fluid is time independent.
3. The flow pattern is constant along the capillary.
4. The flow is isothermal.
5. The material is incompressible.
6. The flow properties are independent of hydrostatic pressure.

Within the normal range of processing conditions raw elastomers
appear to conform with the above assumptions reasonably well, provided the
critical shear stress for melt fracture (nonlaminar flow giving a distorted
extrudate) is not exceeded. However, rubber mixes, which are microcom-
posites, behave in a much more complicated manner. The consequences of
deviations from the basic assumptions above for the analyses which follow
will be indicated.

The equilibrium of forces for flow through a capillary illustrated in
Figure 2.5 results in the following expression[10]:

$$\tau_w 2\pi R L = \Delta P \pi R^2 \tag{2.27}$$

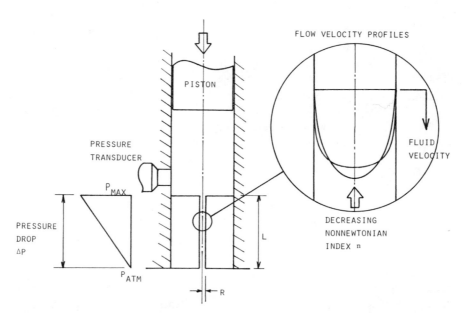

FLOW VELOCITY PROFILES

PISTON

PRESSURE
TRANSDUCER

P_{MAX}

FLUID
VELOCITY

PRESSURE
DROP
ΔP

P_{ATM}

L

R

DECREASING
NONNEWTONIAN
INDEX n

FIGURE 2.5. Configuration of a capillary (ram extrusion) rheometer showing flow
velocity profiles in the capillary.

giving the shear stress at the capillary wall, irrespective of material flow behavior, as

$$\tau_w = \frac{\Delta P R}{2L} \tag{2.28}$$

For a pseudoplastic material,

$$\tau = \eta_0 \dot{\gamma}^n = \eta_0 \left(\frac{du}{dr}\right)^n \tag{2.29}$$

which can be manipulated[10] to give the velocity of the fluid, in the z-direction, at the center of the capillary:

$$u_{max} = -\frac{n}{n+1}\left(\frac{\Delta P}{2\eta_0 L}\right)^{1/n} R^{(n+1)/n} \tag{2.30}$$

The velocity at any radius $u(r)$ is given by

$$u(r) = u_{max}\left(1 - \frac{r}{R}\right)^{(n+1)/n} \tag{2.31}$$

enabling the velocity profile to be determined. For a Newtonian fluid ($n = 1$), the velocity profile is parabolic in form, as shown in Figure 2.5, and the shear rate at the capillary wall is given by

$$\dot{\gamma} = \frac{4Q}{\pi R^3} \tag{2.32}$$

This enables the volumetric flow rate from the capillary to be expressed by the Poisseuille equation, which is obtained by substituting Eqs. (2.28) and (2.32) in $\tau = \eta_N \dot{\gamma}$:

$$Q = \frac{\pi \Delta P R^4}{8\eta_N L} \tag{2.33}$$

The simple relation between $\dot{\gamma}_w$ and Q expressed by Eq. (2.32) does not give the true wall shear rate of a pseudoplastic material, since the velocity profile assumes a "plug-like" shape as the non-Newtonian index n decreases, as illustrated in Figure 2.5. However, Eq. (2.32) is often used and the resulting quantity is termed the apparent wall shear rate $\dot{\gamma}_{wa}$. For a pseudoplastic material the wall shear rate is given by

$$\dot{\gamma}_w = \frac{3n+1}{n}\frac{Q}{\pi R^3} \tag{2.34}$$

and the volumetric flow rate now becomes

$$Q = \frac{n}{3n+1} \left(\frac{\Delta P}{2\eta_0 L} \right)^{1/n} \pi R^{(3n+1)/n} \tag{2.35}$$

The validity of the preceding relations for practical flow measurement depends on the test material conforming with the assumptions stated at the beginning of this section. The behavior of many rubber mixes shows significant deviations from this ideal behavior, particularly with respect to slippage, time dependence, and isothermal conditions.[11] Isolating the relative magnitudes of the contributions from these effects is difficult due to the lack of control of shear history or total shear, which is a function both of capillary length and radial position within the capillary.

Assuming that the temperature rise during flow down the capillary, resulting from viscous dissipation, is negligible and that the thixotropy is also negligible, giving time independence and a constant velocity profile, the contribution of wall slip to the total volumetric flow rate is expressed by

$$Q = \frac{n}{3n+1} \left(\frac{\Delta P}{2\eta_0 L} \right)^{1/n} \pi R^{(3n+1)/n} + \pi R^2 V_s \tag{2.36}$$

The slippage velocity V_s can be obtained from a plot of $4Q/\pi R^3$ vs. $1/R$, giving a straight line of slope $4V_s$.[12] This plot is derived from a series of capillary experiments performed using a standard pressure drop ΔP and a range of dies of different radius but constant length-to-radius ratio.

The transition from the reservoir or barrel of the rheometer to the capillary results in a region of convergent flow being formed, which must be eliminated or accounted for, to enable the pressure drop ΔP resulting from the fully developed laminar flow in the capillary to be determined. There is also a transition region at the capillary exit which must be taken into account. Two methods are available for making these "end corrections." The first and simpler method involves using two dies of similar radius but different length, under similar conditions of volumetric flow rate and temperature, enabling the shear stress to be determined from

$$\tau_w = \frac{(\Delta P_1 - \Delta P_2) R}{(L_1 - L_2)^2} \tag{2.37}$$

However, this two-point method cannot give any indication of the compliance of the material to the general assumptions for capillary flow. The technique due to Bagley[13] is more reliable and useful in this respect, since a number of dies of similar radius but differing lengths are used. If the general assumptions are valid, plotting ΔP vs. L/R gives a straight line with a negative intercept B on the L/R axis (X-axis). The shear stress is then obtained from

$$\tau_w = \frac{\Delta PR}{2(L/R + B)} \qquad (2.38)$$

where B is the extra length of die, in terms of its radius, that is needed to give the same pressure drop as the entrance region. The correction factor B can then be used with dies of any L/R ratio, as indicated by Eq. (2.38) and with any volumetric flow rate within the laminar flow region. However, if significant wall slip, thixotropy, or viscous dissipation effects are present, the ΔP vs. L/R plot will not conform to a straight line, indicating that the capillary rheometer is unsuitable for the material.

It must be noted that end corrections are not necessary for the determination of slip velocity, since the entrance and exit effects in each die are similar and do not influence the slope of the graph.

Elastic effects occur in both the convergent flow from the reservoir to the capillary and at the capillary exit, where the lateral expansion and longitudinal shrinkage of the extrudate, termed "die swell" or "extrusion shrinkage," occurs. The significance of both these effects is dealt with in detail in Section 4.5 on die design.

2.4. THERMAL AND HEAT-TRANSFER PROPERTIES

2.4.1. Thermal Conductivity

Thermal conductivity is defined as the quantity of heat passing per unit time normally through a unit area of material of unit thickness with unit temperature difference between the surfaces. The thermal conductivity of solid rubbers is about $1–2 \times 10^{-10} \, \mathrm{W \, m^{-1} K^{-1}}$ (watts per meter per degree Kelvin), which places rubber in a region of low conductivity where accurate measurement is difficult because of heat losses.

The methods of measuring thermal conductivity can be divided into steady-state methods and transient methods. Steady-state methods are mathematically more simple to handle and are more widely used. In the steady state, when the temperature at any point in the material is constant with time, conductivity is the parameter which controls heat transfer. It is then related to the heat flow q and the temperature gradient dT/dx by

$$q = -KA \frac{dT}{dx} \qquad (2.39)$$

where K is the thermal conductivity and A is the surface area of the test piece. The low thermal conductivity of rubbers can result in steady-state tests,[14] which involve the use of expensive equipment, being very time consuming; thus the transient methods have experimental advantages once the mathematical treatment has been worked out. An enclosed method of measuring conductivity has been described by Hands and Horsfall.[15] This method is capable of maintaining the geometry of the test piece at elevated temperatures, enabling thermal conductivity to be determined at typical processing temperatures.

2.4.2. Specific Heat

Specific heat is the quantity of heat required to raise a unit mass of the material through 1 K. The principal specific heats are those at constant pressure c_p and at constant volume c_v. However, the former is the quantity normally measured and the difference between the two values is usually small enough to be ignored. A typical figure of specific heat for a carbon-black-filled rubber mix is $1500 \, J \, kg^{-1} K^{-1}$.

Except where the very highest precision is required, when an adiabatic calorimeter would be used, it is usual to measure specific heat by a comparative method using differential scanning calorimetry (DSC) or differential thermal analysis (DTA); both are widely used for characterizing the properties of polymers. In these techniques heat losses to the surrounding medium are allowed but are assumed to be dependent on temperature only. The heat input and temperature rise for the material under test are compared with those for a standard material of known specific heat. In DTA the two test pieces are heated simultaneously under the same conditions and the difference in temperature between the two is monitored, while in DSC the difference in heat input to maintain both samples at the same temperature is recorded. Of the two methods, DSC is preferred for specific heat determination.

2.4.3. Thermal Diffusivity

Thermal diffusivity is the parameter used to determine the temperature distribution through a material in non-steady-state conditions, that is, when the material is being heated or cooled. It is a function of thermal conductivity, specific heat, and the material density:

$$\alpha = \frac{K}{\rho c_p} \qquad (2.40)$$

Thermal diffusivity can be obtained directly from a simple test and is easier to measure than conductivity, but the results of the necessarily transient test methods require a fairly complicated mathematical treatment. Methods of measurement have been reviewed by Hands[16] and test apparatus developed at the Rubber and Plastics Research Association (RAPRA) is described by Hands and Horsfall.[17]

Values of thermal diffusivity ranging from $7.5 \times 10^{-8} \, m^2/s$ for a natural rubber gum stock to $1.5 \times 10^{-7} \, m^2/s$ for a mix highly loaded with carbon black have been reported.[18] Hands reports that thermal diffusivity is a function of temperature and that for accurate prediction of temperature distributions in rubber products a relationship between thermal diffusivity and temperature of a second-order polynomial can be established by regression analysis.

2.4.4. Convective Heat-Transfer Coefficient

During many heating operations the rate of heat transfer to a body may be restricted so that the surface of the body rises in temperature for some time after the commencement of the heating cycle. For example, a cold solid suspended in a hot fluid cools the fluid in immediate contact with its surface and, if the fluid is steam, condensation may occur. A similar situation occurs in cooling operations; and a film of rust or scale on the inner surfaces of the fluid circulation channels of an item of processing equipment can seriously limit the efficiency of heat transfer to the fluid.

The convective heat-transfer coefficient is defined as the quantity of heat flowing per unit time normal to a surface across unit area of the interface, with unit temperature difference across the interface; or by

$$q = hA(T_w - T_\infty) \tag{2.41}$$

where h is the heat-transfer coefficient, T is the surface temperature, and T_∞ is the bulk fluid (rubber) temperature. If $T_w = T_\infty$, then h is infinite; but for practical purposes T_w can be assumed to equal T_∞ if $h \geqslant 500 \, W/m^2 K$.

Despite the importance of the heat-transfer coefficient for oven, autoclave, and continuous vulcanization, few values have been published. Apparently only those for air to rubber, water to rubber, and media to rubber in a fluidized bed have been reported[19]; these are 13–90 $W/m^2 K$, 570 $W/m^2 K$, and 590 $W/m^2 K$, respectively.

2.5. VULCANIZATION CHARACTERISTICS

2.5.1. Changes in Physical Properties due to Cross-Linking

The essential change in physical properties which occurs when chemical cross-links are inserted between adjacent molecular chains is from predominantly viscous behavior to predominantly elastic behavior. Cross-links confer dimensional stability by forming a three-dimensional network of rubber molecules and also result in the deformational properties becoming temperature insensitive over a broad and useful range. A further result is to render a rubber insoluble in liquids which were effective solvents for the unvulcanized rubber, although gross swelling of the rubber and loss of useful physical properties can still occur. Swelling by a solvent can be used to determine the cross-link density of a rubber,[20] provided that there are no compounding ingredients present which confuse the result.

Most commercial instruments used to measure the changes in cross-link density which occur during vulcanization do so by monitoring a physical property known to vary in direct proportion to cross-link density. The shear modulus is usually chosen for this purpose since[20]

$$G = NkT \qquad (2.42)$$

where T is absolute temperature, k is the Boltzmann constant, and N is the number of molecular chains per unit volume, which is directly proportional to cross-link density.

2.5.2. Prediction of Scorch or Onset of Cross-Linking

Since the requirement of scorch testing is the detection of the onset of cross-linking, when the material can no longer be subjected to laminar flow, rheometers and viscometers are suitable test instruments, in addition to the curemeters to be discussed later.

In the Mooney viscometer and the Monsanto oscillating disk curemeter, the scorch time is defined as that time for 5 Mooney units or 2 lbf. in^{-1}. (pounds-force inches) rise in torque above the minimum recorded value, respectively, at a given temperature. These are single-point measurements and for quality-control purposes a standard temperature is normally used. The Mooney instrument is usually set at 130°C, giving a reasonable compromise between testing at a typical processing temperature and the duration of the test. When a curemeter is used the scorch and cure tests are generally integrated, requiring that a higher temperature is used. For quality-control tests this temperature is usually in the region of 180°C, which, in

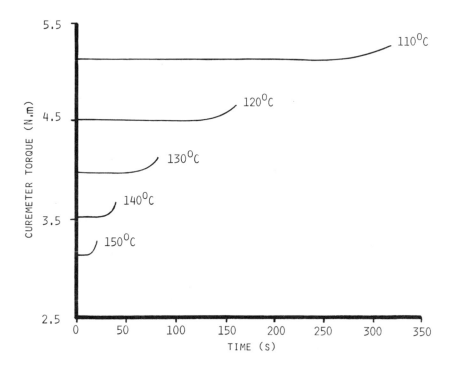

FIGURE 2.6. A family of idealized scorch curves (viscosity reduction during sample heating is not shown) taken over a range of temperatures.

most curemeters, results in the onset of cross-linking at the surfaces of the specimen long before a uniform temperature distribution has been achieved. Thus the test is dominated by heat-transfer effects and discrimination between mixes is very poor.

For process development and problem solving, single-point tests are often inadequate and the scorch characteristics are required over the range of temperatures encountered in the processing operations which constitute the manufacturing route. Testing within this range will generate a family of curves such as those shown in Figure 2.6. The information may then be transformed into the more useful form of the time – temperature – percent scorch (TTS) chart shown in Figure 2.7. This enables a quantitative assessment of the adequacy of processing safety (freedom from onset of cross-linking) of a rubber mix to be made, provided that an estimate can be made of the heat history of the rubber mix in each process. This is a function of the residence time and can be represented by a temperature – time profile, such as the one shown in Figure 2.8.

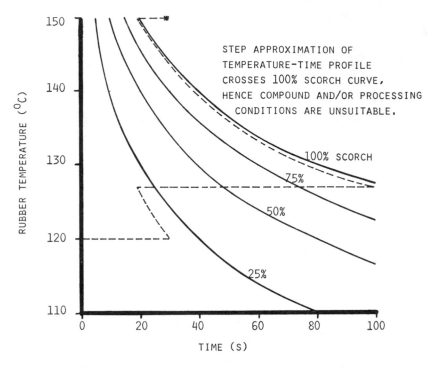

FIGURE 2.7. Time–temperature–percent scorch plot derived from scorch curves in Figure 2.6 and showing temperature–time profile transferred from Figure 2.8.

For transfer of the heat history to the TTS chart it is first necessary to approximate the former by a series of steps. Taking the first step of the temperature – time profile approximation, the time represented by this can be plotted on the TTS chart parallel to the time axis, starting from zero time and at the appropriate temperature. The increment to the next temperature takes place in zero time, therefore no progress is made towards the onset of cross-linking. For this reason the increment is drawn parallel to the nearest contour of constant percent scorch. After this a further line is drawn parallel to the time axis to represent the time spent at the next higher temperature, starting from the end of the contour line representing the increment from the lower temperature. This procedure is repeated until all the heat history has been transferred. Reducing the step size and increasing the number of steps in the approximation of the heat-history profile will obviously increase the accuracy of the result. In fact, the whole procedure can incorporated into a computer program.

When constructing the heat-history profile it must be remembered that

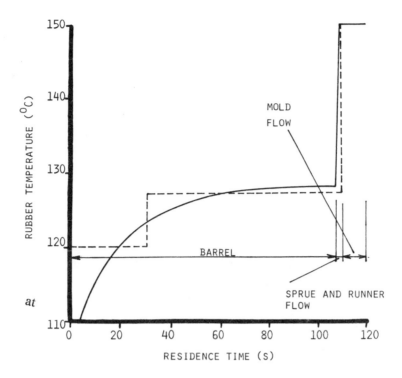

FIGURE 2.8. Rubber temperature/residence time profile for injection molding.

some processes, such as extrusion, can give a broad residence time distribution, requiring that the longer times in the distribution are used instead of the mean residence time. The practice of recycling scrap from processes prior to the vulcanization stage must also be taken into account to arrive at a profile which represents realistically the manufacturing route. If the transferred-heat-history profile then crosses the scorch line on the TTS chart, either the cure system, the processing conditions, or the manufacturing route requires alternation. Also, to avoid variations in the cure system combining with a "worst-case" heat history to give scorch, it is desirable to include a safety margin.

If a cure simulator of the type produced by Göttfert is available then the construction of a TTS chart or the development of a computer program can be avoided. It is possible to generate the temperature – time profile directly at the platens of a curemeter, and check, from the resulting torque *vs.* time plot, that the scorch criterion is not exceeded.

2.5.3. Curemeters and Interpretation of Results

Referring to Eq. (2.42) we see that a curemeter should be designed to measure the elastic properties of a rubber mix, requiring that the deformation of the test sample is quite different from that in a viscometer or rheometer. For viscous flow measurement, continous deformation is necessary. For measurement of elastic properties which are changing with time, a cyclic deformation is necessary. The amplitude of the cyclic shear stress is limited by the need to avoid slippage between the test sample and the instrument and to avoid rupture of the newly formed cross-links. Conversely, for particulate-filler reinforced mixes, a lower limit is set by the need for the stress amplitude to be sufficiently large to induce breakdown of the secondary filler–rubber links, which would otherwise mask the influence of the primary cross-links on the shear modulus.

The practical range of oscillatory frequencies is bound at the upper limit by the need to avoid significant viscous dissipation of energy, so that the set temperature can be maintained accurately; and at the lower limit by the need to gather enough information to permit the recording apparatus to construct an accurate curve. The latter constraint arises because the measurements must be made at the same points in each cycle, for comparison with preceding and succeeding measurements. This gives a maximum of two points per cycle, normally measured at the peak positive and negative values of amplitude.

The elastic shear modulus may be monitored by two alternative methods during a cure test. The values of the stress at the peaks of a constant strain amplitude waveform can be measured or, alternatively, the peak values of the strain cycle needed to maintain a stress waveform of constant amplitude can be measured. The former method is preferable because it eliminates the contribution of viscous effects to the stress measurement.[21] Instruments based on the latter principle have to operate at lower frequencies to minimize the influence of the viscous contribution to the measurement.

The constant strain cycle principle, which adapts well to modern instrumentation and recording methods, is embodied in the somewhat misnamed Monsanto oscillating disk rheometer, the Göttfert Elastograph, and the recently introduced Wallace isothermal curemeter. The former instrument, which has been standardized in BS 1673 Part 10[22] and in ASTM D2084-75,[23] has a design consisting of a biconical rotor in an enclosed die cavity whereas the latter two are rotorless instruments, using the oscillatory motion of the lower half of the die cavity to deform the test sample, as shown in Figure 2.9.

The general form of cure trace obtained from each of the instruments is

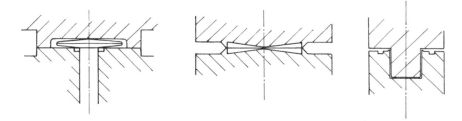

FIGURE 2.9. Test cavity configurations of curemeters. (a) Monsanto oscillating disk rheometer (b) Göttfert Elastograph. (c) Wallace isothermal curemeter.

similar, with torque, which is directly proportional to shear modulus, being plotted against time, enabling the interpretation of results suggested in the Standards[22,23] to be used in each case. However, the rotorless instruments have substantial advantages over the biconical-rotor design with respect to sample heating times. The unheated rotor acts as a heat sink, giving a heating time which is long in comparison with the normal range of cure times obtained at characteristic test temperatures (~180°C). Figure 2.10[24] shows cure curves for which the torque readings are converted to percent cure obtained from an isothermal method, in which the sample heating time is negligible in comparison with the cure time; from a Wallace-Shawbury Curometer, a constant stress cycle instrument in which the sample is small and the paddle deforming it has a low thermal mass; and from a Monsanto oscillating disk curemeter. It can be expected that the Göttfert Elastograph would give sample heating characteristics similar to those of the Wallace Shawbury Curometer, although it is a superior instrument in many other respects. The fast cure in Figure 2.10 emphasizes the differences in heating times between the instruments. These differences would be somewhat reduced for longer cure times, although they would never become insignificant.

The influence of sample heating times on scorch and cure measurements has important practical consequences for process monitoring and development. It has previously been mentioned that the scorch-time measurement can be dominated by heating time when testing at high temperature. The cure time and shape of the cure trace will be similarly influenced in instruments with long heating times, reducing discrimination in routine cure testing and producing inaccurate information for establishing vulcanization process conditions. For the curing of products with thicknesses significantly greater than the curemeter sample, the procedures described in Sections 6.2 and 6.3, involving heat-transfer calculations, are required to determine the conditions for optimum productivity. If essentially isothermal

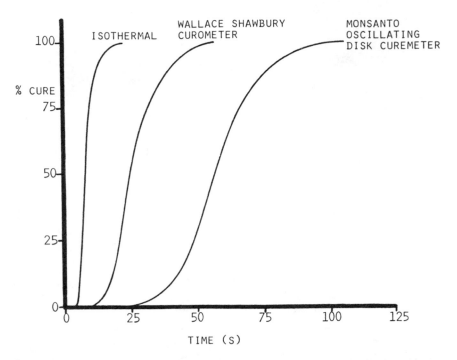

FIGURE 2.10. Comparison of results of isothermal method, Wallace Shawbury Curometer, and Monsanto oscillating disk curemeter at 200°C. (Courtesy Rubber and Plastics Research Association of Great Britain.)

conditions are not achieved in the cure test, the product cure time predicted will be longer than necessary. Also, in the cases of transfer and injection molding, where mold filling must be complete before the onset of cross-linking, the prediction of a spuriously long safe processing time can give serious problems.

In each of the constant strain cycle instruments referred to, the torque necessary to maintain the strain cycle is measured with a strain gauge load cell, amplified and plotted against time on a chart recorder. Torque is directly proportional to shear modulus, although the conversion to shear modulus is unnecessary for most purposes. The sketch cure traces of Figure 2.11 show the three main types of curve which are observed in practice. From each of these it is possible to obtain a cure time and a cure rate. For a plateau-type cure the cure time t_c' is defined as the time in minutes for the torque to increase to

$$(Y/100)(M_{HF} - M_L) + M_L \tag{2.43}$$

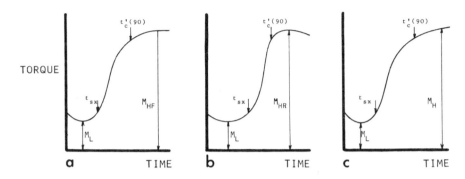

FIGURE 2.11. Typical curemeter traces. (a) Plateau cure. (b) Reverting cure. (c) Marching cure.

For a cure trace which shows reversion after reaching a maximum torque, M_{HF} may be replaced in Eq. (2.43) by M_{HR}; but if a pronounced degree of "marching cure" is encountered, then the calculation of a cure time is arbitrary in that the derived cure time will depend on the specification of M_H. The quantity Y in Eq. (2.43) is the percentage of full cure required; this percentage is usually 90% for a "best technical cure," where the best compromise for a number of cure-related properties is assumed to occur. This emphasizes the fact that cure testing only optimizes the cure time with respect to shear modulus. Other properties, such as stress relaxation, strength, and fatigue life will probably have different optima, which should be taken into consideration when selecting vulcanization methods and setting conditions.

For routine testing there are four basic parameters which need to be determined and checked against tolerance bands for a go/no go decision. These are:

1. Scorch time.
2. Cross-link insertion rate.
3. Cure time.
4. Maximum — minimum torque.

The minimum torque M_L is used in the determination of cure time, scorch time, and the increase in shear modulus during vulcanization (item 4); but it is not quoted as an independent parameter since it only measures the low-strain elastic behavior of the test sample and is not a useful prediction of processing (flow) behavior. Another testing machine, such as the Sondes Place Research Institute TMS rheometer, should be used to assess processing behavior. The cross-link insertion rate is defined by the slope of the cross-link insertion curve, taken at its steepest point. BS. 1673 Part $10^{(22)}$ presents an expression which gives the cross-link insertion rate in terms of "change in percent cure per unit time," but for routine testing torque units can be used. The cure time of a compound having a marching cure can also be defined by slope. Although the torque will continue to increase with time, its slope will decrease progressively, enabling a slope to be specified at which the test will be terminated.

Calculation of the four parameters and comparison with their upper- and lower-limit values can easily be achieved with a microcomputer, which will then provide a go/no go decision. By this method the need for the provision of chart paper and interpretation of a graphical trace is avoided, except when a "no go" decision is given. At this point a printout of the cure trace is required for decisions on disposal and corrective action by a technologist.

REFERENCES

1. Stevenson, J. F., *Am. Inst. Chem. Eng. J.* **18**, 540 (1972).
2. Trouton, F. T., *Proc. R. Soc.* London, Ser. A **77**, 426 (1906).
3. Clegg, P., Contribution to *The Rheology of Elastomers*, ed. by P. Masos and W. Wookey, Pergamon, London (1957).
4. White, J. L., and N. Tokita, *J. Appl. Polym. Sci.* **10**, 1011 (1966).
5. Turner, D. M., M. D. Moore, and R. A. Smith, Contribution to *Elastomers: Criteria for Engineering Design*, ed. by C. Hepburn and R. J. W. Reynolds, Applied Science Publishers, London (1979).
6. ASTM D3346-74, "Processability of SBR with the Mooney Viscometer" (1974).
7. Moore, M. D., Private Communication (1979).
8. Moore, M. D., and D. M. Turner, Paper given at the Plastics and Rubber Institute/British Society of Rheology Conference, "Practical Rheology in Polymer Processing," Loughborough University, U.K. (March, 1980).
9. Fenner, R. T., and F. Nadiri, Private Communication (1979).
10. Lenk, R. S., *Polymer Rheology*, Applied Science Publication, London (1978).
11. Targiel, G., "Practical Rheology in Polymer Processing," Paper given at Joint Plastics and Rubber Institute/British Society of Rheology Conference, Loughborough University, U.K. (March, 1980).
12. Lupton, J. M., and J. W. Regester, *Polym. Eng. Sci.* **5**, 235 (1965).
13. Bagley, J., *J. Appl. Phys.* **28**, 626 (1957).
14. BS 874, "Determining Thermal Insulating Properties" (1973).
15. Hands, D. and F. Horsfall, *J. Phys. E. Sci. Inst.* **8**, 687 (1975).
16. Hands, D., *Rub. Chem. Technol.* **50**, 480 (1977).
17. Hands, D. and F. Horsfall, *Rub. Chem. Technol.* **50**, 253 (1977).
18. Hills, D. A., *Heat Transfer and Vulcanisation of Rubber*, Elsevier, London (1971).
19. Griffiths, M. D., and R. H. Norman, *RAPRA Members J.*, July/August, p. 87 (1976).
20. Treloar, L. R. G., *Physics of Rubber Elasticity*, Oxford Univ. Press, London (1975).

21. Freakley, P. K., and A. R. Payne, *Theory and Practice of Engineering Design with Rubber*, Applied Science Publishers, London (1978).
22. BS 1673 Part 10, "Measurement of Pre-Vulcanising and Curing Characteristics by Means of Curemeters" (1977).
23. ASTM D2084-75, "Vulcanisation Characteristics Using the Oscillating Disk Curemeter" (1975).
24. Hands, D., and F. Horsfall, *Kaut. und Gummi Kunst.* **33** (6), 440 (1980).

3

Principles of Mixing and Internal Mixers

3.1. INTRODUCTION

Mixing, being the first step in a sequence of operations, determines the efficiency with which subsequent processes may be carried out and exerts a considerable influence on product performance. Adequate and consistent mixing is a prerequisite for successful manufacture.

The type of system used for the mixing of rubber depends on both the form of the raw materials and the scale of the operation. The majority of rubber is supplied in bale form, dictating the use of batch mixing. In medium- and large-scale mixing, systems based on the internal mixer are used throughout the industry, while small-scale mixing is usually carried out on a two-roll mill.

The mixing of rubber is a complex operation and is very difficult to quantify. This complexity is a result of the viscoelastic behavior of the rubber and of the nature of the materials with which it is required to be mixed. Particulate fillers are not masses of simple particles but consist of groups of particles called agglomerates, which must be broken down and uniformly distributed throughout the rubber during mixing. Carbon blacks are pelletized, making the breakdown more difficult, and also consist of primary aggregates—clusters of particles which survive the mixing operation and influence the behavior of the finished product.

Liquids, in the form of oils, waxes, and plasticizers, are also added to rubber mixes and generate their own problems for mixer operation.

This chapter is primarily concerned with the internal mixer. However, the principles of mixing rubber with particulate materials and liquids are similar for all rubber mixing systems and common stages in the conversion of the raw material to a finished mix can be identified. In addition, most mixing systems consist of an internal mixer and another machine—either a two-roll mill or an extruder/continuous mixer. The whole system must then be considered when setting the conditions necessary to produce an adequately mixed material.

TABLE 3.1
Francis Shaw Intermix Internal Mixers

Machine size		K0	K1	K2	K2A	K4	K5	K6	K7	K8	K10
Approximate capacities in liters											
Chamber volume with standard rotors											
Useful		1.0	3.0	11	27	50	80	112	190	285	550
Total		1.8	5.3	18.8	45.5	84	132	189	315	475	915
Chamber volume with new rotors 1											
Useful		1.33	4	14.6	36	66.5	106	149	253	380	732
Total		2.0	6	22	54.5	100	160	225	383	574	1100
Chamber volume with new rotors 2											
Useful		1.15	3.45	12.65	31	5715	92	129	218	328	633
Total		1.82	5.5	20	49	91	146	205	346	520	1005
Approximate net weight in tonnes excluding motor	Sturdigear drive	2.5	4.5	5	11.5	16.5	22.5	32	—	—	—
	Unit drive	—	—	7	14	20	28	43	65	78	105

Note: Variable-speed dc drives are usually fitted to the above machines, with power ratings according to application.

Outline specifications of internal mixers from two major manufacturers of these machines are given in Tables 3.1 and 3.2. The smallest machines in each range are laboratory models, used mainly for compound development and mixing of test batches. General-rubber-goods manufacturers usually have mixers with chamber volumes in the range 40–250 liters while tire manufacturers, due to their larger volume requirements, tend to use machines in the range 250–700 liters. It should be noted in Tables 3.1 and 3.2 that a chamber volume is given for the Farrel Bridge machines; but the volume of material which can be mixed is specified for the Francis Shaw machines, which is 60–75% of the chamber volume.

3.2. THE MECHANISMS OF MIXING

The mixing of rubber is a composite operation, involving a number of different mechanisms and stages. These can be resolved into four basic processes:

TABLE 3.2
Farrel Bridge Banbury Machines—'F' Series

Machine size	F.80	F.80UD[a]	F.120	F.120UD	F.160	F.160UD	F.270	F.270UD	F.370UD	F.620UD
Mixing chamber volume (liters)	80	80	120	120	160	160	270	270	370	620
Approximate batch weight at specific gravity of 1.0kgs	60	60	90	90	120	120	202	202	277	465
Standard mixing speeds (rpm)	35 70 105	35 70 105	30 60 90	30 60 90	30 40 60	30 40 60	20 30 40	20 30 40 60	20 30 40 60	40 50 60
Typical powers at above speeds (kW)	130 260 390	150 300 450	168 336 504	225 450 675	225 300 450	375 560 750	260 400 520	500 750 1000 1500	550 850 1100 1700	2200 2000 3000
Normal maximum power per rpm (including motor and gearbox)	3.7	6.3	5.6	9	7.5	12	13	24.5	37	51.5
Length (meters)	5.5	5.8	6.0	6.0	6.2	6.2	7.0	7.3	6.4	11.0
Width (meters)	3.0	3.0	3.5	3.5	3.7	4.0	4.0	4.3	4.6	4.6
Height	4.3	4.9	5.0	5.0	4.9	5.2	5.5	6.0	6.4	6.7
Approximate weight excluding motor (tonnes)	16.6	20.3	22.8	28.0	23.0	35.0	32.8	48.0	82.0	115.0
Approximate cooling water requirement (liters per minute)	340	340	400	450	460	575	680	680	920	920

[a]UD = unit drive.

1. Viscosity reduction.
2. Incorporation.
3. Distributive mixing.
4. Dispersive mixing.

Each of these can occur simultaneously and each can be the main rate-determining process, which will control the mixing time. The mixing time will depend on the type of compound being mixed and the mixing conditions.

When a charge of highly elastic rubber is fed into a mixer it must be rapidly converted to a state in which it will accept particulate additives. This stage is called viscosity reduction and is achieved by three interdependent mechanisms: temperature rise, chain extension, and mastication. Because of rubber's high viscosity and elastic stiffness, the initial deformation of the rubber in a mixer requires considerable mechanical energy, which is converted to heat causing a rapid temperature rise and viscosity reduction. A concurrent viscosity reduction also occurs purely by chain extension and the freeing of untrapped chain entanglements[1]; which is the origin of the pseudoelastic or power-law behavior described in Chapter 2. The dependence of apparent viscosity on both temperature and rate of deformation is described by Eq. (2.7). In some rubbers, notably natural rubber, the viscosity is irreversibly changed by chain scission, whereas the changes due to a rise in temperature and chain extension are recoverable. Consequently, viscosity reduction from chain scission (mastication) influences mixed-compound behavior and must therefore be controlled. Despite adding yet another variable to the mixing process, mastication provides the opportunity of achieving uniform mixed-material properties from a variable feedstock.

As the viscosity and elasticity of a rubber are reduced the rubber can be caused to flow around additives, incorporating and enclosing them in a matrix of rubber. The efficiency of incorporation is dependent upon free-surface "folding" flows being induced in the rubber by the mixer, overlapping and enclosing volumes of additives.[2]

Incorporation and distributive mixing generally proceed simultaneously, the latter commencing as soon as incorporated additives are available for distribution. These two types of mixing are accompanied by subdivision, in which the size of the volumes of additives is progressively reduced. Incorporation, subdivision, and distribution are largely due to exponential mixing mechanisms. Folding flows provide exponential mixing, as does the separation and recombination of flow streams in different patterns. In contrast, laminar mixing is extremely inefficient, as shown in Figure 3.1. The complex geometries of practical mixers reflect the requirement for exponential mixing,—to achieve an acceptable uniformity of distribution of

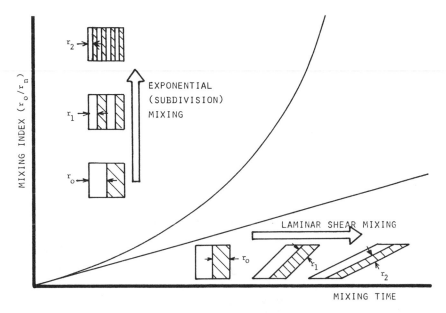

FIGURE 3.1. Comparison of efficiencies of exponential mixing and laminar shear mixing, as measured by reduction of striation thickness of a two-component blend from an intial thickness r_0.

additives throughout a rubber mix in a short mixing cycle. Expressing the effect of exponential mixing in a simple equation,

$$\text{index of distributive mixing} = \text{const} \times e^N \qquad (3.1)$$

where N is the number of revolutions of the rotors or screw of the mixer. Given the form of the expression, the mixing index will rise very quickly to a value at which the distribution of additives will be adequate for practical purposes.

Exponential mixing is also necessary for the absorption of liquid additives. The penetration rate of a liquid into a rubber is proportional to the square root of time. Hence the depth of penetration in the time scale presented by typical mixing cycles is very small, leaving the generation of large interfacial surface areas between the liquid and the rubber as the controlling factor. A rapid incorporation of liquids can often be achieved by absorbing the liquid into the interstices of filler-particle agglomerates, although liquids so absorbed must be largely replaced by rubber for acceptable mix properties.

During incorporation, subdivision, and distributive mixing the rubber flows around filler–particle agglomerates and penetrates the interstices between particles in the agglomerates.[3,4] This action has two effects. First, due to the "wetting out" of the filler by the rubber and reduction of voids, the rubber mix becomes less compressible and its density increases. Second, the rubber which has penetrated the interstices becomes immobilized and is no longer available for flow. Medalia[5] refers to this as occluded rubber and points out that immobilization reduces the effective rubber content of the mixture. This reduction has the effect of increasing viscosity; the incompressibility of the mixture now allows high forces to be applied to the particle agglomerates, causing them to fracture. This action is termed dispersive mixing, and will continue while the forces being applied to particle agglomerates, of both fillers and minor additives, are sufficient to cause fracture. However, these forces are progressively reduced, by the release of occluded rubber due to agglomerate fracture and by the reduction in agglomerate size. If a rise in temperature accompanies mixing, the mixture viscosity will be further reduced and the efficiency of dispersive mixing will decrease more rapidly. Surprisingly, increasing the rate of deformation of a mixture, by increasing rotor or screw speed, does little to improve dispersive mixing and can result in a deterioration, by raising mixture temperature. This fact can be attributed to the strongly non-Newtonian behavior of rubber, enabling large increases in rate of deformation to be achieved with small increases in stress.

Distributive mixing occurs concurrently with dispersive mixing, which serves the purpose of separating the fragments of agglomerates once they have been fractured.

Many particulate additives are pelletized or produced in flakes and other forms which give rapid incorporation. In the absence of a mastication stage, mixing time is generally dictated by distributive mixing when large particle size diluent or semireinforcing fillers are used; and by dispersive mixing when reinforcing fillers, particularly carbon blacks, are used. Mohr[6] also points out that the smaller the volume fraction of a minor additive, the more mixing is needed to ensure a uniform distribution of that additive. It is more difficult to mix a small amount into a large amount than it is to achieve an acceptable 50–50 mixture.

3.3. ELEMENTS OF INTERNAL MIXER DESIGN

3.3.1. Rotor and Chamber Designs

There are two basic designs of internal mixer available—those with meshing rotors and those with nonmeshing rotors. As shown in Figure 3.2 the mixing chambers have a similar internal shape in each design, differing only in the distance between the centers of rotation needed to accommodate the two different rotor types.

Both rotor designs take a complex form to give good distributive mixing and have helically set projections (rotor wings or nogs) to give axial transfer of the material in the mixing chamber for this purpose. The main difference appears in the method of achieving the high forces in the rubber necessary for effective dispersive mixing. In the nonmeshing design this is carried out in the clearances between the chamber wall and the rotor, in the areas approaching and at the rotor tip. The efficiency of dispersive mixing of this type of design depends on both the angle of lead-in to the rotor tip and the tip clearance; and service life is determined by wear in these areas. With intermeshing rotors, dispersive mixing is also achieved in the region between the rotors. As Figure 3.3 shows, the projections or nogs mesh with the smaller radius of the rotor body. Since both rotors turn at the same angular velocity this gives a difference in the surface speed of each rotor and results in a "friction ratio" similar to that of a two-roll mill, with similar high forces being generated in the material.

A number of designs for nonmeshing rotors are available; those of the Farrel Bridge Company show an interesting progressive development. Initially, this company installed in their machines a two-wing rotor of the type shown in Figure 3.4(b). Later the four-wing rotor of Figure 3.4(a) was developed to increase the area dedicated to dispersive mixing, with a view to reducing mixing times and/or improving the uniformity of the resulting rubber mixes. By increasing the dispersive mixing in this way heat generation is also increased, which requires the new rotor design to give improved extensive mixing and to improve the efficiency of transfer of material to the cooling surfaces of the chamber and rotors.

The rate of heat extraction from the rubber mix is a limiting factor on mixing speed due to the very poor thermal conductivity of rubber. The only way to achieve effective cooling in an internal mixer is to arrange the mixer geometry and operating conditions so that the flow paths effectively and continuously expose new rubber surfaces at or very near the chamber and rotor surfaces. This factor applies equally to all mixer designs and the mode of heat transfer is termed forced convection.

3.3.2. Ram and Door Configurations

As mixing times are shortened more attention must be given to the loading and discharging of materials in order to obtain a correspondingly reduced "dead time" between active mixing stages. This requires that the entry to and exit from the mixer be large. A rapid and repeatable loading of materials is also essential with short mixing cycles to obtain consistent mixing. In fact, considerable increases in the size of the hoppers, throats, and

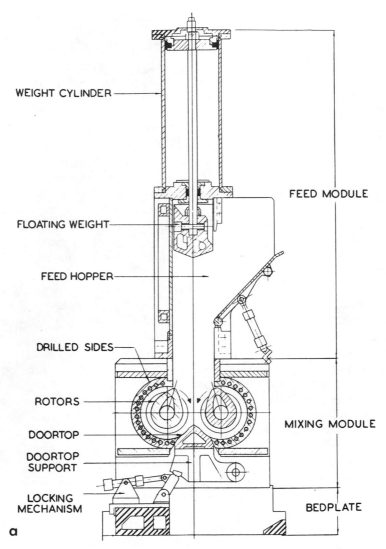

WEIGHT CYLINDER

FLOATING WEIGHT

FEED HOPPER

DRILLED SIDES

ROTORS

DOORTOP

DOORTOP
SUPPORT

LOCKING
MECHANISM

a

FEED MODULE

MIXING MODULE

BEDPLATE

FIGURE 3.2. Cross sections of internal mixers. (a) Farrel Bridge Banbury showing tangential rotors. (b) Francis Shaw Intermix showing interlocking rotors.

discharge doors of mixers have been made in recent years. These changes of size have led to the ram and discharge door (see Figure 3.2) forming a much larger part of the mixing chamber.

As ram and door sizes have increased, increased importance has been given to shaping them to give preferred flow paths and to eliminate stagnant

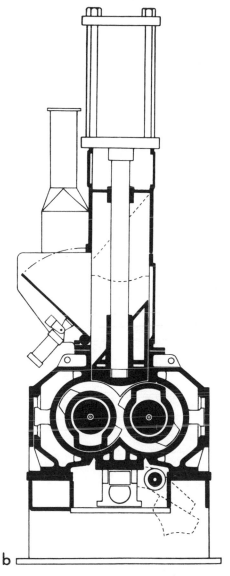

FIGURE 3.2 (continued)

areas, and to the size of the pneumatic cylinder needed to maintain the ram pressure on the material being mixed. The forces which the discharge door must withstand have also increased as a result of its larger area. The use of the drop door in preference to the earlier sliding door used on Banbury machines stems partly from mechanical-engineering considerations but

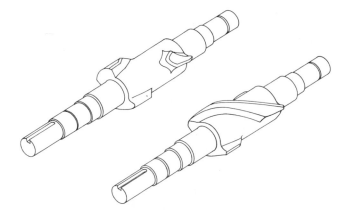

FIGURE 3.3. Francis Shaw Intermix rotors.

mainly from operational requirements. The contamination associated with "hangup" of material and with oil from the slideways of the sliding doors has been eliminated by the use of the drop door. The configurations of the doors used with machines having meshing- and nonmeshing-rotor types are slightly different; but both designs need to ensure good flow over the mix thermocouple which is mounted to project from the doortop, and to avoid forces of a magnitude which would damage the door mechanisms.

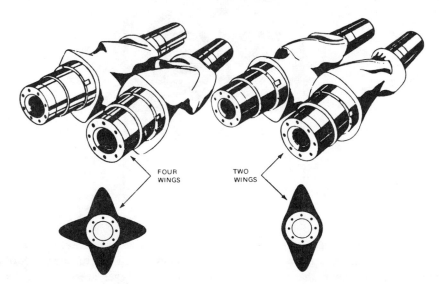

FIGURE 3.4. Farrel Bridge Banbury rotors. (b) Two-wing rotors. (a) Four-wing rotors.

3.3.3. Rotor Shaft Seals

The alternative name of dust seals defines the major function of shaft seals—they prevent the escape of particulate ingredients from the mixing chamber. The seals are a critical factor in mixer performance since they involve small clearances, where considerable heat buildup can occur due to the ingress of rubber. The generation of local hot spots around the seals has often led to cured lumps forming during high-speed mixing of batches containing curatives, thus placing an upper limit on the productivity of the machine.

Shaft seals are now available which generate little frictional heat. The Farrel Bridge machines are fitted either with the SSA type, which is generally adequate up to rotor speeds of 40 or 50 rpm (for an 11 D or equivalent machine), or the FYH type, which is designed for high-speed operation.

The Francis Shaw Intermix machines, which run at rotor speeds of up to 66 rpm, use a lubricated U-ring restrictor-type seal. The SSA seal relies on the pressure of the mix to provide the sealing force and also to force a small quantity of the rubber mix into the cavity. This rubber, which is softened by an oil feed to limit the temperature rise, prevents the entry of powders into the seal. The lubricated U ring also works on a controlled leaking principle to prevent the ingress of powders. Type FYH seals are hydraulically loaded positive contact seals which compensate for the small axial movements of the rotor during mixing.

3.3.4. Temperature-Control Facilities

Temperature control of an internal mixer usually refers to the maximum rate at which heat can be extracted from the process, although improvements in mixing resulting from accurate temperature control of the water circulated for cooling have been reported.[7] In either case the efficiency of the system is dependent on the transfer of heat from the metal surfaces in contact with the rubber mix to the circulating water and vice versa.

The temperature-control problems associated with high-speed mixing can be acute and are often the limiting factor on mixer performance. Every available contact surface between the rubber and the machine is utilized to achieve a high rate of heat transfer. The rotors and the curved surfaces of the mixing chamber are invariably used. The ram and hopper door have water passages and the end frames are sometimes cooled to alleviate the problem of heat buildup in the shaft seals. The modes of water circulation for the mixing chamber of the Farrel Bridge Banbury machines have developed progressively from a spray technique, where water was sprayed onto the outer surfaces of the chamber walls, through the jet method, where water was circulated in channels at the outer surfaces of the mixing chamber, to drilled passages, which represent the latest development. Francis Shaw Intermix machines generally have flood cooling, with substantial reservoirs of water adjacent to the outer surfaces of the mixing chamber; but they offer drilled passages as an alternative for conditions giving high heat generation. The drilled passages and flood-cooling reservoirs are shown in Figures 3.2(a) and 3.2(b), respectively where the proximity of the drilled passages to the inner surfaces of the mixing chamber can be seen clearly. It is this proximity, coupled with a high rate of water circulation to give turbulent flow, which makes drilled passages the most effective for heat transfer.

The rotors of the Banbury and Intermix have very different constructions. The Banbury rotors are cast as a hollow unit with a large internal volume, while the Intermix rotors are heat shrunk onto a machined shaft, giving an annular water passage of a much smaller volume. The smaller volume makes it easier to ensure an active circulation to all the metal surfaces. Special measures are required in the cast hollow design to avoid stagnation of flow in the areas of the rotor tips due to the inertial forces caused by rotation. This stagnation of flow is obviously a greater problem at high rotor speed and a spray pipe design is now available (Figure 3.5) which directs a high-velocity jet of water into the regions of potential stagnation. The passages in the ram and door for water circulation are shown in Figure 3.2 to be of the flood-cooling type. Again, drilled passages may be used if the process requires the improved heat transfer. The Francis Shaw Company reports an improvement of heat-transfer rate in the region of 33% going from a machine with flood cooling to all surfaces to one with drilled passages.

Most mixers are supplied with facilities for water cooling or steam heating. Water-tempering systems are a relatively recent innovation.[7] Where these systems are used they usually take the form of the system of Figure 3.6, and a separate circuit is used for the chamber, the rotors, and the door (where the latter is used). In Figure 3.6 the circuit has a pump which circulates the water through the passages of the mixing chamber; a ther-

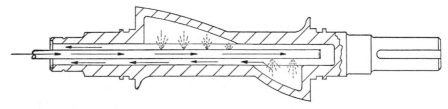

FIGURE 3.5. Spray pipe arrangement for cooling Farrel Bridge Banbury rotors.

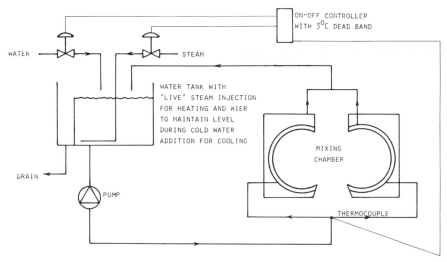

FIGURE 3.6. A typical water-tempering circuit.

mocouple detects the temperature of the water into the chamber. If the water is too cold, steam is admitted via a diaphragm valve until the temperature setpoint of the controller is reached. Water-tempering systems were first introduced to overcome the problem of condensation forming on rotor and chamber surfaces as a result of the very effective cooling occurring between mixes when drilled water passages are used. These systems also render a mixer independent of the effects of day–night and summer–winter temperature cycles and can substantially reduce the "first-batch effect" caused by the startup of a cold mixer.

3.3.5. Drive Systems

Internal mixers are normally supplied with single- or two-speed drives and with a number of options available for these speeds. Variable-speed units are uncommon, although the versatility of a variable-speed mixer for a wide range of mix types is generally acknowledged.

The gearing arrangements are of two main types, both involving substantial gearboxes. For high torques a unit drive is used, having a separate drive shaft from the gearbox to each rotor. For lower torques a simpler arrangement is possible, with a single shaft from the gearbox to one rotor and transmission between rotors by gear wheels mounted directly on the rotor shafts. Oil coolers are often fitted to new main drives to permit them to work in a soundproof enclosure when legislation regarding noise levels is enforced.[7]

3.4. PRACTICAL MIXING VARIABLES

3.4.1. Criteria for Material Addition and Batch Discharge

There are three main criteria used to determine when an additive or filler should be charged into an internal mixer or when mixing should be terminated. These are mixing time, batch temperature, and mixing energy. In process-control terms mixing time is an open loop, providing no compensatory adjustment for the influence of relatively uncontrolled variables, such as mixer temperature and feedstock properties, on mixer performance. Batch temperature, measured from the thermocouple set in the mixer drop door or end frame, is related both to the amount of energy expended in mixing and the rate at which it is extracted. Despite being influenced strongly by variations in mixer temperature,[8] it is sensitive to mixing performance and provides a good indication of state of mix. Mixing energy, which is invariably taken to be the electrical energy delivered to the motor, minus that required to run the mixer in an empty condition, is measured by a power integrator. This criterion again provides a reasonable indication of mixing performance, although it does not take into account the rate of energy input or power. This rate is proportional to rotor torque in constant speed mixing and is therefore directly related to the forces applied to particle agglomerates during dispersive mixing.

Mixing time and energy are often used to determine when materials should be charged into the mixer during conventional or sequential addition mixing, whereas temperature is normally used only as a dump or discharge criterion. Additionally, the electrical-power requirement of the mixer motor is often used as a criterion for material additions and particularly for oil additions, usually immediately after the power peak due to dispersive mixing has been reached.

All of the mixing criteria identified here are indirect indicators of state of mix and therefore have to be related to the material properties required in the mixed batches, for downstream processing and product performance. Their value depends on their sensitivity for predicting changes in material behavior during mixing and the extent to which they can be influenced by other variables. Ideally, a mixing criterion should only be sensitive to mixed material properties. These attributes are best displayed by mixing energy; and commercial instruments are available to generate mixer-control signals when preset energy levels are reached. It has also been pointed out that the first-batch effect, whereby the first four or five batches in a production run are dissimilar from those following, is substantially reduced by using the mixing-energy criterion.

3.4.2. Material Input Sequence

The choice of a material input sequence to a mixer can exert a profound influence on the efficiency of mixing; and hence on productivity and the properties of the resulting mix. As a general guide, the number of material additions in a mixing cycle, which involve raising and lowering the ram, should be minimized, due to their time content. With this in mind, the requirements of different types of compounds can be examined.

General-rubber-goods compounds normally have diluent or semireinforcing fillers which present few problems for dispersive mixing, resulting in distributive mixing being the rate-determining stage. Unless a natural rubber requiring mastication is being used, or the compound includes a high proportion of oil, it is preferable to add the ingredients with the rubber at the start of the mixing cycle, to commence their distributive mixing as early as possible. Provided that mixer cooling is effective and a suitable rotor speed is chosen, the curatives and accelerators can usually be added without danger of scorch. Referring again to Mohr's statement that "the minor ingredients are more difficult to distribute than major ones,"[6] the discharge or dump criterion can be set by the in-batch variation of the cure behavior. The value of the dump criterion can be established where the standard deviation s or the variance s^2 of the maximum curemeter torque or the cure time indicate that their distributions fall between acceptable limits. Using

$$s^2 = \frac{1}{n-1} \sum_{i=1}^{n} (x_i - \bar{x})^2 \tag{3.2}$$

where n is the number of samples taken from a batch mixed to a specified value of the dump criterion, x_i is the value of the ith sample, and $\bar{x}$ is the average value of n samples, for a number of batches mixed to different dump-criterion values, a plot of the type shown in Figure 3.7 can be constructed.

For mixes having substantial proportions of oil, the main problem is to prevent the oil from lubricating the surfaces of the rotors and chamber wall, causing gross slippage and preventing effective mixing. This problem is often overcome by adding the oil and bulk filler together, prior to the rubber, in an upside-down mixing method and allowing time for the oil to be absorbed onto the filler surface and into the interstices between the particles in agglomerates. The amount of free oil is then sufficiently reduced to enable effective mixing to start immediately after the rubber is added.

For tire and conveyor-belt type compounds containing substantial quantities of reinforcing carbon black, dispersive mixing is usually the rate-determining stage. The material input sequence should therefore be chosen to

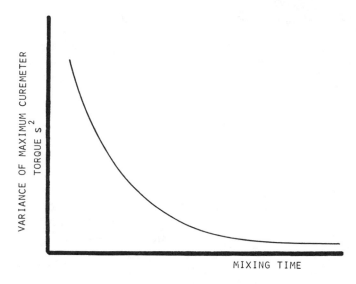

FIGURE 3.7. Practical determination of in-batch uniformity as a function of mixing time.

maximize the forces acting on filler agglomerates. This maximization is best achieved by withholding oils, waxes, and fatty acids from the early stages of the mixing cycle and charging the mixer with only the rubber and bulk filler, in addition to any other particulate additives appropriate to the mixing temperatures. These should all be charged into the mixer at the same time, unless mastication of a natural rubber is required. When adequate dispersive mixing has been achieved or its efficiency has been much reduced by a temperature rise in the rubber, the oil and other viscosity-reducing ingredients can be added. For constant-speed mixing this point is usually in the region where the power required by the mixer motor is decreasing rapidly after the dispersive mixing peak or plateau, as shown in Figure 3.8. The mixing cycle can be terminated when the oils, waxes, and fatty acids, in addition to any other ingredients withheld to minimize their residence time at an elevated temperature, are adequately distributed. Alternatively, some highly reinforced compounds may need additional mixing, enabling temperature-sensitive additives to be included at a later stage. These techniques will be dealt with in Section 3.4.7.

3.4.3. Rotor Speed

Dispersive and distributive mixing generate conflicting requirements with respect to rotor speed. The rate of distributive mixing is a function of

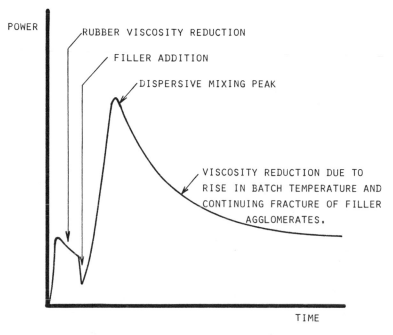

POWER

RUBBER VISCOSITY REDUCTION

FILLER ADDITION

DISPERSIVE MIXING PEAK

VISCOSITY REDUCTION DUE TO
RISE IN BATCH TEMPERATURE AND
CONTINUING FRACTURE OF FILLER
AGGLOMERATES.

TIME

FIGURE 3.8. Power–time trace for dispersive mixing.

rotor speed, proceeding more rapidly as speed is increased; but to retain a high viscosity in the rubber for dispersive mixing it is desirable to run a mixer slowly, to minimize the rise in batch temperature. This restriction on batch temperature is also necessary when mixing temperature-sensitive additives, such as curatives, placing a practical limit on the rotor speed for distributive mixing. Using both the mixing-energy and batch-temperature dump criteria, the mixed-material properties which depend on filler dispersion, such as hysteresis and fatigue life, are generally improved as rotor speed is reduced, at the expense of longer mixing times. However, low rotor speeds, by improving filler dispersion, may enable the requirement for a second dispersive mixing operation to be avoided and, due to the low batch temperature, allow curatives to be added, considerably reducing the total cost of mixing and the associated batch handling.[9]

3.4.4. Fill Factor

Fill factor defines the proportion of the mixing chamber volume occupied by the finished mix, that is, the material volume calculated from the weights and densities of the compound ingredients.

Underfilling of the mixing chamber is essential for efficient mixing[2] and fill factors in the range 0.65–0.85 are generally used, depending on mix type. Highly reinforced mixes, such as tire tread and conveyor-belt-type compounds, mix more successfully toward the lower end of the range, while moderately reinforced general-rubber-goods compounds, particularly those having a large proportion of oil, can be mixed effectively at the upper end of the range.

Within the range 0.65–0.85 it is possible to retain the batch in regions where active mixing occurs. Very low fill factors are obviously uneconomic and excessively high fill factors result in material remaining in the "throat" of the mixer and not taking part in the mixing. Underfilling of the mixer also results in voids forming in the rubber mass behind the rotor wings or nogs, providing effective mechanisms for exponential distributive mixing, by free-surface folding flows and by flow stream division. As the fill factor is increased there is a transition from exponential mixing to laminar mixing, with a consequent increase in the mixing time needed to achieve an adequate distribution of additives throughout the rubber.

The efficiency of heat transfer from a mix is also controlled by the mechanisms which govern distributive mixing. Free-surface folding flows provide an excellent means of transferring material to the cooling surfaces of the rotors or chamber for heat exchange, utilizing the mode of heat transfer known as forced convection. As the fill factor is increased the mix tends to flow in closed streamlines, resulting in conductive heat transfer becoming predominant over forced convection; thus very inefficient cooling in a poor conductor of heat such as rubber results. The practical consequence of this influence of fill factor on heat transfer is that batch temperature, measured at either a constant mixing time or energy, increases substantially as fill factor is increased, over the whole range of practical fill factors.[9]

3.4.5. Ram Pressure

It has been noted that in modern internal mixers the ram forms a substantial part of the total surface area of the mixing chamber, giving it a greater influence on mixing performance in comparison with older machines. The force applied to the ram should ensure that materials charged into the mixer engage rapidly with the rotors and be sufficient to prevent the subsequent upthrust of the batch from displacing it upwards, producing a stagnant region similar to that resulting from an excessively high fill factor. The upthrust is strongly dependent on fill factor, requiring high compressed-air-line pressures and large pneumatic cylinders to provide the ram force necessary for working at high fill factors. Often, available ram force places a practical limit on fill factor.

During normal operation upthrust on the ram is cyclic, varying as the rotor wings or nogs sweep past it, resulting in a *small* amount of cyclic movement. Whitaker[10] shows, by a series of trials on a Francis Shaw K2A Intermix, that the effect of increasing ram pressure follows a trend of diminishing returns. Once sufficient pressure is available to limit ram travel to a few centimeters, little is gained by applying more pressure. However, a substantial reduction in ram pressure causes a radical drop in mixing efficiency; and many mixers are protected against overload by an automatic ram pressure release coupled to the motor ammeter.

3.4.6. Circulating Water and Batch Temperature

The temperature of the mixer exerts a strong influence on the characteristics of mixing. This is demonstrated by the first-batch effect, where the physical properties of batches produced immediately after startup are substantially different from those subsequently produced when the mixer has achieved its "operating temperature," particularly when mixing time or batch-temperature dump criteria are used. During startup heat transfer from the batch is extremely efficient, due to the considerable mass of metal in the rotors and chamber acting as a heat sink, delaying the rise in batch temperature. The first-batch effect and subsequent variations in the properties of mixed batches are reduced by controlling the temperature of the water circulated through the chamber, rotors, and drop door for cooling. Traditionally, mixers have been fed with cooling water at a relatively uncontrolled "ambient" temperature, giving day–night and summer–winter cyclic variations, as well as less regular fluctuations due to local weather.

Water-tempering systems, which usually enable the circulating-water temperature to be controlled in a range extending from the feed-water temperature up to approximately 80°C, have brought substantial improvements in the batch-to-batch uniformity of mixing. However, it must be remembered that only the temperature of the circulating water is being controlled directly. The temperature gradients through the chamber and rotors, and the batch temperature, will also depend on other mixing variables.

When mixing to a constant energy, raising the circulating-water temperature has the effect of slightly decreasing the mixing time but causing a small deterioration in mixed-material properties.[9] Mixing to a batch-temperature dump criterion magnifies these trends, due to the strong dependence of batch temperature on circulating-water temperature.

For most mixes the best results are obtained with water temperature set points in the region of 30°C,[9,11] although in short-run mixing the significant first-batch effect associated with such a low temperature may dictate moving as high as 50°C. In all cases the set point should be one that can be consistently maintained, dictating that it should be above the maximum expected feed-water temperature. It should also be above the dew-point temperature, at which condensation will begin to form on the surfaces of the chamber and rotors.

With a mixer having a variable-speed drive the batch temperature can be controlled directly, by adjusting the rate of mechanical energy input to the batch, using rotor speed, to a level which is appropriate to the required batch temperature and the rate of heat transfer from the batch. In simple terms, if the mechanical energy input via the rotors is equal to the heat extracted from the batch, its temperature will remain constant. The advantages of variable-speed mixing for batch temperature control were reported as early as 1964 by Perlberg,[12] who suggested a simple reduction of rotor speed in the latter stages of a mixing cycle, to limit the batch-temperature rise and to allow the addition of curatives. Using modern control technology this can be considerably improved—by utilizing the signal from the batch-temperature thermocouple to provide the feedback for automatically adjusting rotor speed in a closed-loop system.

3.4.7. Mixing Sequences

This section is concerned with the operations which have to be performed in order to achieve a mixed batch having the properties required for downstream processing and product performance. In an ideal mixing sequence, all the ingredients of a compound would be charged into an internal mixer together and mixed adequately, without danger of scorch, in a single cycle; and then the ingredients would be converted to a form required for the downstream processes (strip or granular), in a dump extruder sited directly below the mixer.

General-rubber-goods compounds can often be mixed by the ideal route just described, by utilizing the advantages of a variable-speed drive or, in the case of single-speed mixers, by selecting gearing to give a rotor speed sufficiently low to avoid a batch-temperature rise which would preclude the addition of curatives. The increased mixing time needed to achieve an adequate distribution of ingredients is usually compensated for by avoiding the requirement for a labor-intensive and operator-dependent addition of curatives on a two-roll mill, following internal mixing.

For compounds containing large quantities of reinforcing fillers and requiring substantial dispersive mixing, three-stage mixing sequences are commonly used, with the following sequence[13]:

Masterbatch

↓

Remill

↓

Final mix

Each of the three stages is carried out in an internal mixer and, unless the output requirement is very large, the same mixer is used for each stage. In the masterbatch stage the rubber and the reinforcing filler are mixed at a high rotor speed (40–60 rpm, depending on mixer size), together with any particulate additives which are not temperature sensitive. A substantial temperature rise occurs during this stage, and the batch is dumped when the viscosity reduction is judged to preclude further effective mixing, using time, temperature, or energy dump criteria. The final batch temperature is often in the region of 140–160°C. Following discharge from the internal mixer, the batch is either sheeted out on a two-roll mill, cooled, and cut into slabs or is extruded and pelletized. It is then refed to the internal mixer for further dispersive mixing, which can now continue due to the initial low temperature and high viscosity of the batch. Dispersive mixing is substantially completed during this stage, before the temperature rise again renders it ineffective. After sheeting or pelletizing and cooling, the compound is ready for the addition of curatives in the final mix stage. Due to the temperature limitations now imposed, this stage is carried out at a low rotor speed, to give a final batch temperature in the region of 90–105°C. After this the compound is again sheeted or pelletized and cooled, for downstream processing.

There are many variations on the sequence just described. For moderately reinforced compounds the remill stage may not be necessary or, as an alternative to the internal mixer, the two-roll mill can be used for the remill stage, due to the absence of a problem with loose filler. It is also a common practice to add curatives on a two-roll mill that are premixed with rubber to avoid powder loss.

The necessity for multistage mixing should be examined closely in terms of the factors which control dispersive mixing, which is dependent on the forces which can be applied to filler agglomerates. These forces are related to the torque on the mixer rotors and dependent on the batch viscosity. Although high torques can be generated by high rotor speeds in the early stages of a mixing cycle, they are rapidly reduced by the resulting temperature rise, as well as by the progress of dispersive mixing. Prolonging the period of effective dispersive mixing by batch-temperature control can preclude the need for a remill stage and, with some compounds, enable the curatives to be added during dispersive mixing. While the mixing cycles thus obtained will be considerably longer than for multistage sequences, the overall cost of mixing should be substantially lower.

3.5. FLOW INSTABILITIES

With some compounds and mixing conditions there is a tendency for the rubber to undergo melt fracture, breaking up into "crumbs" and becoming coated with particulate additives, which act as effective partitioning agents. When this occurs effective mixing ceases and is very difficult to recover, although an oil addition will sometimes remedy the situation, by wetting the loose filler and providing a nucleus from which viscous flow can recommence.

Crumbing of a batch is a phenomenon which is usually encountered only with narrow molecular-weight distribution rubbers of low green strength.[14] It is associated with the region 3 behavior[15] discussed in connection with the two-roll mill in Section 5.2.3 and is thought to occur as the rubber passes over the rotor wing or nog and into the void which forms behind it. Reference to the conditions for a transition from region 3 to the stable region 4 observed with a two-roll mill[16] indicates that an increase in both circulating-water temperature and fill factor, coupled with a decrease in rotor speed, if possible, may be used to effect a similar transition in the internal mixer.

3.6. LABORATORY SIMULATION OF FULL-SCALE MIXING

The problem of scaleup from a laboratory mixer to one of factory size has been tackled by a number of workers using mathematical modeling techniques, without being successfully resolved. Adopting a more practical approach, the problem can be viewed as one of approximately simulating full-scale mixing in a laboratory machine, so that the properties of the batches produced are similar to those which would be obtained from full-scale machines. This simulation is vital for compound development and setting material specifications, to avoid devising compounds which do not give the required performance under production conditions or specifying values of material properties which cannot be achieved in production.

Gunberg et al.[17] introduced the concept of unit work (mixing energy per unit volume of material) and showed that a common processing profile could be constructed of a property of the mix against unit work for a series of mixers of different sizes. In Figure 3.9 Mooney viscosity is plotted against unit work; and Myers and Newell[18] have also used relaxation time, obtained using a dynamic stress relaxometer, to show that the principle is applicable to viscoelastic properties. The first stage of simulating factory mixing must therefore be the use of a value of unit work similar to that which would be used in the factory machine. The mixing energy for the laboratory machine E_L is obtained from

$$E_L = \frac{V_L}{V_F} E_F \tag{3.3}$$

where V_L and V_F are the batch volumes of the laboratory mixer and the factory mixer, respectively, usually expressed in m^3, and E_F is the mixing energy used for the factory mixer, usually expressed in joules. Unit-work values are normally in the range 400–1200 MJ/m^3, depending on compound type.

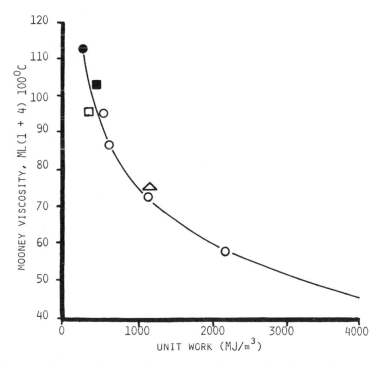

FIGURE 3.9. Scaleup: Mooney viscosity as a function of unit work for a passenger tread masterbatch mixed using (○) Brabender plastograph, (□) BR Banbury, (△) IA Banbury, (●) 11 Banbury, and (■) 27 Banbury.[17]

Practical trials[9,11] have shown that the relationship between unit work and mixed-material properties can be influenced by other mixing variables, such as rotor speed, fill factor, ram pressure, and cooling efficiency. Fill factor and ram pressure are simply simulated. A fill factor similar to that for the factory mixer should be used and an equivalent specific ram pressure. Specific ram pressure P_s is defined as the pressure exerted by the ram on the material in the mixing chamber, and is obtained from

$$P_s = \frac{A_P}{A_T} P_L \tag{3.4}$$

In Eq. (3.4) A_p is the area of the piston in the pneumatic cylinder or actuator used to move the ram, A_T is the area of the throat of the mixer, in which the ram moves, and P_L is the compressed-air-line pressure. The rotor speed and cooling efficiency are rather more complicated to deal with.

A commonly quoted scaling rule for rotor speed is equivalent shear rates at the rotor tips or nog surfaces for the laboratory and factory mixers; but this gives laboratory-mixer rotor speeds which are far too high. A better approach can be made through the use of equivalent unit power. Unit power is defined as "the unit work divided by mixing time in seconds" and has units of MW/m^3. Its purpose is to give approximately equivalent dispersive mixing characteristics. Taking the unit power from the mixing of an appropriate compound in the factory machine, a short series of trials with the laboratory mixer can be used to determine the rotor speed for an equivalent unit power.

The cooling efficiency of laboratory mixers is substantially superior to that of factory mixers. This is simply an effect of size and presents the problem of adjusting the cooling of the laboratory mixer to enable the temperature profile obtained with a similar compound in the factory mixer to be simulated. Although any solution to this complex problem will be somewhat imprecise, it is preferable to use an equivalent circulating-water temperature and adjust the flow rate to give the necessary simulation. In the trials needed to determine an appropriate water flow rate the first-batch effect should be avoided, by taking the temperature profile from the fourth or fifth batch in a sequence.

The procedure described here for setting the conditions for simulation is somewhat lengthy and involved but need not be repeated for each compound investigated. Compounds tend to fall into groups which have approximately equivalent mixing behavior. It is generally adequate to establish and use a single set of conditions for each group. Finally, when the mixing conditions have been established, the mixing procedures and sequences which will be used in the production situation must be followed during laboratory mixing.

REFERENCES

1. Borentski, F. J., Paper presented at the ACS Rubber Division's 111th Meeting, Chicago, Illinois (May 1977).
2. Freakley, P. K., and W. Y. Wan Idris, *Rub. Chem. Technol.* **51**, (1), 134 (1979).
3. Boonstra, B. B., and A. I. Medalia, *Rub. Age* **92** (6), 892 (1963).
4. Boonstra, B. B., and A. I. Medalia, *Rub. Chem. Technol.* **36** (1), 115 (1963).
5. Medalia, A. I., *Rub. Chem. Technol.* **45** (5), 1171 (1972).
6. Mohr, W. D., in *Processing of Thermoplastics Materials*, ed. by E.C. Bernhardt, Van Nostrand Reinhold, New York (1959).
7. Ellwood, H., Paper presented at the second Annual National Conference of the Inst. Rub. Ind., Blackpool, U.K. (1974).
8. Johnson, P. S., Paper presented at ACS Rubber Division's 116th Meeting, Cleveland, Ohio (October 1979).
9. Ebell, P. C., Ph.D. Thesis, Loughborough University, U.K. (1981).
10. Whitaker, P. J., *J. Inst. Rub. Ind.* **4** (4), 153 (1970).
11. Freakley, P. K., Paper given at International Rubber Conference, Harrogate, U.K. (1981).
12. Perlberg, S. E., *Rub World* **150** (2), 27 (1964).
13. Bristow, G. M., *NR. Dev.* **12** (3), 45 (1981).
14. Tokita, N., and I. Pliskin, *Rub. Chem. Technol.* **46** (4), 1166 (1973).
15. Erwin, L., *Polym. Eng. Sci.* **18** (13), 1044 (1978).
16. White, J. L., and N. Tokita, *J. Appl. Polym. Sci.* **9**, 1929 (1965).
17. Gunberg, P. F., S. B. Turetzky, and P. R. Van Buskirk, *Rub. Chem. Technol.* **49** (1), 1 (1976).
18. Myers, F. S., and S. W. Newell, *Rub. Chem. Technol.* **51** (2), 180 (1978).

4

Screw Extrusion and Continuous Mixing

4.1. INTRODUCTION

Extruders are widely used in the rubber industry in a variety of applications. In large mixing systems, dump extruders are used to accept the batch of material from an internal mixer and to give it a shape suitable for further operations. Again in the mixing system, mixing extruders or continuous mixers are used to incorporate and distribute particulate additives. Further down the production line, extruders are used to preform rubber for further operations and to form finished products. All these applications generate their own machine performance requirements, and the wide range of extruder designs available reflects this.

Extruders may be categorized in two ways. First, extruders may be identified by the temperature of the feedstock necessary for successful operation. Traditionally, hot-feed extruders have been used by the rubber industry, where the feedstock is prewarmed in a prior operation. For conventional hot-feed extrusion a two-roll mill is usually used for prewarming. Cold-feed extruders, taking strip or granulated rubber at ambient workshop temperature, are a more recent introduction, probably resulting from the advances in extruder design for the plastics industry. Second, extruders may be identified by application. Many companies require an "undedicated" machine which is capable of operating successfully, if not efficiently, with a wide range of rubber mix types. Here the emphasis in design is to minimize the time taken to change a die and return the extruder to useful operating conditions, and to achieve efficient self-purging to minimize cross-contamination from mix changes. When an extruder is to be used for long runs with rubber mixes having a narrow range of flow properties, the screw, head, and die can be designed to give both high output rates and good dimensional control. Also, the feed and haul-off equipment and the control system may be selected to ensure that the good dimensional control is maintained, despite minor variations in the feed material.

The major physical difference between hot- and cold-feed extruders lies in the length-to-diameter ratio of the screw. For hot-feed machines, where a considerable portion of the input of energy to the rubber mix for heating and preplasticizing is carried out on a two-roll mill, the functions of the extruder screw are simply those of conveying and pressurizing. This has resulted in "short" machines having screw lengths, in terms of their diameters, of 3D to 5D. In addition to conveying and pressurizing, the screw of a cold-feed extruder must input to the rubber all the mechanical work necessary to raise it to the desired temperature for smooth flow through the die. This requires screws having lengths in the region of 9D to 15D, and for some applications longer screws than these may be used.

Cold-feed extruders have largely replaced hot-feed types in production lines where long runs are achieved and where good dimensional accuracy is required, and have made considerable inroads into the "undedicated" area with improvements in versatility resulting from design development and operating "know-how." However, hot-feed extruders are widely used, and Iddon[1] points out that the capital cost and energy consumption of conventional cold-feed extruders increase rapidly when the screw diameter exceeds approximately 150 mm.

4.2. ELEMENTS OF EXTRUDER CONSTRUCTION

4.2.1. The Modular Machine

The modern extruder may be viewed as a modular machine where interchangeable components are assembled into a complete extruder to the customers' requirements. Not all manufacturers offer this versatility, but it does present a good method of dealing with the mechanical design of extruders. Here the available mechanical components, their service life and performance, will be discussed with a view to providing a guide for the prospective purchaser of an extruder.

4.2.2. Drives

A number of motor types are available to the extruder manufacturer. The choice of type is made on the basis of the speed range and torque characteristics required.

Constant-speed ac motors, including pole-changing motors, are only suitable for very simple operations, such as extrusion of blanks or straining of mixes, even when coupled with a speed-change-type gearbox. The high starting torque, associated particularly with small-diameter screws, usually requires the provision of a clutch. Also, the use of a mechanical variable-speed gearbox is not advisable for extruders which are to be fed with "tough" material of large cross-sectional area, particularly for machines having screws of 90 mm or larger.[2]

Variable-speed ac motors are inexpensive and easy to maintain but do not develop maximum torque at low speeds. This characteristic requires the provision of an oversize motor if low-speed operation is required. Also, the electronic constant-speed control system, which should always be provided with motors of this type, cannot always respond rapidly to the short-term variations in torque and speed which occur as a result of variations in the feeding of the screw. The larger motor overcomes this problem by reducing the small variations in extrudate dimensions resulting from the fluctuations in screw speed due to uneven feeding.

Direct current motors are preferable for applications requiring a speed range of more than 4.4:1 and for products which demand good dimensional control. With thyristorized speed control, load variations of 100% can be accommodated, while the speed is maintained within 0.5% of the set value. However, the current-limiting devices used do not permit short-term overloading and here again a larger-size motor should be chosen if sharp loading peaks are expected in order to avoid "tripping-out" of the electricity supply. Two-speed gearboxes are sometimes used in conjunction with smaller motors if mixes requiring high torques are to be extruded. Slow speeds are then used to limit heat buildup in the rubber. For the same reason, dc motors should have combined armature and field requlation so that they develop maximum power at approximately two-thirds of their maximum speed.[2]

4.2.3. Transmissions, Gearboxes, and Bearings

V-belt drives are almost universally used to transmit power from the motor to the gearbox because of their smooth starting and overload capacity.

Gearbox selection is mainly dependent on the maximum torque to be transmitted. Anders[2] states that a continuous overload of 25% should not reduce the normal gearbox life of approximately 40,000 hours. For speed ranges of greater than 4.4:1 and ratings over about 150 kW, a gearbox should be provided with a separately driven oil pump. For high-power drives, filters and an oil cooler are necessary and there should be oil flow and pressure monitors fitted to provide automatic protection from lubrication failure.

Self-aligning roller thrust bearings have been proven to be very satisfactory for withstanding the axial thrust resulting from extrusion pressures on the screw, remaining evenly loaded despite misalignment between the screw and barrel. A suitable bearing should have a service life of about 30,000 hours at maximum load and maximum speed.

4.2.4. Feed Zone

The normal commercial practice is to use a feed roller or a spiral undercut in the feed pocket; in modern extruders both are often provided. The feed roller ensures continuity of feed and gives some preliminary plasticization, while the undercut provides uniformity of feed and good filling of the screw.

The material may be fed in strip or granule form. The latter is usually only successful with "harder" rubber mixes that do not require excessive amounts of partitioning agent to prevent agglomeration during storage and difficulties in feeding. Particulate rubbers present some exceptions to this and will be discussed in the sections dealing with continuous mixing. If granule feeding is possible, reliable metering can be obtained with an automatic feed-level detector placed a short distance above the screw. Strip feeding is advantageous for companies producing their own mixes, since the strip may be produced directly from a two-roll mill or extruder at the end of the mixing line. Alternatively zig-zag cutting of a continuous slab may be used where batch cooling after the final two-roll mill or extruder precludes the direct production of a strip.

4.2.5. Barrel and Screw

For high output an extruder barrel should, in all cases, have a "wet liner" so that the heat generated in the material by the action of the screw can be removed effectively. For this reason it is important that the heat-transfer fluid should be as close as possible to the screw—about 15–30 mm, according to screw diameter. Anders[2] recommends that nitriding steel grade 8550 be used for the liner, providing a bore hardness of at least 900 Vickers and conferring very good wear resistance.

The screw material must have a high yield point and toughness and should be heat treatable so that the "lands" can be further hardened to resist the high surface pressure and abrasion produced by the rubber mix. Here again heat nitriding has been found to be satisfactory, and recent experience[2] has also shown ionizing nitriding to be very effective. However, as the nitrided layer is very thin with the latter process, the screw should only be polished and not ground.

4.2.6. Extrusion Heads

Even though there are many head designs, each engineered for specific applications, there are a number of essential requirements common to all. Head temperature should be controllable so that even at high material temperatures (e.g., 120°C) the metal surface temperature can be maintained at a level low enough (e.g., 70°C) to prevent scorching on the head and die walls, where the flow rate is slower and the residence time correspondingly longer. For many mixes it may also be necessary to heat the lips of the die in order to obtain a good surface finish.

Since it is generally necessary to clean out the head when changing mixes, processers should estimate as accurately as possible how often mixes will be changed and how long this will take with different type heads. Though the initial cost may be higher, it is almost always possible to find a way of reducing downtime to a minimum. Where screen packs are essential, automatic or continuous screen changers may be used to avoid interrupting a long run to replace a clogged screen pack. The opening of a head can also be simplified in a number of ways. The provision of two heads on hinges and with bayonet locking mechanisms can completely eliminate the time taken for die changing and cleaning from the total downtime. Tire-tread extrusion heads that open upwards and the split sheeting die of a roller head extruder held closed by a swan-neck press are further examples of means of reducing the time and effort involved in the changing of dies and mixes. Consideration should also be given to minimizing material wastage during die changes.

4.2.7. Temperature Control

The modern extruder is divided into a number of zones for temperature control, each zone being capable of independent control. Normally the barrel is divided into three or four zones; the head forms a separate zone; and the die temperature is controlled independently of the head. In many extruders the screw is bored out for temperature control. The instrumentation and strategies of control will be discussed in later sections.

Two types of systems are generally used to control temperature; both systems are of the fluid-circulation type. For high cooling capacity, a "direct" system is used in which cold water is added directly to the cooling line to the temperature control zone, provided that a supply of clean and relatively lime-free water is available and that temperatures will seldom exceed 70°C. The high cooling capacity of the direct system is desirable in extruders of 150 mm or greater screw diameter; but the more expensive indirect system must be used if conditions are unsuitable for the former. The "indirect" system feeds cold water to a heat exchanger connected in parallel to the cooling line, giving freedom of choice for the heat-transfer fluid fed to the extruder. According to Anders[2] the total heat-exchange capacity should be sufficient to ensure that not less than about 60% of the power output of the extruder drive motor can be removed in the form of heat. For initial heating, either electric elements or coils through which steam may be passed are incorporated into the heat-exchanger vessels.

4.3. HOT-FEED EXTRUDERS

4.3.1. General-Purpose Machines

In addition to the costs and energy consumption of cold-feed extruders rising rapidly above screw diameters of approximately 150 mm,[1] Evans[3] suggests that practically there is very little to choose between hot- and cold-feed extruders. Clearly, hot-feed extruders cannot simply be dismissed on the grounds of "outmoded technology" and deserve a detailed appraisal in order to identify possible areas of use.

The following advantages are usually claimed[3] for cold-feed extruders in comparison with hot-feed machines:

1. Lower capital cost of equipment—no mills, etc.
2. Reduced labor costs.
3. Better temperature control.
4. Better dimensional control of the extrudate.
5. Capability for handling a wider range of rubber mix types.

When considering both capital cost and labor costs, the equipment and time required in the mixing department to support a cold-feed extruder must be added into the total costs.

The superior temperature control of a cold-feed extruder is essential to offset the temperature rise associated with the long residence times. The short residence times of hot-feed extruders do not create such a severe problem. However, the capability for precise temperature control with the cold-feed machine does result in advantages with regard to the dimensional control of the extrudate. Extrusion rate and die swell are primarily dependent on head pressure and material temperature. The hot-feed extruder, having a short residence time, relies largely on the two-roll mill for temperature control, where the potential for precise temperature control under the non-steady-state conditions of extruder feeding is limited. The uniformity of "processing history" that a cold-feed extruder confers upon a rubber mix thus proves to be its main advantage.

Moving to item (5), although cold-feed extruders are capable of handling a wider range of rubber mix types by invoking specialist screw design, the range of materials which may be successfully extruded by a "general-purpose" machine is narrower than for the equivalent hot-feed extruder. Also, Evans[3] states that the technical controls necessary for mixes that are to be extruded using a cold-feed machine need to be much tighter, particularly with respect to material storage time.

The available facts indicate that the cold-feed extruder is preferable when long runs are planned and a limited range of mixes is to be extruded. If versatility is required and short runs are envisaged, the hot-feed machine would appear to offer some advantages, provided that dimensional control is adequate.

4.3.2. Dump or Batch Extruders

This class of hot-feed extruder, designed to accept the mix discharged from an internal mixer, possesses features which differentiate it from the normal hot-feed machine. In order to accept and feed successfully the mass of material from the internal mixer, the screw is of large diameter in the hopper section, tapering to the discharge diameter just forward of the hopper. An inclined pneumatically operated pusher in the hopper section forces the rubber down into the flights of the screw. A plain slit die may be fitted to produce slab; alternatively, a roller die may be used to produce dimensionally accurate and flaw-free thick sheet. A third option is the use of a pelletizing head for the production of granules for subsequent cold-feed extrusion. In fact, single-stage internal mixing, coupled with discharge to a batch extruder equipped to produce granules or strips for a number of cold-feed extruders, presents a very efficient manufacturing route.

4.4. COLD-FEED EXTRUDERS

4.4.1. Developments in Screw Design

The advent of the cold-feed extruder has resulted in a series of developments in screw designs. Screws for hot-feed extruders are simple conveying elements of two-start design, sometimes having a single start section in the hopper region. Typically, flight depths of 0.2 D and a pitch of 1 D are used. These proportions are valid for all sizes of a hot-feed extruder screw, giving an output that increases as a function of the cube of the diameter ratio of the screw at a specified speed as a simple scaleup rule.

Following the introduction of the cold-feed extruder, it was found that simple screws of the hot-feed type gave inadequate homogenization of the rubber mix. This results from the rotating laminar planes which generate around a simple conveying screw, with little exchange of material between planes; and from the poor heat conductivity of rubber. The material in the center of the screw channel is therefore insulated from the warm plasticized outer layers, giving a dimensionally and visually unacceptable extrudate.

A series of changes were then initiated to overcome this difficulty. Lengthening the screw and decreasing the pitch increased the residence time and laminar mixing; and decreasing the flight depth alleviated the heat-transfer problem. The limitations of this approach arise from the increasing heat generation in the rubber as the flight depth is reduced and from the definite limit set on residence time by the cross-linking characteristics of a rubber mix. A more significant development in design has been the introduction of mixing sections to the extruder screw. At the same time, the development of mixing screws, where practically the whole screw serves a mixing function in addition to conveying, was proceeding. These mixing screws are discussed in Section 4.7, but are equally applicable to the extrusion of premixed material.

Mixing sections were initially sited at the die end of the screw, resulting in the screw channel being filled at all times, independent of die resistance. On the other hand, it was found that there was more than normal wear and that mixes tended to overheat when working against a high die resistance. Conversely, a low die resistance resulted in cold spots due to inadequate homogenization in the mixing section. Thorne[4] states that if the appearance of cold spots in the material at the die is to be avoided, homogenization must take place where the material has uniform viscosity, that is, near the hopper section. Troester screws now follow this practice and have a mixing section consisting of several intersecting flights.

Mixing screws and screws having mixing sections generally have a common disadvantage—they only operate efficiently with a limited range of rubber mixes. A wide range of mixes can be processed only if the output rate is reduced. Even so, lengthening the residence time may lead to premature cross-linking. Trials carried out with the recently developed pin extruder indicate that high output rate can be maintained with a wide range of rubber mixes.[5] The pin extruder, the principle of which is illustrated in Figure 4.1, presents a method of achieving the flow division necessary to good homogenization without recourse to specialist screw design or the creation of areas of intensive heat generation in the screw. The latter capability is particularly important for large extruders where heat transfer is usually a limiting factor.

Similar capabilities and attributes appear to be possessed by the RAPRA cavity transfer mixer (CTM).[6] This device takes the form of a screw and barrel extension for a conventional extruder and so carries with it the additional advantage of retrofitting. It works on the principle of flow stream division and reorientation, to give effective exponential distributive mixing. The design of the mixer, which is shown in Figure 4.2, involves exchange of material between the hemispherical cavities machined into both the screw and barrel extensions. Cavity transfer mixers with three to six circumferential rows of cavities have been found to give a high quality of uniformity of both temperature and mix composition, producing considerable improvements in extrudate appearance and dimensional stability in comparison with the

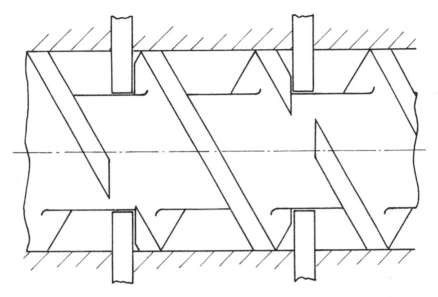

FIGURE 4.1. Schematic section of pin extruder with two-start screw, showing screw flights cut away at pin–plane positions.

unmodified extruder. The pressure drop and temperature rise resulting from adding the mixer to an extruder is reported to be negligible, indicating that an output similar to that from a conventional screw extruder can be maintained, without an increased danger of scorch being incurred.

4.4.2. Vented Extruders

The need for vented extruders arose with the development of low-pressure continuous vulcanization techniques. The vented extruder is designed to extract trapped air, moisture, and other volatile components from a rubber mix, thus reducing the porosity resulting from their expansion during low-pressure vulcanization. Vented extruders are also used for improving extrudate quality, even for sponge rubber.[4] In the latter case, greater control of the cell structure can be achieved if the cells result primarily from the blowing agent and not from trapped air or moisture.

The first section of a vented extruder is essentially a cold-feed extruder and all the factors influencing cold-feeder extruder performance which have been previously discussed are relevant to it. At the end of this first section the material is extruded over a dam—a narrow annular gap providing a thin film with a large surface area—and passes into a low-pressure zone created by a large flight depth. At this point the barrel is vented and extraction is

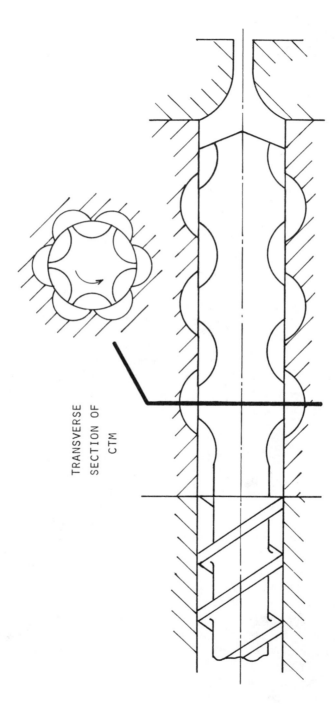

TRANSVERSE
SECTION OF
CTM

FIGURE 4.2. Schematic section of extruder fitted with a RAPRA cavity transfer mixer (CTM), showing hemispherical cavities in screw and barrel extensions. (CTM is a trade name of the RAPRA of Great Britain).

usually assisted by a vacuum pump. A filter system is employed here if the vapors and gases vented are suspected to be toxic. In the vent region the material has to be conveyed without pressure; otherwise extrusion from the vent would result. In the final section of the extruder, pressure is again built up to overcome die resistance; this final section has proportions similar to those of a hot-feed extruder.

The main prerequisite for the proper operation of a vented extruder is that the second stage of the screw should be able to extrude at least as much material as the hopper section is able to feed. Otherwise, the vacuum zone would be filled and venting would become inadequate due to the small surface area of material exposed. The extreme case is extrusion from the vent mentioned previously. Screws for vented extruders are designed to avoid this; but die resistance and extrusion conditions must be carefully determined. Most manufacturers supply alternative second stages for screws, to cater for a wide range of die resistances.

4.5. DESIGN OF EXTRUDER HEADS AND DIES

4.5.1. Pressure Drop Due to Screen Packs

A considerable number of extrusion operations require the use of a screen pack to remove undesirable particles from a rubber mix. If it is intended that a screen pack should be used, the pressure drop caused by its presence must be added to that due to the die in order to determine the"operating point" of the extruder.

In most cases the screen is supported by a "breaker plate," which has a large number of closely spaced holes to permit the passage of the rubber mix. The pressure drop due to the breaker plate may be determined using the capillary flow equation for power-law fluids:

$$\Delta P = 2\eta_0 L \left(\frac{Q}{\pi R^{(3n+1)/n}} \frac{3n+1}{n} \right)^n \tag{4.1}$$

where Q is the volumetric output rate of the extruder divided by the number of holes in the breaker plates. Entrance effects for both breaker plate and screen pack are negligible since comparatively little convergent flow occurs in comparison with a die or capillary.

Carley and Smith[7] derive the following expression for pressure drop across a screen:

$$\Delta P_3 = c_n \eta_0 (\dot{\gamma}_0)^{1-n} \left(\frac{W}{\rho D_s^2} \right)^n \frac{d}{m^{2n} D_0^{3n+1}} \tag{4.2}$$

where

$$c_n = 2^{n+3} \left(3 + \frac{1}{n}\right)^n \left(\frac{\pi}{4}\right)^{-n} \tag{4.3}$$

In Eq. (4.2) η_0 is the viscosity at the reference shear rate $\dot{\gamma}_0(1\ s^{-1})$; W is mass flow rate; D_s is the diameter of the screen; d is the diameter of the wire used in the screen; m is the mesh number (for square woven screens there are m^2 openings per unit area); and D_0 is the average minimal opening between adjacent wires. Other symbols have their usual meaning. Equation (4.2) refers to the pressure drop across a single screen. Carley and Smith state that the pressure drops of individual screens are additive, regardless of the mix of meshes or of relative alignment of screen wires.

4.5.2. Elements of Die Design

In contrast with the dies used in capillary rheometry (Section 2.3.2), practical extrusion dies are normally short in comparison with their aperture size and are often required to impart a complex shape to the extrudate. These differences result in a marked dependence of the possible extruder output rate on the extensional flow characteristics in the die entry region and on the achieving of a constant velocity around the exit periphery of a die of complex shape. An abrupt lead-in taper to a die, or a design giving a nonuniform exit velocity distribution, will severely limit the output rate which can be achieved before the onset of melt fracture.

A precise definition of melt fracture is difficult; it includes both die entry and exit effects and in appearance can range from a slight, but unacceptable, roughness on the surfaces of an extrudate to an extreme distortion. However, it does, in many cases, set a very definite limit on the extrusion rate which may be achieved. For effective die design its causes must be clearly understood; and the techniques for increasing the critical output rate, at which the onset of melt fracture occurs, need clear definition.

The criteria for melt fracture must be considered in conjunction with the requirement for a pressure drop over the die that is compatible with both the extruder performance and with the need to limit the temperature rise of a rubber mix during extrusion. Die design is then concerned with the choosing of a die geometry that will lead to a prescribed output and extrudate shape while working within definite limits of pressure drop and temperature rise.

Die-design problems may be divided into two broad classes:

1. Dies in which the melt flow is one-dimensional, that is, where the velocity is changing in only one direction.
2. Dies in which the flow is two- or three-dimensional.

One-dimensional-flow dies are an obvious extension of the capillary-rheometer-type case and include dies having circular, annular, and thin slit cross sections. Two- and three-dimensional-flow dies include all those which cannot be approximated to circular, annular, or slit cross sections. In this case the problem is to determine an internal die geometry to give the uniform exit velocity already mentioned. Failure to do this results in the setting up of forces which tend to distort the shape of the extrudate from that of the die. This falls within the definition of melt fracture, since the severity of the distortion is a function of the viscoelastic response of the rubber. If the extruder is run slowly enough, sufficient time may be allowed for relaxation of the stresses causing the distortion.

A practical one-dimensional-flow die consists of an aperture having a constant cross section of length L_c (the land), which imparts the required shape to the extrudate, and a further length L_1, in which the cross section changes progressively from that of the die to that of the extruder head. The function of this lead-in, which may be constructed separately from the primary die section, is to create the necessary conditions for streamline convergent flow, without the occurrence of "dead areas" or circulatory flows which increase local residence times and hence give rise to premature vulcanization. The pressure drops across these two sections will be additive.

If the normal die construction, with both lead-in and land sections, gives an unacceptably high-pressure drop, resulting in low output and high material temperatures, it may be necessary to opt for a tapered die. In this construction there is a direct taper from the barrel diameter to the die exit, substantially reducing the pressure drop needed for extrusion. While tapered dies can often be used to alleviate material-temperature-rise problems, their design is subject to a number of uncertainties, due to forces which tend to distort the shape of the extrudate from that of the die not being dissipated by stress relaxation in the land region.

4.5.3. Lead-In Sections and Tapered Dies

Both lead-in sections and tapered dies take two general forms: conicylindrical and wedge shaped. Cogswell[8] presents analyses for both these forms, using a power-law model to describe shear flow behavior and assuming that the tensile component of the convergent flow can be characterized by an extensional viscosity λ which is independent of stress. This extensional viscosity can be obtained by inserting the results of a capillary rheometer experiment using a zero-length (knife edge) die into the following expression:

$$\lambda = \frac{9(n+1)^2}{32\eta_a} \left(\frac{P_0}{\dot{\gamma}}\right)^2 \tag{4.4}$$

In Eq. (4.4) η_a is the apparent viscosity for the shear rate $\dot{y}$ in the knife-edge die, n is the power-law index, and P_0 is the pressure drop across the knife-edge die.

The conicylindrical and wedge-shaped forms are shown in Figure 4.3, with the nomenclature used in the following analyses. For the conicylindrical form the pressure drop due to extensional flow is given by

$$\Delta P_E = \frac{\lambda \tan\theta \, \dot{\gamma}_1}{3} \left[1 - \left(\frac{r_1}{r_0}\right)^3 \right] \tag{4.5}$$

where

$$\dot{\gamma}_1 = \frac{4Q}{\pi r_1^3} \tag{4.6}$$

For shear flow the pressure drop is given by

$$\Delta P_s = \frac{2\tau_1}{3n \tan\theta} \left[1 - \left(\frac{r_1}{r_0}\right)^{3n} \right] \tag{4.7}$$

where τ_1 is the inlet shear stress, determined from the power-law equation. The extensional and shear flow pressure drops may now be added together to give the total pressure drop over the tapered section.

Cogswell and Lamb[9] indicate that nonstreamline flow (onset of extrudate distortion, surface roughness, or melt fracture) occurs when the extensional

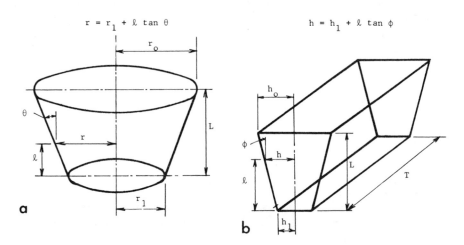

$r = r_1 + \ell \tan\theta$ $h = h_1 + \ell \tan\phi$

a

b

FIGURE 4.3. Tapered die forms. (a) Conicylindrical lead-in section. (b) Wedge-shaped lead-in section.

stress σ exceeds some critical value. This value can also be obtained from a capillary rheometer experiment with a knife-edge die, using

$$\sigma \text{ (critical)} = \tfrac{3}{8}(3n + 1)\, P_0 \text{ (nonlaminar flow)} \qquad (4.8)$$

The identification of the pressure drop P_0, which gives an unacceptable extrudate, is entirely dependent on the quality required in the product for which the die-design exercise is being carried out. During flow through a conicylindrical die, the critical stress may have two maxima—at the die entrance, due to convergence from the head of the extruder, and at the die exit, due to convergence within the die. The critical region of a lead-in section is usually at its exit. The expression for the critical flow rate at the entry takes the form

$$\sigma \text{ (critical)} = \frac{1}{2^{1/2}} \left(\frac{3n+1}{n+1}\right) \frac{4Q}{\pi r_0^3} (\eta_a \lambda)^{1/2} \qquad (4.9)$$

where η_a is the apparent viscosity at $\dot{\gamma} = 4Q/\pi r_0^3$. Equation (4.9) assumes free convergence, which will occur if the angle of lead-in from the head diameter of the extruder to the die entrance is greater than the natural angle of convergence, given by

$$\tan \theta_0 = \left(\frac{2\eta_a}{\lambda}\right)^{1/2} \qquad (4.10)$$

The critical flow rate at the exit may be determined from

$$\sigma \text{ (critical)} = \left(\frac{3n+1}{n+1}\right) \lambda \left(\frac{2Q}{\pi r_1^3}\right) \tan \theta$$

$$= \left(\frac{3n+1}{n+1}\right) \lambda \frac{\dot{\gamma}_1}{2} \tan \theta \qquad (4.11)$$

Values of $\tan \theta$ which give the maximum output prior to the onset of melt fracture can be determined from Eq. (4.11) and checked with Eqs. (4.5) and (4.7) for the pressure drop. Dies and lead-in sections with angles greater than the free-convergence angle θ_0, defined in Eq. (4.10), should be avoided, due to their low output capabilities and the recirculatory flow which occurs outside the boundary of the conical free-convergence zone.

For wedge-shaped dies or lead-in sections, referring to Figure 4.3(b), the pressure drop due to extensional flow is given by

$$\Delta P_E = \frac{4}{3}\dot{\varepsilon}\lambda \left[1 - \left(\frac{h_1}{h_0}\right)^2\right] \qquad (4.12)$$

where

$$\dot{\varepsilon} = \frac{\dot{\gamma}}{3} \tan \phi \tag{4.13}$$

For shear flow, the pressure drop is given by

$$\Delta P_s = \frac{\tau}{2n \tan \phi} \left[1 - \left(\frac{h_1}{h_0} \right)^{2n} \right] \tag{4.14}$$

for which

$$\dot{\gamma} = \frac{3Q}{2Th^2} \tag{4.15}$$

enabling τ in Eq. (4.14) to be determined and the relationship between pressure drop, volumetric output, and the lead-in or taper angle ϕ to be established.

The expression for the critical flow rate at the entrance to the section, assuming free convergence of flow from the head diameter, takes the form

$$\sigma \text{ (critical)} = \frac{Q}{Th_0^2} \lambda \left(\frac{\eta_a}{\lambda} \right)^{1/2} \tag{4.16}$$

and the critical flow rate at the exit may be determined from

$$\sigma \text{ (critical)} = \frac{2}{3} \lambda \frac{Q}{Th_1^2} \tan \phi \tag{4.17}$$

for which the value of σ (critical) can again be found from a capillary rheometer experiment and Eq. (4.8).

The equations for wedge-shaped sections ignore the influence of the width-to-depth ratio on the flow pattern. As the T/h ratio approaches unity, the accuracy of the equations will deteriorate, although they can still be used to provide qualitative guidance. As with the conicylindrical section, Eqs. (4.16) and (4.17) can be used to determine the lead-in or die angles for maximum flow rates and checked with Eqs. (4.12) and (4.14) for the associated pressure drops.

4.5.4. Parallel Die Sections

These include simple plate dies and the land sections of dies having tapered lead-ins. Normal practice in the plastics industry suggests that a land length to diameter, or other critical cross-sectional dimensional ratio, of a least 10:1, is necessary to establish a stable velocity profile[10] and to ensure that the forces generated in the material during convergent flow, which tend to distort the shape of the extrudate from that of the die, are minimized. In rubber extrusion, where viscosities are generally an order of magnitude higher than in plastics extrusion, such long land lengths may give excessive die resistance (pressure drop) and material temperature. However, when shorter land lengths are used, the die shape may need adjustment to compensate for the distortion.

Simple plate dies do not have lead-in sections, resulting in free-convergent flow from the extruder head diameter to the die, with an angle of convergence dependent on the flow properties of the rubber, as described by Eq. (4.10) for a circular die and by[8]

$$\tan \phi_0 = \frac{3}{2} \left(\frac{\eta_a}{\lambda} \right)^{1/2} \tag{4.18}$$

for a slit die (rectangular cross section). The pressure drops for free-convergent flow are given by

$$\Delta P_0 = \frac{2^{1/2} 4}{3(n+1)} \dot{\gamma}_0 (\eta_a \lambda)^{1/2} \tag{4.19}$$

for a die of circular cross section, where η_a is the apparent viscosity at the shear rate in the die $\dot{\gamma}^0$, and by

$$\Delta P_0 = \frac{2^{1/2} 4}{3n+1} \dot{\gamma}_0 (\eta_a \lambda)^{1/2} \tag{4.20}$$

for a die of slit form. The critical tensile stress will again determine the maximum output which can be attained prior to the onset of melt fracture, and is defined by Eq. (4.8). The maximum output can be estimated from Eqs. (4.9) and (4.16).

The following expressions[11] describe the fully developed flow in the land region of dies amenable to one-dimensional analysis. For a die of circular cross section, the volumetric flow rate of a power-law fluid is given by

$$Q = \frac{n}{3n+1} \left(\frac{\Delta P}{2\eta_0 L} \right)^{1/n} \pi R^{(3n+1)/n} \tag{4.21}$$

including a correction for the non-Newtonian flow profile. The analysis for a slit die is actually for flow between parallel plates, since edge effects are not taken into account. The expression for flow rate then takes the form

$$Q = \frac{3n}{2n+1}\left(\frac{h\,\Delta P}{2L\eta_0}\right)^{1/n}\frac{Th^2}{6} \tag{4.22}$$

where T is the width of the die, L is the length (along the flow path), and h is the depth, being the distance between the plates or die lips. The tube die is a variant of the slit die, where the slit is "rolled up" to form an annulus. For a die having an outer radius r_0 and an inner radius r_i, with a length L, the flow rate is given by

$$Q = \frac{\pi}{6}(r_0 + r_i)(r_0 - r_i)^2 \frac{3n}{2n+1}\left(\frac{(r_0-r_i)\Delta P}{2\eta_0 L}\right)^{1/n} \tag{4.23}$$

The slit-die analysis can also be extended to T sections, U sections, and more complex cases, provided that the T dimension is substantially greater than the h dimension, giving small edge effects.

4.5.5. Wall Slip and Die Swell Behavior in Simple Dies

The correspondence between the cross-sectional shape of an extrudate and the die orifice which produced it, is generally accepted as a natural and inevitable result. However, the extrusion of a Newtonian liquid obeying the common assumption of zero slip gives the situation shown in Figure 4.4, where the extruded material tends to accumulate as a more or less spherical mass of a much bigger cross section than the die orifice. Continuing with the assumption of zero slip and examining the case of power-law fluids suggests that the trend toward a "plug-like" velocity profile as the power-law index n decreases also results in a trend toward a more conventionally shaped extrudate.

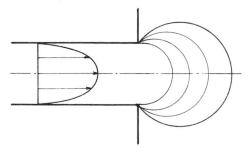

FIGURE 4.4. Globular mass of material forming at die exit due to equalization of flow velocities in the absence of wall slip.

If wall slip occurs, the contribution to the overall volumetric flow rate is simply additive, as demonstrated by the following equation for a circular die:

$$Q = \frac{n}{3n+1} \left(\frac{\Delta P}{2\eta_0 L} \right)^{1/n} \pi R^{(3n+1)/n} + \pi R^2 V_s \qquad (4.24)$$

where the wall-slip velocity V_s is related to the wall shear stress by a power-law function[12]:

$$V_s = k\tau^m = k \left(\frac{\Delta P R}{2L} \right)^m \qquad (4.25)$$

Although the shear deformation of the rubber in convergence to the die and passage through it will be reduced by the presence of wall slip, the tensile deformation will remain largely unaffected. Melt fracture, due to exceeding a critical tensile stress, will still conform to the analyses developed by Cogswell[8] and presented in Section 4.5.3; but it is likely that increased slippage will result in improvements in extrudate smoothness, due to the reduction of shear stresses.

The gross effect of slippage at the die exit is to prevent the situation shown in Figure 4.4. Slip and die swell are further related by the relative magnitudes of the viscous flow and slip components of the total output from a die. For a given output rate, increasing the slip velocity will result in reduced strain rates; and die swell, or extrusion shrinkage as it is more correctly called, is a result of the elastic recovery of strains set up during flow through the die assembly. Hence an increase in slip will give a decrease in die swell.

For dies which approximate to the one-dimensional flow case, die swell can be simply defined by a swelling ratio B, where

$$B = \left(\frac{\text{cross-sectional area of extrudate}}{\text{cross-sectional area of die}} \right)^{1/2} \qquad (4.26)$$

and may be obtained directly from a capillary rheometer experiment. Extrusion from a capillary die with a similar length to cross-sectional dimension ratio as the projected production die, over a suitable range of temperatures and shear stresses, will give the necessary swelling ratio information. The die cross-section dimensions can then be scaled using appropriate swelling ratios; but it must be borne in mind that production extrudates are invariably subjected to draw-down, by the application of tension from the haul-off. Due allowances should be made for the reduction of cross-section size from this source.

4.5.6. Elements of Die Design for Complex Extrudates

The design of dies which cannot be approximated to the one-dimensional cases may be resolved into the determination of the die form to give the required extrudate dimensions and the prediction of the pressure drop across the die. The extrudate cross section depends on both the flow velocity distribution in the die and on the elastic recovery which takes place on exit from the die. It follows that the die shape is required to provide compensation for the forces acting to distort the extrudate shape from that of the die. This involves the consideration of die swell as an integral design factor, since it influences the required die form.

Solid extrudates include cross sections, such as the square, the rectangle, the triangle, the ellipse, and other simple geometrical shapes. Composite shapes are made up from combinations of these sections with each other and with those shapes which can be treated as one-dimensional cases.

The problem of extruding solid "massive" shapes is well illustrated by considering the flow of fluid through a square channel. The contours of constant fluid velocity (isovels) plotted onto the die cross section in Figure 4.5 show that the stable velocity distribution does not conform to the shape of the channel and tends to assume a circular configuration well away from the channel boundary. If the fluid velocity along lines parallel to the channel wall is plotted (line A-A in Figure 4.5), it can be seen that it will decrease from a maximum at the center of the side to a minimum at the corner. [13]

The effect of this velocity distribution at the channel exit will be a "rounding-off" of the extruded profile: this rounding-off will be further

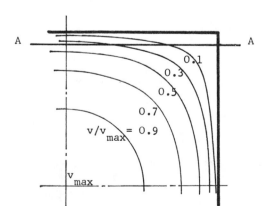

FIGURE 4.5. Isovels (contours of constant flow velocity) in a square-section die. Only one quadrant of the die cross section is shown due to the symmetry of flow.

accentuated by the greater elastic recovery occurring at the corners as a result of the greater velocity gradient.

A further problem resulting from the elastic memory of rubber is encountered in practical dies. If the velocity distribution in the lead-in section of the die is significantly different in form that in the land section, the stress distribution, and hence the die swell, will be different from that predicted from the land section alone. This will depend on the time of retention of elastic memory, which is analogous to the relaxation time Γ of the rubber, where

$$\Gamma = \eta_0/G \qquad (4.27)$$

Röthemeyer[13] shows that the elastic memory may be used to advantage by including a preforming section in a die between the lead-in and the land. The function of this preforming section is to set up a stress pattern which, when evoluted to that due to the cross section of the land, achieves a

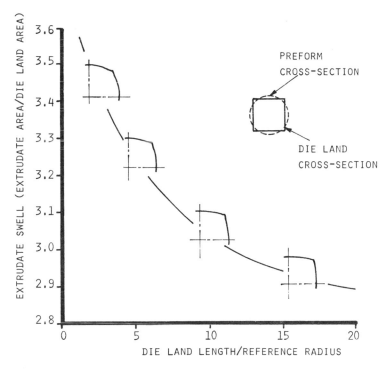

FIGURE 4.6. Influence of die land length on the cross-sectional shape and swell of a polyethylene extrudate, using a circular section preform.[13]

more uniform die swell over the whole extrudate cross section than would have otherwise been obtained.

In conjunction with a square cross-section land, Röthemeyer used two preforms to demonstrate the effect of preform shape. Also, by varying the land length of the die, the time for relaxation of the stress pattern was systematically changed. Figures 4.6 and 4.7 show the cross sections of the preforms, their orientation with respect to the land cross section, and the influence of land length on the extrudate, expressed as a ratio of the reference radius r_v of both the preform and the land (they are equal), where

$$r_v = \frac{2 \times \text{cross-sectional area}}{\text{circumference}} \qquad (4.28)$$

Although the results shown are for polyethylene, the concepts are equally valid for the extrusion of rubber.

Some practical dies combine the function of preform and land in a

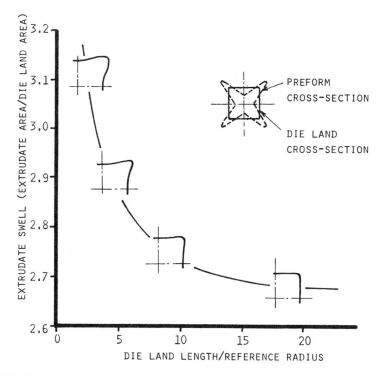

FIGURE 4.7. Influence of die land length on the cross-sectional shape and swell of a polyethylene extrudate, using a "star-shaped" preform.[13]

single section to reduce the total pressure drop over the die, resulting in a die exit crosssection significantly different from that of the final extrudate cross section. However, this can result in high stresses being set up in the extrudate, leading to melt fracture and low output rates.

The problem of specifying a die to produce a required "massive" crosssection is one that can, except in a few cases, only be approached using approximations based on investigations such as that of Röthemeyer[13] and on one-dimensional die design coupled with practical development trials. In the cases where quantitative methods have been successfully applied, the mathematical techniques involved are beyond the scope of this text. However, the determination of pressure drop and flow behavior in channels of simple geometric shape has been effectively resolved.

Although the fundamental determination of flow through such channels is complex, Miller[14] has resolved the problem into one of relating the flow in the required channel to that in a simple case, such as a circular capillary, via a geometric shape-factor. The pressure drop in a channel of arbitrary cross section may be expressed by

$$\frac{\Delta P}{L} = \frac{2}{r_v} \bar{\tau} \tag{4.29}$$

where $\bar{\tau}$ is the average wall shear stress. Miller then postulates that the average wall shear stress $\bar{\tau}$ and the apparent wall shear rate $\bar{\gamma}$ are, to a good approximation, independent of duct geometry, such that

$$\bar{\tau} = \bar{\tau}(\bar{\gamma}) \tag{4.30}$$

Ignoring, for the moment, the correction for non-Newtonian flow in a capillary and assuming a power-law-type fluid, Eq. (4.30) may be written as

$$\tau = \eta_0 \left(\frac{Q\Omega}{4Ar_v}\right)^n \tag{4.31}$$

for a channel of arbitrary cross section. Substituting $\Omega = 16$ (Table 4.1) in the right-hand side of Eq. (4.31) yields the familiar expression $(4Q/\pi R^3)$ for the shear rate at the wall of a channel, of circular cross section. Using Eq. (4.30) to relate the flow in the channel to that in a square-section channel yields

$$\bar{\tau}_c(\bar{\gamma}) = \bar{\tau}_s(\bar{\gamma}) \tag{4.32}$$

where the subscripts c and s refer to the circular and the square channels, respectively.

TABLE 4.1
Values of Ω for Various Cross-Sectional Geometries

Geometry	Parameter	Ω
Circular cylinder[15]		16.0
Parallel plates[15]		24.0
Isosceles triangles[16]	$2\alpha = 10°$	12.5
	$= 20°$	12.8
	$= 30°$	13.1
	$= 40°$	13.15
	$= 50°$	13.25
	$= 60°$	13.35
	$= 70°$	13.25
	$= 80°$	13.2
	$= 90°$	13.15
Rectangular ducts[17]	$b/a \leqslant 1.0$	$\dfrac{24.0^a}{(1 - 0.351b/a)(1 + b/a)^2}$
	$b/a = 0.0$	24.0
	$= 0.5$	15.7
	$= 1.0$	14.3
Annular ducts[15]		$\dfrac{16(1 - K)^2}{(1 + K^3) + (1 - K^2)/\ln K}$
	$K = R_1/R_2$	
	$K = 0.0$	16.0
	$= 0.1$	22.4
	$= 0.3$	23.4
	$= 0.5$	24.0
	$= 0.7$	24.0
	$= 1.0$	24.0
Star-shaped conduits[18]	Number of points	
	3	6.50
	4	6.61
	5	6.63
	8	6.64
	11	6.63
Regular polygonal conduits[18]	Number of sides	
	3	13.33
	4	14.25
	5	14.74
	6	15.05
	8	15.41

[a]Approximate.

(Table continued)

TABLE 4.1. (continued)

Geometry	Parameter	Ω
Rhombic conduits[19]	$b/h = 1$	14.22
	$= 2/3$	14.00
	$= 1/2$	13.62
	$= 1/3$	13.1
	$= 1/4$	12.75
	$= 1/10$	12.15
Rounded rectangular ducts[20]	$b/a \leqslant 1.0$	$\dfrac{24.0^a}{1 + 0.9b/a - 0.4(b/a)^2}$
	$b/a = 0.0$	24.0
	$= 0.5$	17.8
	$= 1.0$	16.0
Half-rounded rectangular ducts[20]	$b/a \leqslant 1.0$	$\dfrac{24^a}{1 + 1.24b/a - 0.63(b/a)^2}$
	$b/a = 0.0$	24.0
	$= 0.5$	16.4
	$= 1.0$	14.9
	$= 2.0$	14.8
Elliptical ducts:[17]	$b/a \leqslant 1.0$	$19.75[1 - 0.310b/a + 0.120(b/a)^2]$
	$b/a = 0.0$	19.75
	$= 0.25$	18.25
	$= 0.667$	16.4
	$= 1.0$	16.0

Miller shows that the use of a geometric-shape factor to correlate flow in channels of dissimilar cross section provides solutions comparable to those derived from more fundamental treatments. However, Miller's technique does not permit the flow velocity distribution in a channel of unusual cross section to be determined. For this determination, reference must be made to the fundamental analyses, which are identified in Table 4.1 against the appropriate channel shape, or to the experimental method proposed by Fisher and Malsen.[21] This method involves stretching a thin membrane, such as a thin rubber sheet, across the channel exit and applying pressure to one side of the membrane.

4.5.7. Composite Extrudate Shapes

The frequent requirement for an extrudate having a composite cross section imposes on the die designer the problem of equalizing the velocities across the exit of a die, which may be made up of a number of geometric shapes of dissimilar form and size. If the land length for a die of this type were everywhere equal, the thicker sections would show a much more rapid exit velocity than the thinner ones. Therefore distortion of the extrudate, due to an unbalanced stress field, and melt fracture would result.

Squires[22] suggests that the pressure drop should be approximately equalized in all sections of a composite die. This has the effect of equalizing the shear stresses and may be accomplished by making all the L/r and L/h ratios constant for all branches of the die. The average velocity in each branch can then be determined by dividing the volumetric flow rate Q for each branch by its cross-sectional area A. If the haul-off rate is then taken as being equal to the highest of the calculated velocities, a "draw ratio," defined as the ratio of the haul-off speed to the average exit velocity, can be determined for each branch. Squires states that for thermoplastics a maximum draw ratio of less than 3:1 is required for a die to work successfully; but the applicability of this statement to rubbers requires experimental verification. If any of the draw ratios exceeds the critical draw ratio, the equality in the L/r and L/h ratios must be adjusted in order to equalize them more effectively.

The use of a geometric-shape factor to relate flow in channels of dissimilar cross section can be employed to good effect in the determination of pressure drop and average exit velocity in the branches of a composite die. Miller[14] also suggests that flow through channels of cross sections different from those in Table 4.1 can be determined approximately by taking an "average" shape factor based on the most similar shapes in the table. The relatively small differences between the geometric-shape-factor values should confer a reasonable degree of accuracy to this procedure.

4.6. DETERMINATION AND CONTROL OF EXTRUDER OPERATING CHARACTERISTICS

There is extensive literature on the mathematical modeling of plastics extruders, and a substantial proportion of it is applicable to rubber extrusion. However, the models which have been developed mainly benefit the designers of extruders and are unsuitable for the precise determination of specific operating characteristics, due to the simplifications and assumptions made in the models' development. Models are only useful to the processor for providing a conceptual understanding of the primary factors which determine extruder performance.

The flow of material in the barrel of a single-screw extruder is a result of four flow mechanisms. The first two of these—drag flow and transverse flow—are obtained by resolving the rotational velocity of the screw surface V into a longitudinal component $V \cos\phi$, parallel to the screw flight, and into a transverse component $V \sin\phi$, perpendicular to the screw flight. The drag flow results in the material being conveyed toward the die, while the transverse flow, although not contributing directly to the extruder output, results in a circulatory flow which is important for heat transfer and mixing.

Both the flow restriction caused by the die and the design of the screw result in the creation of a pressure gradient along the screw channel. For most screw designs, the maximum pressure occurs at the die, resulting in a pressure flow in the opposite direction to the drag flow. The pressure gradient also results in leakage flow, through the clearance between the screw flights and the barrel. Hence, the volumetric output Q of the extruder is given by

$$Q = Q_D - Q_P - Q_L \tag{4.33}$$

where the subscripts D, P, and L refer to drag, pressure, and leakage flow, respectively.

The amount of leakage flow will depend largely on the clearance between the screw flights and the barrel, and will increase as wear occurs. This leakage is undesirable for three reasons:

1. It causes reduced output, as described by Eq. (4.33).
2. It can give an extruder undesirable operating characteristics.
3. It will increase the residence time of some of the material passing through the extruder, giving an increased danger of scorch.

Items (1) and (2) are coupled together—as wear and leakage flow increase, the penalty of reduced output can be accepted or the extruder screw speed can be increased to compensate. If the latter course is taken, the extruder power requirement and material temperatures generally increase. The best solution to the problem is to refurbish a screw before its influence on performance becomes troublesome. The pumping characteristics of an extruder can be checked at intervals by the practical method suggested in the following paragraphs for matching extruder performance with die design.

The conditions under which an extruder will operate can be determined by equating the screw and die characteristics. If both are plotted on a single graph of volumetric output *vs.* pressure difference, an "operating point" can be identified at the intersection of the two curves.[23] Points A, B, C, and D

in Figure 4.8 identify four operating points. Referring first to the screw characteristics, the deeper screw shows more sensitivity to back pressure than the shallow one, resulting in the output rate being more dependent on the consistency of material flow properties. Also, this sensitivity to back pressure results in the output from the small die being less than for the shallow screw, even though the pumping capacity is greater when the large die is used.

   The relationship between pressure drop and volumetric output for a die can be determined with reasonable accuracy from the procedures developed in Section 4.5; but, as already indicated, rubber extruders have very complex performance characteristics which preclude accurate theoretical predictions of the influence of die resistance on output. Extruder operation curves, such as those shown in Figure 4.8, must be obtained experimentally. It can now be seen that die design involves matching the theoretically determined characteristics of a die with the experimentally determined characteristics of an extruder, enabling the operating point and the associated output performance to be estimated in advance of die manufacture. This gives the die designer

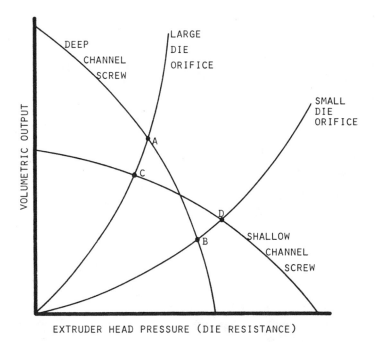

FIGURE 4.8. A simplified extruder operation diagram, showing the influence of the interaction between screw and die characteristics on performance.

much greater freedom to examine alternative die forms, without incurring the expense and time involved in the trial-and-error approach.

The extruder operation curve will obviously depend on the flow and wall-slip behavior of the compound being used; but it can also be strongly influenced by the set temperatures for the barrel zones, head, die, and screw. Surprisingly, haul-off tension has been found to have little effect on volumetric output,[24] enabling this variable to be disregarded. The influence of temperature levels and profiles on the operation curve can be explored using the experiment design and analysis methods described in Section 7.3.

Experimental determination of an operating curve requires a special die fitted with an adjustable flow restrictor and a pressure transducer. The die form is not important, so it can be of a simple circular cross section with a conical lead-in. In small dies, the flow restrictor can take the form of a screw, which can be progressively adjusted to partially block the land region of the die. For larger dies, a restrictor plate, which can be adjusted to partially block the die exit, is preferable. The pressure transducer must be sited in a position where it detects pressure prior to the start of convergent flow to the die exit. Transducers sited in the extruder head for monitoring and control are ideal for this purpose; otherwise, it is necessary to design the special die so that a preconvergence pressure can be measured.

During an extruder trial it is then necessary to determine the influence of the back pressure (die resistance), as measured by the pressure transducer and adjusted by the die restrictor, on output and extrudate temperature, to give an operating graph such as the one shown in Figure 4.9. The extrudate temperature can be simply measured by inserting a handheld needle pyrometer into it.

When the extruder characteristics have been determined, the relationships between output and pressure drop for the die design being investigated can be calculated, using rheological data for a number of temperatures within the regions of interest for high productivity. These relationships can then be plotted onto Figure 4.9 where they are shown as the curves starting from the origin. The operating points of the extruder are now defined by the triple intersections of output rate, pressure, and temperature, identified by the symbol $\otimes$. By drawing a curve through the intersection points, the relationship between screw speed, output, and extrudate temperature is defined, for the particular set-point temperatures used.

The characteristics of a number of alternative die designs can be plotted onto the extruder operating graph, to enable a selection to be made and to indicate the direction which should be taken for improvements in design. When a satisfactory die form has been achieved, it can be manufactured and subjected to a practical trial, after which some adjustments to the form will

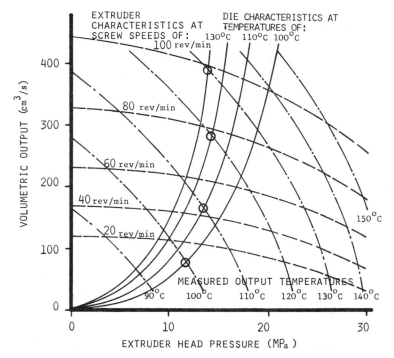

FIGURE 4.9. An operating diagram obtained from a practical extrusion trial on a 90-mm cold-feed extruder and a theoretical die design. The diagram shows a range of operating points at the triple intersections of the screw and die characteristics with material temperatures.

probably be necessary, due to the assumptions and simplifications made in obtaining the die-design equations. However, this route will give dies of superior performance to those produced by the traditional trial-and-error methods. It will also generally shorten the introduction time, particularly when an extrudate of unfamiliar cross section is required.

When a die has been proved to give satisfactory results, it is passed to the production group, who then have the task of controlling the extrusion process so that it continues to give satisfactory results. It is obvious that the flow properties of the feedstock should fall within reasonable tolerance bands, which requires good control of mixing and compound storage time. Moving to the extruder, one of the major causes of variable output is variable input. Rubber extruders are very sensitive to variations in feed rate, and for high-precision extrusion, the feed strip should be of a consistent width and thickness. It is also helpful to subject the feed strip to a constant tension, using an arrangement similar to the one shown in Figure 4.10, with

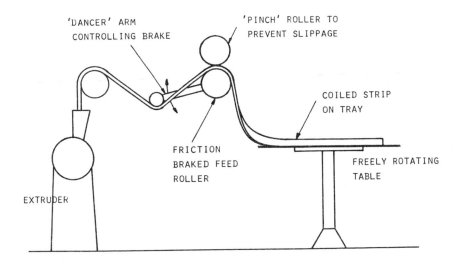

FIGURE 4.10. A simple system for control of strip feeding to an extruder.

a "dancer" arm controlling a simple friction brake. Power-feed devices designed for the purpose of assuring a uniform feed are also available. These usually take the form of feed rollers in a two-roll-mill configuration sited above a wide feed pocket. Francis Shaw and Company use a slipping clutch principle to ensure the uniform feeding, while Scheisser employs a pressure transducer in the feed pocket to control the motion of hydraulically driven rollers. However, the latter device has been used successfully in the slipping clutch mode by setting the hydraulic pressure release valve at an appropriately low level.

Progressive deviations from the operating point defined in Figure 4.9 will be caused by the progressive clogging of screen packs. The deviation will be in the direction of reduced output and increased extrudate temperature. If a pressure transducer is sited in the extruder head, between the screen pack and the die, it can be used in a control loop to set the screw speed. Maintaining the head pressure constant by adjusting the screw speed will help to maintain the output rate at a constant level, although there will be some change due to the extrudate temperature. In a more sophisticated system the dependence of output on temperature could be included in the control algorithm, to give a more consistent output rate.

When the temperature rise from its normal value reaches an upper set limit, a manual or automatic screen pack change can be initiated. For continuous wind-through screen packs, the rate of wind-through can be determined from the extrudate temperature, to ensure that it is only changed when necessary. The control techniques needed to implement these systems are dealt with in Section 7.5.

Although the measures described in the previous paragraphs can minimize variation in the extrudate cross-section dimensions, it is unreasonable to expect them to eliminate it. Random variations will occur due to changes in flow properties within the tolerance bands and through imprecision in the feed, in addition to imperfect compensation of the effects produced by screen-pack clogging. The influence of all these effcts on the extrudate dimensions can be reduced by control of the haul-off tension or speed in response to continuous automatic weighing or measuring of the extrudate. Again Section 7.5, in conjunction with Section 7.4, describes the methods by which this can be achieved.

4.7. CONTINUOUS MIXING

4.7.1. Feedstock Form and Preparation

Continuous mixing involves the incorporation of additives into rubber, followed by distributive and dispersive mixing, which differentiates it from mixing in extruders discussed in Section 4.4.1, where only further distributive mixing of a previously mixed compound is involved. Also, whereas extruders are generally operated with a strip feed, continuous mixing normally demands that solids are fed in a particulate form.

From the processers' point of view there are two routes to obtaining particulate rubber: to buy it in particulate form from a supplier or to granulate it in-house. Suppliers use a number of methods of producing particulate rubber, which influence its particle size, particle shape, and handling behavior. The latter is the main problem of particulate rubber, due to its tendency to creep under self-weight in storage. Although the partitioning agents used to prevent interparticle adhesion are effective, a mechanical interlocking of particles occurs over a period of time, due to flow, precluding the use of simple bulk powder transport, storage, and handling methods. It is possible to use live storage hoppers, with augers or other devices, to break up the interlocked powder mass and initiate discharge; but the additional cost is substantial and must be added to the price premium already paid for choosing particulate rubber instead of bale rubber.

Despite the bulk delivery, storage, and handling route for particulate rubber being currently barred by technical and economic factors, the concept of continuous mixing is still very attractive. Attention has recently been focused on in-house granulation of bale rubber,[25,26] enabling the storage

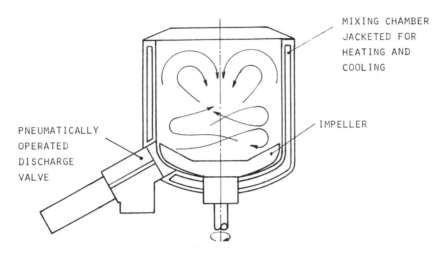

FIGURE 4.11. High-speed intensive powder blender.

time of the granulated rubber to be minimized and the problem with mechanical interlocking of particles thus avoided. It has also been shown that the particle size can be as large as 10 mm without impairing the mixing characteristics, and that the 1-mm particle size of powdered rubber does not confer any significant advantages for incorporation or distributive mixing. This is extremely important, as the energy requirement and cost of granulating a bale of rubber increases in approximate inverse proportion to the square of particle diameter.

When feeding to continuous mixers, particulate rubber is usually preblended with the other particulate and pelletized ingredients of the rubber compound, so that the cost and technical difficulty of metering a number of separate ingredients into the continuous mixer can be avoided. The powder blenders suitable for this test are of two types and are both batch-operated machines:—impeller blenders and trough blenders.

Intensive impeller blenders have variously been called intensive dry blenders, turbo blenders, and high-speed intensive blenders. The construction of these machines, as shown in Figure 4.11, consists of a stainless-steel bowl, jacketed for heating and cooling, with an impeller which rotates at a high speed set in its base. The mixing action depends on all the materials in the blender being lifted in a recirculating vortex by the impeller. For this to occur the fill factor must be in a certain region, avoiding overfilling, when all the material cannot be raised into the vortex and underfilling, when the vortex is unstable and tends to collapse. A typical impeller blender would have a batch weight of 120 kg and be driven by a 40/60 kW motor, giving impeller speeds of 400 and 800 rpm, with mixing times in the region of three to six minutes. During mixing, a temperature rise to 40–60°C is normal, due to the energy dissipated in collisions of particles with each other and with the walls of the mixing vessel. These collisions can present problems for pelletized material, fracturing them into their individual particles and creating handling problems and, if the system is not totally enclosed and sealed, a dust hazard. Small-particle-size ingredients can also cake on the walls of the mixing vessel, if there is moisture present. For this reason the mixing vessel must be maintained above the dew point, to avoid condensation. Despite this, experience has shown that it is possible to add oils and plasticizers into the blender, by injecting them through a rose giving a number of fine streams into the vortex. It has been claimed that up to 40 parts of liquid can be added in this way, without destroying the free-flowing behavior of the powder blend. [27]

Intensive impeller mixers are acceptable if the whole system can be enclosed and if automatic weighing and handling of all ingredients takes place. For companies with a large range of ingredients this is rarely possible, leading to a labor-intensive and inefficient requirement for a large number of manual weighings, and the creation of a dusty environment in the region of the blender which can be substantially worse than that of an internal mixer. Ellwood [25] suggests that the problem can be alleviated by using trough blenders.

Trough blenders are also known as ribbon blenders, due to the form of the blending blades; these are narrow and set helically on a horizontal drive shaft at the longitudinal axis of the trough, so that they sweep the surface of the trough with a small clearance. Each blade then displaces material axially and circumferentially, giving an effective blending action. The low speed of these machines and their gentle blending action gives few problems with pellet attrition and dust hazards. These machines are available in a very wide range of sizes, from one liter for a bench top model to 14 cubic meters. Ellwood [25] suggests that a machine with a batch size which will provide approximately one hour's feed for the continuous mixer is appropriate. This reduces the frequency of weighing to an economically attractive level and allows sufficient time for this low-speed machine to achieve a uniform blend. This slow mixing speed also requires that the rate of addition of oils and plasticizers be slow, to avoid overwetting a small proportion of the blend and creating a slurry.

Feeding of continuous mixers from batch-blending machines requires intermediate storage of the blended materials. The main problem during this storage is segregation of the ingredients, which results in the composition of the feedstock deviating from that required for the compound. Segregation occurs when blends of particles of different sizes and densities are used. Rubber, with a density in the region of $1000 \, kg/m^3$ and particle sizes of up to 10 mm, blended with zinc oxide, having a density of 5500 kg/m^3 and particle sizes less than 0.1 mm,presents a serious segregation problem. For this reason "live" intermediate storage is preferable to a simple hopper, enabling the uniform distribution of ingredients produced during blending to be maintained. A wide range of live hopper designs are available; but a very-low-speed ribbon blender provides an effective and compact solution to this storage problem, and has a reasonably low-energy requirement.

In Figure 4.12 a complete system for supplying a continuous mixer with a blended feedstock is shown. Bulk filler storage, conveying, weighing, and feeding equipment is dealt with in Sections 8.2.3 and 8.3.1 together with oil or plasticizer weighing and injection units.

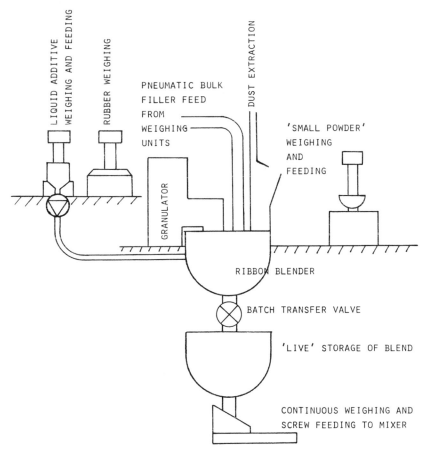

FIGURE 4.12. Weighing, blending, and feeding system for the continuous mixing of particulate rubber.

The bulk density of the blend supplied by the metering equipment of Figure 4.12 is usually in the region of $500 \, kg/m^3$, in comparison with $1200 \, kg/m^3$ for a typical finished mix. For some continuous mixers, particularly the single-screw type, a precompaction operation is necessary, to ensure that the channels of the mixing screw are adequately filled. Compactors generally take the form of conical extruders,[28] of light construction and low-power requirements.

4.7.2. Single-Screw Continuous Mixers

Single-screw continuous mixers have been developed by Werner and Pfleiderer, Berstorff, Iddon, and others. These are essentially cold-feed extruders equipped with mixing screws designed to input adequate distributive and dispersive mixing to a particulate preblend, through flow division and the generation of localized high stresses. They have the objective of producing an extrudate of a quality similar to that achieved by a cold-feed extruder.

Most practical mixing screws are variants on the barrier screw design, of which the Werner and Pfleiderer Contimix screw shown in Figure 4.13 is a good example. At each set of dams in the screw channel, the flow stream is divided and both streams pass through the high-stress region created by the small clearances between the top of the dam and the barrel. The distributive mixing is influenced strongly by the length-to-diameter ratio of the screw and

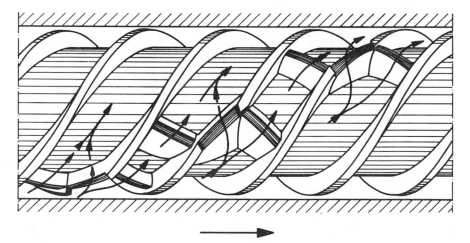

FIGURE 4.13. Barrier screw design for the Werner and Pfleiderer EVK (Contimix) continuous mixer.

the dispersive mixing by the size of the clearances between the dams and the barrel.

Lehnen[28] reports that a 90-mm screw-diameter machine having an L/D ratio in the range 20:1 to 24:1 requires a motor of at least 45 kW and is capable of outputs between 140 kg/h and 170 kg/h. However, single-screw continuous mixers must be considered as dedicated machines, for which the processing "window," created by the combination of screw geometry, operating variables, and compound flow behavior, will only produce a viable output rate of adequately mixed compounds shaped into an extrudate with acceptable dimensions for a narrow range of compound types. It is the strong interaction of the conditions for mixing efficiency with those for extrudate dimensions, coupled with temperature constraints, which reduces the versatility of these machines. This identifies them as being best suited to production lines where long runs and a narrow range of compounds are used.

Methods of determining operating conditions similar to those described for cold-feed extruders in Section 4.6 can be used; but the combinations of screw speed and back pressure which are of interest for their output rate and extrudate temperature will be constrained by the need for the output to be adequately mixed, as described in the previous paragraph.

Finally, a number of machine manufacturers produce mixing screws for high-precision extrusion of previously mixed compounds, and do not claim that these screws will also perform dispersive mixing. Among these are the NRM Plastiscrew, used in Farrel Bridge extruders, and the Iddon high-intensity mixing scroll.

4.7.3. Integrated Mixing Systems

If the single-screw continuous mixers are "dedicated" machines, optimized to operate at high efficiency with a limited range of compounds and extrudate cross sections, there exists a requirement for undedicated machines, which are capable of mixing and extruding a wide range of materials. Farrel Bridge, Stewart Bolling, and Baker Perkins have each achieved such a machine by separating the mixing and extrusion functions. The Farrel Bridge MVX (mix–vent–extrude) is the most highly developed of these machines, while the Baker Perkins MPC/V has been mainly used for thermoplastic and thermoset compounding, although its design seems well suited to rubber mixing. In each of these machines a twin rotor or screw mixing section is closely coupled with an extruder, as shown in Figure 4.14 for the Farrel Bridge MVX. The following discussion will concentrate on this machine, information on the other two being sparse and undetailed.

The MVX presents a formidable problem for optimization of operating conditions, although establishing conditions under which it will operate

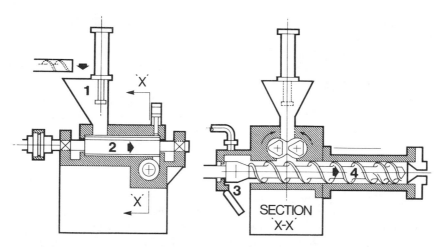

FIGURE 4.14. Farrel Bridge MVX (mix-vent-extrude) integrated continuous mixing system.

successfully appears to present few problems, indicating that it has a very wide operating window. Its performance is dependent on a number of strongly interacting variables. Starting with the mixing section, the residence time, which is equivalent to mixing time, is controlled by the rate at which the extruder accepts material. The temperature rise in the material passing through the mixer will be a function of the residence time, the temperature of the water circulating through the rotors and jackets, the fill factor, and the rotor speed. Incorporation, distributive mixing, and dispersive mixing will also depend on these variables, emphasizing the similarities with internal mixing.

The extruder characteristics can be determined using methods similar to those described for a cold-feed extruder in Section 4.6, with the exception that the feed temperature is now an important variable. In fact the extruder is a hot-feed machine, but is under much more precise feed control than is achieved with a two-roll mill strip-feed arrangement.

To reduce the number of variables which have to be adjusted during a setting-up exercise and to improve the consistency of performance, Farrel Bridge offer microprocessor control of the MVX, with a number of control options. These go beyond the usual options on local automatic control of machine functions and include closed-loop control of variables, such as the temperatures of the material in the mixing and extrusion sections and the extruder head pressure, using the techniques described in Sections 7.4 and 7.5. However, as mentioned previously, optimization is still a formidable task, and is seen to be essential to maximize productivity when long runs are envisaged. The techniques described in Section 7.3 provide an ideal tool for dealing with this complex multivariable process.

In conclusion, Ellwood[25] points out a number of technical and economic advantages to be gained from using continuous mixing in preference to batch mixing. Some of these apply to continuous mixing in general and some are specific to the MVX. He envisages the MVX producing feedstock for downstream processes, to permit a direct and simple comparison with batch internal mixers.

It is a well-known fact that continuous processes, running under equilibrium conditions, are inherently capable of producing a more consistent product than batch mixers operating under cyclic conditions. In addition, mixes with lower rubber content can be produced using continuous mixers, in compounds where the rubber is only required as a binder. This follows the internal mixer practice of "seeding" a batch with a portion of the preceding one, to initiate incorporation; but, apparently, continuous seeding is more efficient.

Ellwood[25] shows, by a cost analysis comparing a 240/175 MVX system with an F80 Banbury dumping into a two-roll mill mixing a typical GRG (general rubber goods) compound, that:

1. The installed capital cost of the MVX is 8% greater than that of the F80 Banbury, assuming the provision of in-house granulation facilities for the former.
2. The plant cost of the MVX, obtained by dividing the installed cost by output capacity, based on a 24-hour working period, is 72% of that for the F80 (16.07 tonnes for the F80 in comparison with 24.48 tonnes for the MVX).
3. The energy usage of the MVX system, including the granulator, is 89% of that for the F80 Banbury system.
4. The labor cost of the MVX system is 40% of that for the F80 Banbury system, assuming that the trough blender takes sufficient material to keep the MVX supplied with pre-blend for one hour.

4.7.4. Continuous Mixers in Batch Mixing Systems

Continuous mixers have been used for the second stage of two-stage mixing systems employing an internal mixer for the first stage. In this position they can be supplied with the material from the first-stage mixing in pellet form, to be metered in the required proportion with the curatives and accelerators. Both Whitaker[29] and Peakman[30] point to the use of the Transfermix machine, developed by Uniroyal, for this purpose; and the latter also refers to the Farrel Continuous Mixer (FCM). The primary requirement of both of these machines in this application is effective distributive mixing; neither of them are capable of producing a finished extrudate.

The Transfermix is a single-screw machine which operates on the so-called enforced order principle. The screw flights vary in depth, becoming shallow where opposite-hand screw flights cut in the barrel increase in depth. Material is exchanged between the screw and barrel flights, giving distributive mixing of exponential characteristics with a claimed narrow residence time distribution.

In concept the FCM is a continuous internal mixer. Contrarotating screws feed material into the mixing zone which has rotors with a form resembling those of a Banbury internal mixer. The output rate can be controlled by a hydraulically operated discharge orifice, enabling the residence time to be controlled. The fill factor is determined by the balance of feed and discharge rates, as for the mixing section of the Farrel–Bridge MVX. In fact there are many operational similarities and the FCM has been closely coupled with an extruder in the MVX configuration. However, it should not be inferred that their efficiencies as mixers are similar.

REFERENCES

1. Iddon, J. M., *Eur. Rub. J.* **158** (3), 18 (1976).
2. Anders, D., *Rub. World* **166** (1), 31 (1972).
3. Evans, C. W. *Rub. Age* **105** (9), 33 (1973).
4. Thorne, B. A., Papers given at Plastics and Rubber Institute Symposium, "Optimisation of Quality and Productivity in Rubber Processing," Burton-upon-Trent, U.K. (November 1978).
5. Harms, E. G., *Eur. Rub. J.* **160** (5), 33 (1978).
6. Hindmarch, R. S., and G. M. Gale, Paper given at Meeting of ACS Rubber Division, Philadelphia, Pennsylvania (May, 1982).
7. Carley, J. F., and W. L. Smith, *Atlanta Gazette*, May, p. 594 (1975).
8. Cogswell, F. N., *Polym. Eng. Sci.* **12** (1), 64 (1972).
9. Cogswell, F. N., and P. Lamb, *Trans. Plast. Inst.* **35**, 809 (1967).
10. Powell, P. C., Paper given at Joint British Rheology Society and Plastics and Rubber Institute Conference, "Rheology in Polymer Processing," Loughborough University, U.K. (1975).
11. Brydson, J. A., *Flow Properties of Polymer Melts*, 2nd ed., George Godwin, London (1981).
12. Fenner, R. T., and F. Nadiri, Private Communication, Imperial College of Science and Technology, London, (1979).
13. Röthemeyer, F., *Kunststoffe* **59** (6), 333 (1969).
14. Miller, C., *Ind. Eng. Chem. Fundam.* **11** (4), 524 (1972).
15. Knudsen, J. G., and D. L. Katz, "Fluid Dynamics and Heat Transfer," McGraw-Hill, New York, (1958), p. 97.
16. Sparrow, E. M., *A. I. Chem. E. J.* **8** (5), 509 (1962).

17. Bougher, D. F., and G. E. Alves, in J. H. Perry (Ed.), *Chemical Engineers Handbook* (4th ed.), Section 5, McGraw-Hill, New York (1963).
18. Shih, F. S., *Can. J. Chem. Eng.* **45**, 285 (1967).
19. Shih, F. S., Paper given at 61st Annual Meeting of American Institute of Chemical Engineering, Los Angeles, California (December 1968).
20. Lahti, P., *J SPE* **19**, 619 (1963).
21. Fisher, F. G., and W. A. Malsen, *Brit. Plast.* **31** (7), 276 (1958).
22. Squires, P. H., Contribution to *Processing of Thermoplastic Materials*, E. C. Bernhardt (Ed.), Van Nostrand Reinhold, New York (1959).
23. Tadmore, Z., and C. G. Gogos, *Principles of Polymer Processing*, Wiley, New York (1979).
24. Kakouris, A., M. Phil Thesis, Loughborough University, U.K. (1983).
25. Ellwood, H., Paper given at the International Rubber Conference, Harrogate, U.K. (1981).
26. Wheelans, M. A., Paper given at the International Rubber Conference, Harrogate, U.K. (1981).
27. Fielder, T. K., Paper given at the Plastics and Rubber Institute Conference, "Powdered Rubbers," Southampton University, U.K. (1978).
28. Lehnen, J. P., Paper given at the Plastics and Rubber Institute Conference, "Powdered Rubbers," Southampton University, U.K. (1978).
29. Whitaker, P., *J. Inst. Rub. Ind.* **4** (4), 153 (1970).
30. Peakman, M. G., *J. Inst. Rub. Ind.* **4** (1), 35 (1970).

Calendering and Milling

5.1. INTRODUCTION

Calendering and milling are sufficiently similar to enable them to be treated together. Both have been used for many years in the rubber industry, resulting in a considerable fund of practical expertise for their operation. However, technical studies of both the two-roll mill and calender yield information which can be put to practical use in any company.

Residence time may be used to differentiate conveniently between calendering and milling. Two-roll milling involves a substantial residence time and many passes through a single rolling nip. This is necessary for mixing or for raising the temperature of premixed material to that required by subsequent processes. Calendering is essentially a shaping operation where the work done on the material is required to produce a change of shape, not a change of state. This involves only a small number of passes between rolling nips, but does require prewarming and homogenization of the feed material—on a two-roll mill, for example. For continuity of production, calendering operations which use more than one rolling nip have to be performed on machines with the requisite number of rolls to form these nips.

5.2. THE OPERATING CHARACTERISTICS OF TWO-ROLL MILLS

5.2.1. Factors Influencing Mill Capacities

For effective mixing and homogenization in a rolling nip, a bank of material must be formed above the nip. This reservoir of material ensures that the nip is adequately fed and that effective flow work is being done on the material. The capacity of a two-roll mill is determined by the setting of the nip, which fixes the volume of material banded round one roll, and by the need for the bank of material to be in continual motion, that is, a rolling

TABLE 5.1
Two-roll Mill Capacities

Roll length (m)	Roll diameter (m)	Mix capacity (kg)	
		Minimum	Maximum
0.75	0.35	9	13.5
1.1	0.4	13.5	23
1.2	0.45	20	32
1.5	0.55	34	57
1.8	0.6	57	91
2.1	0.65	68	114
2.1	0.7	80	136

bank. If the bank of material is too large, stationary volumes of material will form which take no part in the milling operation. Table 5.1 gives typical mill sizes and capacities.

5.2.2. Streamlines and Velocity Profiles in Rolling Nips

The configuration of material in a rolling nip is shown in Figure 5.1. The rolls act upon a thin flowing wedge of material which is simultaneously compressed and forced to flow between the rolls. The minimum distance between the rolls $2h_0$ is the nip. Using hydrodynamic relations due to Gaskell[1] and developed by Bergen,[2] it is possible to construct streamline flow patterns and velocity profiles for symmetric and nonsymmetric milling. These expressions refer to milling or calendering with the rolls forming the nip running at even spreads or at differential speeds, respectively. The ratio of roll speeds is termed the "friction ratio."

The velocity profiles and corresponding streamlines for a symmetrical mill show the material flowing more or less parallel to the roll surface, which is also a streamline. Toward the center of the upstream region of the wedge backflow occurs, which results in the formation of two closed vortices. This effect can be observed in the rolling bank of a mill with equal roll speeds. The material enclosed in these vortices does not enter the nip, so that there exists a fraction of the material which is never subjected to high shear forces and mixing is unsatisfactory. The upstream point beyond which these closed vortices form is termed the stagnation point.

In practice nonsymmetrical milling is used to improve the mixing action of the rolling nip. Unequal roll speeds serve to warp the velocity profile and will direct the backflow toward the slower roll. Bergen[2] shows that the streamlines now all lead through areas of finite shear deformation beyond the

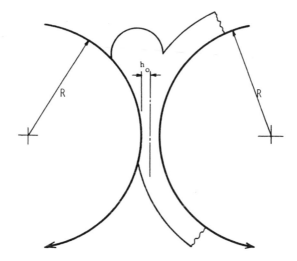

FIGURE 5.1. Band formation on a two-roll mill or calender.

stagnation point; this results in increased shear deformation in comparison with symmetrical milling.

In spite of the improvement brought about by nonsymmetrical milling, the mixing efficiency is still very poor due to the presence of closed streamlines and the absence of material exchange along the length of the nip. These conditions must be disturbed by an operator cutting and cross-blending or by a mechanical stockblender.

5.2.3. Analysis of Banding and Bagging on a Two-Roll Mill

The prediction of material behavior on a two-roll mill is important for the identification of machine conditions which give trouble-free milling. In a series of experiments with various grades of SBR, EPR, and BR, Tokita and White[3-5] identified four distinctive regions of flow behavior. In region 1 there is essentially no flow between the rolls due to the force required to cause flow through the nip being greater than the force of adhesion of the rubber to the roll surface. Hence there exists a critical stress which defines the transition between region 1, where no flow occurs, and region 2, where it is possible to form a band around the front roll.

Since rubber is a viscoelastic material the critical stress is dependent on the time scale of the milling operation compared with the relaxation time of the material. The time scale of the operation and hence the appearance of region 2 behavior can therefore be controlled by increasing the nip setting and/or reducing the roll speed of the mill. Alternatively or additionally, a measure of control is possible by increasing the material temperature and thus reducing the relaxation time. However, it must be noted that the coefficients of friction of most rubbers decrease with increasing temperature. The simplest model which exhibits the essential viscoelastic behavior is the Maxwell model, for which the relaxation time Γ is

$$\Gamma = \eta_N/G \qquad (5.1)$$

where η_N is Newtonian viscosity, G is shear modulus, and Γ is the time for the stress to relax to 63.2% of its initial value in a stress relaxation test.

Milling in region 2 is characterized by a tight elastic band of material clinging to one roll. As the roll speed increases minute tears and surface irregularities form which heal within one revolution. A decrease in nip setting gives a smoother, tight band. If the band is cut, the rubber will retract from the cut because of stored elastic energy.

The tears in the band will only propagate if the stored energy arising from deformation in the nip exceeds the critical tearing energy of the material being milled, and the rate of tear propagation is a function of the stored energy. The upper limit for stable operation is set by a stored-energy criterion. Quantitative determination of the criterion is difficult, but qualitative assessments are possible. The critical tearing-energy density decreases rapidly with an increase in temperature, while the stored energy decreases less rapidly. As the temperature is increased, a limit is reached where tears propagate rapidly and a transition to region 3 behavior is usually observed. In some cases there is a direct transition to region 4.[6]

There is a second criterion of the upper limit of stability in region 2 milling; the stress generated in a single pass, which is a function of mill geometry and operating conditions, must essentially relax in one revolution of the rolls. If the stress does not relax, the material will become oriented along the flow streamlines and shear mixing will decrease. Furthermore, stress will rapidly build up and cause gross slippage. For the stress to relax in one revolution, the relaxation time must be small compared with the period of roll rotation. Increasing the material temperature to increase the rate of stress relaxation and reducing the roll speed will both improve the stability of the region 2 operation.

Region 3 behavior is characterized by crumbling and tearing. The band may be unable to support its weight and hangs as a bag from the roll. Near the upper transition the surface becomes rippled rather than torn and the limit is set by hydrodynamic stability.[8] An increase in temperature, a decrease in roll speed, and an increase in nip setting all promote a transition to the stable region 4. In a series of experiments with constant roll speed and geometry, White and Tokita[3] found that the transition occurred at a temperature corresponding to the same relaxation time for different materials.

Any change in material properties, such as broadening the molecular-weight distribution, which increases the characteristic tearing energy, will improve the milling. In a comparison of two polybutadienes, White and Tokita found that a material with a broad moecular-weight distribution gave good mixing with a broad range of conditions in region 2. A narrow molecular-weight material, showing a similar Mooney viscosity, gave only region 3 behavior, with severe crumbling.

Region 4 behavior is characterized by a stable band, usually around the slower roll, and predominantly viscous flow behavior.

5.2.4. Front–Back Roll Transitions in Band Formation

Normally, rubber is banded on the slow roll of a two-roll mill, which is then designated as the front roll. It is desirable that milling conditions are adjusted so that the band will form preferentially around the front roll. Tokita[9] makes the observation that for a given material at a certain mill temperature there will be a tendency for back-roll band formation if the nip distance is very small. Increasing the nip distance will cause a transition of the band to the front roll at a certain nip setting. Reducing the nip from a wide setting will cause a front–back transition at a similar nip setting, thus establishing a transition point.

Tokita[9] establishes a band-formation index, N_0, which is constant at the front–back transition regardless of the type of mill;

$$N_0 = (f + 1) \left(\frac{2h_0}{R} \right)^{1/2} \tag{5.2}$$

where f is the friction ratio, $2h_0$ is the nip distance, and R is the roll radius. For a factory mill a similar dimensionless number, N^*, can be given which characterizes the operation of the mill in production service:

$$N^* = (f^* + 1) \left(\frac{2h_0^*}{R^*} \right)^{1/2} \tag{5.3}$$

Here f^* and R^* are the friction ratio and the radius of the roll, respectively, of the factory mill and $2h_0^*$ is the nip distance, which must be adjusted to prevent back-roll banding for a material of given N_0. Tokita designates N^* as the "mill operational index." For front-roll band formation,

$$N^* > N_0 \tag{5.4}$$

and from Eqs. (5.2)–(5.4), the nip distance of the mill in the factory for front-roll band formation is

$$2h_0^* > N_0^2 \left(\frac{R^*}{(f^* + 1)^2} \right) \tag{5.5}$$

For a factory mill, N^* may be predetermined by batch size and rolling band formation. This leaves temperature as the controlling variable and may be utilized to obtain $N^* > N_0$. Increasing the milling temperature has the effect of decreasing N_0. If the N_0 values of a rubber mix are measured as a function of temperature, then a temperature can be found at which $N^* > N_0$. Tokita notes that the surface temperature of the rubber on the mill is more important than the bulk temperature for the determination of N_0.

In the case of continuous milling, where the mix has already been banded on the front roll, the relation between the front–back transition, bagging, and the geometrical and kinematic parameters of a mill are different from Eq. (5.2) with respect to the friction ratio. It is generally known from factory experience that an increase of nip distance and a decrease of the friction ratio will suppress the bagging phenomena. Tokita postulates that the point of separation of the material from the back roll advances around the back roll as the friction ratio increases. This gives more time for the material to adhere to the back roll and can eventually result in its separation from the front roll.

The front–back transitions for discontinuous and continuous milling operations therefore depend upon the friction ratio f in opposite ways. Given a low friction ratio, a mix has difficulty in self-banding on the drop mill (discontinuous), but shows less tendency for bagging on a blender mill (continuous). The opposite occurs for a high friction ratio. Therefore the friction ratio for a mill has to be chosen for the type of mill operation;—a higher friction ratio to achieve consistent self-banding on a drop mill and a lower value for a blender mill.

5.2.5. Two-Roll Mill Applications and Operations

The inherently poor control of two-roll mills has already been noted. Coupled with their labor-intensive operation and dependence on operator integrity for consistency of performance, it would appear that two-roll mills seem destined for a rapid demise. However, two-roll mills present three valuable assets: low-temperature running characteristics, versatility, and self-purging. They can operate successfully with temperature-sensitive compounds, and change operation from one compound to another in a rapid and efficient manner, without incurring any waste. In a short-production-run environment, these advantages often outweigh the disadvantages; but when extended production runs can be planned, replacement of two-roll mills by extruders and continuous mixers is often justified by the return on investment.

Two-roll mills have been largely dispossessed of their primary mixing role by the advent of stringent regulations concerning dust hazards. There is no doubt that the incorporation of particulate fillers on a two-roll mill constitutes a dust hazard. Some minor ingredients are still added as loose powders; but it seems that these additions will be phased out, since the minor additives are often the ones which have the most dubious health record. However, addition of curatives and accelerators to a batch of rubber on a two-roll mill, after discharge from an internal mixer, is still required in many parts of the industry. Additives which are pelletized or premixed with rubber are being used to overcome the dust problem. For this to be effective, good distributive mixing is required and many two-roll mills are fitted with a stockblender for this purpose, providing good in-batch uniformity. A schematic of a stockblender is shown in Figure 5.2.

As a shaping operation after internal mixing, to provide a sheet or strip of material for feeding to a cooler and subsequently to downstream processes, the two-roll mill works well, provided that the milling is minimized and standardized. Specialized feeding arrangements which enable the two-roll mill to act like a roller die, producing a sheet of the desired thickness in a single pass, achieve this desired level of performance and are capable of semiautomatic operation. As a consequence, most of the disadvantages of the two-roll mill are eliminated; but its important self-purging capability is retained. However, the single-pass mode of operation obviously precludes the addition of materials on the dump mill.

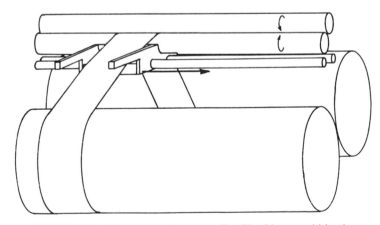

FIGURE 5.2. Schematic of a two-roll mill with a stockblender.

Further down the production line, two-roll mills are used as rubber compound preheating and feeding devices for hot-feed extrusion and calendering. The former application is rapidly being phased out due to the technical and economic advantages of the cold-feed extruder; but the two-roll mill still performs a useful function in short-run calendering, due again to its self-purging characteristics. This is referred to again in Sections 5.6.1 and 5.7.

5.3. MILL AND CALENDER ROLL TEMPERATURE CONTROL

Most mills have relatively primitive temperature-control systems, whereas the precision usually demanded of calenders requires a uniform and well-controlled temperature profile across the surface of each roll. Tempering systems, ranging from simple manual water flood cooling and steam heating to heat-exchanger systems closely controlling the temperature of the circulating fluid, are adequately described in Chapters 3 and 4.

The temperature-control systems for calenders, due to the extremely short residence time, cannot compensate effectively for transient changes in feed material and the main consideration is maintaining temperature stability. For this reason the basic function of a calender temperature-control system is to hold the inlet temperature to the roll constant and to take away the heat developed by normal calendering.[10]

Most calenders (and mills) have cored rolls of the type illustrated in Figure 5.3(a), which are of massive construction due to the need for high bending stiffness. This results in a high thermal capacity, a relatively low rate of heat transfer to and from the roll surface, and a falloff in temperature at each roll edge, where there is an increase in metal thickness and an increase in effective surface area for convection losses and where the journals are effective heat sinks. If the full width of the roll is to be utilized in a calendering operation, it may be necessary to set electrical cartridge heaters in the roll edges to achieve a uniform temperature profile.

A nonuniform temperature distribution, resulting from the cored roll being only partially full of the temperature-control fluid, will occur if the calender rolls are stationary during a warmup period or during a product

FIGURE 5.3. Cooling arrangements for mill and calender rolls. (a) Cored roll. (b) Peripherally drilled roll.

change. In addition to producing a cyclic change in rubber flow properties during each revolution, eccentricity of the roll will result from differential thermal expansion. The effects of these factors on the gauge of the calender product may persist for an hour or more.

The peripherally drilled roll, illustrated in Figure 5.3(b), was developed to improve the rate of heat transfer to and from the roll surface and to remedy the temperature falloff at the roll edges. Both cored and peripherally drilled rolls can be controlled within ± 2°C with standard set-point tempering systems; but this accuracy is a function of calendering speed. Under normal working conditions the surface temperature of cored rolls can be maintained within ± 2°C of the setpoint up to 30–40 m per minute. Above this speed the rate of heat transfer to the circulating fluid is insufficient to prevent an unacceptable temperature rise at the roll surface and peripherally drilled rolls are needed.

5.4. CALENDER CONFIGURATIONS AND OPERATIONS

The choice of calender configuration is determined by the projected applications—each fabric and rubber path through the machine must be considered in conjunction with the required gauge accuracy and the convenience of the configuration for siting ancillary equipment. The majority of calenders installed in general-rubber-goods companies are three-roll units, whereas four-roll machines are normally used by the tire manufacturer. The advantage of the three-roll calender for general goods is its versatility. It can produce a sheet of rubber and friction or skim coat a fabric. When producing a rubber sheet, only a two-roll calender is generally needed and the bottom roll becomes a cooling drum. The four-roll calender has the advantage, for the tire industry, of being capable of double topping. These operations are illustrated in Figure 5.4.

The first and most obvious calender configuration is the vertical or stack type. The faults of this derive from the difficulty of feeding a continuous strip of material to a vertically oriented nip and from the interaction of forces between nips. This interaction influences the bending of the middle roll or rolls and also results in roll "float." The latter effect is a particular problem of vertical calenders having plain journal bearings, which require larger clearances than roller bearings to operate. The roll is able to float up or down, within the bearing clearances, in response to the balance of nip forces and roll weight. The magnitude of this effect will obviously increase with bearing wear.

The disadvantages of nip-force interaction and roll float are practically eliminated by the triangular and horizontal "Z" configurations. The nip

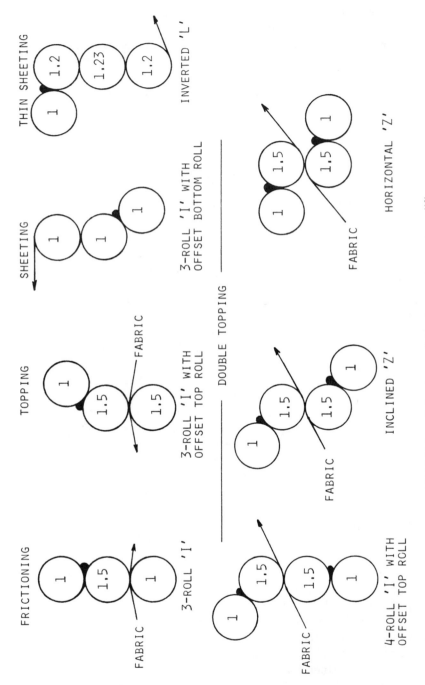

FIGURE 5.4. Calender configurations and operations.[10]

forces act in planes at 90° to each other and roll float is minimal, due to the direction of the resultant of the nip forces being relatively insensitive to changes in these forces.

With the triangular and "Z" calenders, and any others which have adjacent nips set at an angle to each other, preload (zero clearance) devices may be fitted to maintain the rolls in their normal running position when fully loaded and to even out roll movement during load fluctuations. Lee[10] considers preloads to be necessary if close tolerances are to be maintained, particularly at the start and finish of a production run.

Although triangular and "Z" configurations present the best possibility for high-precision calendering, accessibility can be a problem. Offset top roll and inclined Z designs have been developed as a good compromise between accessibility and nip-force interaction.

Perlberg[11] reports that tire cord fabric has been calendered to an accuracy of 0.006 mm using an inclined "Z" calender having a 50° offset on the bottom roll (Figure 5.4) and equipped with preloaded roller bearings.

5.5. ROLL DEFLECTION AND METHODS OF CORRECTION

The deflection of a calender roll due to nip forces can readily be calculated by treating the roll as a beam, if the nip forces are assumed to be uniformly distributed along the length of the nip and the second moment of area of the roll is assumed to be constant.[12] In practice deviations from the assumptions made in the analysis arise from nonuniform temperature distributions along the calender roll (low temperatures at the roll ends), and from the nonuniform cross section of the calender roll. Additionally, the separating forces depend on nip distance and will, therefore, decrease with increasing roll deflection until an equilibrium deflection profile is reached. This equilibrium profile is the so-called "oxbow" contour which is observed on all calenders.

To compensate for the practical factors which result in the oxbow contour the "reverse crown" roll contour has been developed.[13] This is obtained by grinding the roll so that its shape differs from that predicted from the theoretical bending analysis by what might be called a "reverse oxbow." That is, the crown is made lower at those points where the sheet is normally thick. The amount of "reverse crown" required varies with size of roll and calender application. However, if it is adjusted for the thinnest product that is likely to be produced, Gooch[13] states that it will compensate adequately for the oxbow effect over the entire working range of thickness.

To provide active methods of compensation for roll deflection, with adjustments to cater for a wide range of processing conditions, the

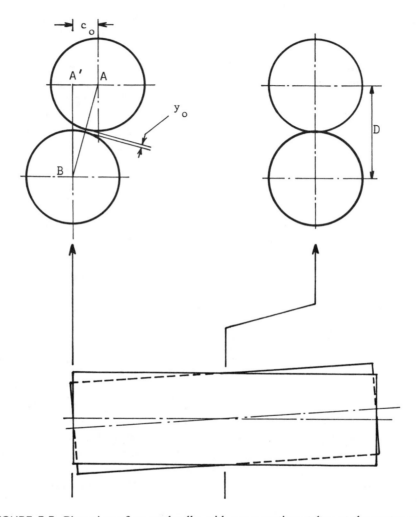

FIGURE 5.5. Plan view of crossed rolls, with cross sections taken at the center and one end of the roll showing the effect of crossing on nip distance.

techniques of roll crossing and roll bending have been developed. Figure 5.5 is a diagrammatic representation of two crossed rolls which are tangent at their longitudinal midpoint and therefore have, for simplicity in this analysis, a zero-clearance nip. The angle of crossing is, in practice, about 1°, and the two end faces are nearly coplanar. Thus, triangle $A'AB$ can be taken to be a right triangle, where A' would be the position of A if the rolls were not crossed. The distance $A'B$ is equivalent to one roll diameter D and AA', the

crossing at the end face, is designated C_0. From Pythagoras's theorem the
nip clearance at the end of each roll y_0 is given by

$$y_0 = (C_0^2 + D^2)^{1/2} - D \qquad (5.6)$$

For the nip clearance at any point Z measured from the end of the roll
towards the center,

$$y = y_0 \left(1 - \frac{2Z}{L}\right)^2 \qquad (5.7)$$

A curve of y vs. Z is termed the contour correction curve of roll crossing.
The preceding analysis assumes that there is no working load to cause roll
deflection.

For the case of roll bending, the theoretical contour (again when there
is no working load) is that produced by two couples acting on a uniform
beam. The bending deflection then produced at the center of the roll is

$$y_0 = \frac{ML^2}{8EI_x} \qquad (5.8)$$

and

$$y = y_0 - \frac{M}{2EI_x}(LZ - Z^2) \qquad (5.9)$$

Then

$$y = y_0 \left(1 - \frac{2Z}{L}\right)^2 \qquad (5.10)$$

This gives the result that the contour correction curves for roll bending
and roll crossing are identical if small second-order errors in the roll-crossing
contour are neglected. Roll crossing requires that each roll has an individual
drive shaft equipped with universal couplings in order to accommodate the
resulting roll misalignment. This increases the cost of roll crossing over and
above that of roll bending, which can be fitted to a bull-gear calender (one
where the drive is transmitted from roll to roll via directly coupled gears).
However, the maximum practical compensation which may be achieved with
roll bending, dependent upon roll size, is about 0.075 mm,[14] due to
maximum allowable bearing loads and to roll neck bending stresses imposed
by both the bending and separating forces. The range of roll crossing is
limited only by the stresses resulting from the separating force and can
provide a roll center to roll end nip clearance difference of about 0.6 mm.[14]

If the full capabilities of roll crossing are to be realized, it must be applied to a pair of rolls which are free to deflect, for example, the top and offset rolls of an inverted-L calender or any roll pair of a Z calender.

5.6. FEEDING; SHEET COOLING, AND BATCH-OFF EQUIPMENT

5.6.1. Feeding Methods

The supply of a material of uniform flow properties to the calender nip at a constant rate is a prerequisite for dimensional accuracy and good surface appearance. A uniform roll-separating force is important and the temperature band in which successful calendering can take place may be quite narrow. The rolling bank size influences both temperature and roll-separating force.

As calender speed is increased, the control of flow properties and feed rate necessary for successful manufacture becomes more critical. There are calendering operations which, due to their slow speed and wide product tolerances, operate successfully with hand feeding and very primitive controls. However, economics usually dictate that a more sophisticated approach be adopted if substantial quantities of calendered products are needed.

For uniform feed rate and even feed across the width of the calender nip, a conveyor feed from the preplasticization machine(s), terminating in a pendulum or wig-wag conveyor head which oscillates back and forth across the width of the nip, is necessary. A conveyor system has the further advantage that, for a given calender speed, the transit time is uniform and the temperature drop during transit is also uniform, although dependent on changes in ambient temperature.

For the preplasticization machinery, a choice must be made between two alternatives—two-roll mills or cold-feed extruders. The comparative advantages of the two methods will depend on calender usage.

To supply a strip of material of uniform flow properties and temperature a two-roll mill must operate under equilibrium conditions, which cannot be achieved if the batch size is progressively changing due to material being removed to feed the calender. To overcome this problem Willshaw[14] recommends that the preplasticization system should be progressional, preferably allowing for working on three separate mills interconnected by strip conveyors.

A cold-feed extruder, particularly one equipped with a mixing section, is capable of giving the calender feed material a uniform shear history and of closely controlling the material temperature. However, whereas the two-roll-mill system can be run at constant speed and changes in calendering rate and gauge accommodated by changing the feed strip width, the extruder output control may be more complicated, whether it is achieved by a variable die orifice or screw speed. Both die resistance and screw speed influence the extrudate temperature. This problem must be faced during installation and commissioning, but thereafter an extruder is capable of a more uniform feed than a two-roll-mill system—hence its use for high-precision operations, such as the production of tire-cord fabric.

A further factor for choice of preplasticization machinery is the convenience of changeover from one compound to another. The speed of changeover will be largely dictated by the time required to change and restabilize the calender roll temperature. Drilled rolls have an advantage over cored rolls here. With the two-roll-mill system the possibility of compound-to-compound contamination is negligible, except by operator error. Also, the new compound can be introduced into the mill progression while the previous run is being finished. The extruder system is less versatile. There will inevitably be some material wastage when the previous compound is purged from the extruder, and there is a definite possibility of contamination, dependent on the extruder design and purging procedure.

5.6.2. Sheet Cooling and Batch-Off Equipment

The common requirement in sheet cooling and batch-off equiment is that the calendered rubber product should be sufficiently cooled so that it does not adhere to the liner—the interleaving material in which most calendered products are wrapped to prevent rubber-to-rubber contact and the subsequent adhesion. The length of the cooling train, which usually consists of water-cooled rolls around which the product is passed, therefore depends on the projected maximum rate of calendering, the projected maximum product thickness, and the adhesive qualities of the compounds to be calendered.

From here separate consideration must be given to calendering operations involving fabrics, which have considerable dimensional stability, and those involving unsupported sheet, which does not. In the latter case a controlled tension, provided by the batch-off or windup equipment, is essential to maintain the accurate gauge imparted by the calender.

When unsupported sheet is taken from the calender nip, it shrinks along its length and increases in width and thickness. This results in a "crown," that is, the sheet is thicker in the center than at the edges, if no prior compensation has been applied at the calender for this effect. Two methods can be used to overcome the problem of shrinkage. One is to chill the sheet quickly and restrain it by wrapping it tightly in a liner. The other is to allow the sheet to shrink freely, or even to force-shrink it before wrapping it in the liner. Sheet wrapped with the shrinkage stresses "frozen in" will shrink lengthways when taken from the liner, especially so when heated. The preferred treatment depends therefore on the ultimate use of the sheeting. For the calendering of fabrics, tension is determined by the need to keep woven fabrics perfectly flat and to ensure uniform cord spacing with cord fabrics. Shrinkage effects are usually minimal.

5.7. DETERMINATION AND CONTROL OF CALENDER OPERATION CHARACTERISTICS

Calender performance is dependent on a number of factors: feedstock temperature; size of rolling bank at the nips; roll temperatures; whether single-, double-, or triple-bank calendering is used; and wind-off tension. Assuming good control of the rheological properties of the feedstock, the maximum practical calender speed is largely dictated by the surface uniformity of the calendered product, which is strongly dependent on the temperature rise of the rubber as it passes through the nips. Two forms of surface fault can generally be identified. V-shaped marks, known as "crows feet," result from calender roll temperatures and/or feedstock temperatures being too low. Blisters result from the calender rolls and/or feedstock being too hot. The tendency of a mix to blister worsens as filler content is reduced and as the required sheet thickness is increased. This can be alleviated, to a certain extent, by double-bank calendering; but the gauges which can be successfully calendered are usually in the range 0.1–1.5 mm. As both roll speed and product thickness are increased, the operating window becomes narrower.

The shape of the operating window is dependent on a number of factors, as indicated in the previous paragraph, and may be difficult to determine using a "prior experience" approach. The techniques described in Section 7.3 for process-capability determination and optimization generally give a good return on the effort and resources invested in them, for high productivity, long-run operations. However, attention should be given to other limitations, such as the relationship between output speed, product thickness, and the calender line cooling capabilities. For frictioning operations the maximum speed may be limited by the physical strength of the fabric being calendered.

The level of sophistication of calender control is largely a matter of economics. Each increment of improvement in control in the climb from total manual control to a fully computerized system gives measurable improvements in speed, accuracy, and material utilization.

The choice of a system can obviously be made by comparing the cost of its purchase, installation, and commissioning with the projected savings and extended capability resulting from its use. However, the value of a control system is a function of the capabilities of the calender to which it is fitted. Realizing the full potential of a new control system can involve the purchase of a new calender! It is also worth reemphasizing that good calender control starts with a consistent feedstock.

The flow in the nips of a calender is transient in character, with high and rapidly changing deformation rates being applied to the rubber in a very short time scale, resulting in it showing a strong viscoelastic response. This is, in contrast to most other processing operations where predominantly viscous flows can be assumed. Consequently, the testing methods used to evaluate and monitor the rheological properties of the feedstock must measure the viscoelastic response. Stress relaxation tests are suitable and may be performed using the Sondes Place Research Institute TMS Rheometer described in Section 2.3.1. After a conditioning run, say at 10 s^{-1}, the rotor is stopped and locked in place, enabling the decay in stress with respect to time to be measured. The results can be fitted to a viscoelastic model, but may be treated in a simpler manner for correlation with calendering performance or checking against the upper and lower limits of tolerance bands. When plotted on a log (shear stress) vs. log (time) graph, the results generally give a good straight line, enabling the viscoelastic response to be empirically characterized by the intercept on the log (shear stress) axis and the negative slope of the line. For routine production monitoring the test results can be transmitted directly to a computer, which will then produce the slope and intercept values using linear-regression analysis, without the necessity to plot, and interpret a graph.

For manual calender control, the main consideration is the instrumentation and ergonomics of the operator's panel or desk. All the instruments which indicate changes in thickness, load, temperature, or line speed should be positioned where the operator can make quick reference and, if necessary, make the required adjustments. To complete the control information, it is necessary that a duplicate set of instruments from the calender feed equipment be installed alongside the calender instruments in the control desk. This will allow the operator to monitor the operation of the feed equipment and to ensure that a uniform feedstock is being supplied to the calender.

Although this type of control is termed manual, each machine variable will be under set-point control. The operator's task is to make adjustments to the set points during product changeovers and in response to small variations in feedstock properties and in the calender operating characteristics. Manual control of this type is adequate for most calenders although closed-loop thickness control may be added by using the signal from the thickness gauges to activate the nip adjustment drive motors.

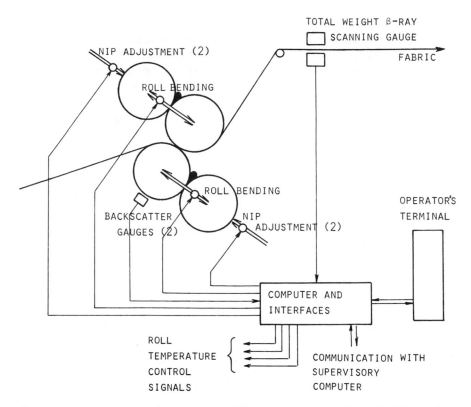

FIGURE 5.6. General schematic of a computer control system for a double-topping operation.

Further sophistication may be necessary if large throughputs are envisaged or if the required product accuracy justifies the use of a calender equipped with roll-bending and/or roll-crossing equipment. Here the number of variables and the precision needed in their control make manual operation inefficient. Microprocessor systems have been successfully applied by Measurex Limited[15] to calenders to meet requirements of this type, with the following claimed improvement over manual control.

60–80%	Improvement in sheet uniformity.
3–8%	Reduction in raw-material costs.
30%	Reduction in startup/mix change times.
1–2%	Increased annual production through reduction of scrap.

A schematic of a typical control system is shown in Figure 5.6. In conjunction with a calender having this level of control, cold-feed extruder material supply to the calender is essential to realize its potential benefits. The volumetric output and feed strip temperature can also be computer controlled, using an operating algorithm analogous to the performance diagram shown in Figure 4.9, to control screw speed and a die restrictor.

REFERENCES

1. Gaskell, R. E., *J. Appl. Mech.* **17**, 334 (1950).
2. Bergen, J. T., Contribution to *Processing of Thermoplastic Materials*, ed. by E. C. Bernhardt, Van Nostrand Reinhold, New York (1959).
3. Tokita, N., and J. L. White, *J. Appl. Polym. Sci.* **10**, 1011 (1966).
4. White, J. L., and N. Tokita, *J. Appl. Polym. Sci.* **9**, 1929 (1965).
5. White, J. L., and N. Tokita, *J. Appl. Polym. Sci.* **9**, 1589 (1965).
6. Funt, J., *Mixing of Rubbers*, RAPRA, Shrewsbury, U.K. (1977).
7. White, J. L., *Rub. Chem. Technol.* **42**, 257 (1969).
8. Pearson, J. R. A., *J. Fluid Mech.* **7**, 481 (1960).
9. Tokita, N., Paper presented at the Second Annual National Conference of the Inst. Rub. Ind., Blackpool, U.K. (1974).
10. Lee, D. E., *Eur. Rub. J.* **18** (9), 15 (1976).
11. Perlberg, E., *Rub. Age* **104** (4), 43 (1972).
12. Seanor, R. C., ASME, Paper No. 56-A-176 (1976).
13. Gooch, K. J., *Mod. Plast.* **29** (2), 165 (1951).
14. Willshaw, H., "Calenders for Rubber Processing," Inst. Rub. Ind. Monograph, London (1956).
15. Brotzel, D., Paper presented at the Plastics and Rubber Institute Symposium, "Optimisation of Quality and Productivity in Rubber Processing," Burton on Trent (November 1978).

Heat Transfer and Vulcanization Methods

6.1. INTRODUCTION

Vulcanization processes divide naturally into two main groups. The first consists of molding methods, all of which involve an integral shaping operation which is completed prior to the onset of cross-linking. The second includes a number of techniques used to cure a previously formed product. For the purposes of analysis, the shaping operations in molding can be considered to be separate from the vulcanization stage, enabling the majority of vulcanization processes to be evaluated using similar techniques.

During vulcanization externally supplied heat flows into the rubber at a rate controlled by the efficiency of heat transfer from the heating medium and by the heat-transfer properties of the rubber. The temperature gradients in the rubber arising from this conductive heat transfer then depend on the temperature of the external heat source, the time of heating, the size and shape of the article being vulcanized, and its initial temperature. Changes in temperature within a rubber product, which occur with respect to both time and position, tend to give a nonuniform state of cure and can result in the properties of the rubber at the surface of a product being quite different from those at the center. One of the main objectives of selection and optimization of vulcanization processes is that of achieving an acceptably uniform state of cure in conjunction with a viable production rate.

6.2. HEAT TRANSFER

6.2.1. Modes of Heat Transfer

Heat transfer within a rubber product during vulcanization is conductive in nature, except in the case of radiation curing processes. However, the mode of supply of heat to the surface of the article can be either conductive or convective. In molding, heating is due to conduction through the mold metal and then across the smooth clean metal surfaces of the mold to the rubber; but cooling of the molded article is usually by free convection, when it is demolded and air cooled. A number of other processes utilize convective heating, immersing the product in a heated liquid, vapor, or gas. This may be either free convection, when thermal currents in the fluid are relied upon to deliver heat energy to the rubber surface, or forced convection, where motion is artificially generated in the fluid to replace the fluid at the rubber surface more rapidly and give a higher heat-transfer rate.

6.2.2. Conduction

Conductive heat transfer in vulcanization is termed unsteady, since the temperature at any point within a product varies with both time and position. This is described by the partial differential equation

$$\frac{\partial T}{\partial t} = \frac{k}{\rho c_p} \frac{\partial^2 T}{\partial x^2} \tag{6.1}$$

where k is thermal conductivity, c_p is specific heat, and ρ is density. The term $k/\rho c_p$ is often referred to as a single parameter called thermal diffusivity, denoted by α; and x is the distance from the surface at which heat is supplied, or from some arbitrary reference surface.

The changes in temperature distribution with respect to time in products having a large surface area in comparison with their thickness, such as sheet, tube, and other thin-walled articles, can be evaluated by assuming that the heat flow is normal to the heated surface and that the influence of edge heating is negligible. This is termed one-dimensional heat flow. For the case of negligible surface resistance to heat transfer, the temperature distribution in a homogeneous rubber slab may be obtained by solving Eq. (6.1) analytically to give the rapidly converging series

$$\frac{T_s - T_x}{T_s - T_0} = \frac{4}{\pi} \sum_{n=1}^{\infty} \frac{\sin[(2n-1)\pi x/2l] \exp[-(2n-1)^2\pi^2\alpha t/4l^2]}{2n-1} \tag{6.2}$$

In Eq. (6.2) T_s is the temperature of the heating medium and the rubber surface; T_x is the temperature at a plane distance x from the slab surface; T_0 is the initial temperature of the rubber; and l is the half-thickness of the slab, assuming that it is heated from both sides. Only the first two terms of this series are of practical importance, the values of successive terms being negligibly small. Also, when $at/4l^2 \geqslant 0.06 - 0.08^{(1)}$ only the first term is needed. Hence

$$\frac{T_s - T_x}{T_s - T_0} = \frac{4}{\pi} \left(\sin \frac{\pi x}{2l} \exp(-\pi^2 at/4l^2) \right.$$

$$\left. + \frac{1}{3} \sin \frac{3\pi x}{2l} \exp(-9\pi^2 at/4l^2) \right) \tag{6.3}$$

and for the center plane of the rubber slab, when $x = l$, Eq. (6.3) becomes

$$\frac{T_s - T_x}{T_s - T_0} = \frac{4}{\pi} \left(\exp(-\pi^2 at/4l^2) - \frac{1}{3} \exp(-9\pi^2 at/4l^2) \right) \tag{6.4}$$

When the least surface dimension of a product (width, circumference, etc.) is less than approximately five times its thickness, the use of the one-dimensional heat flow analysis underestimates the rubber temperatures by a factor too large to ignore for practical vulcanization. For homogeneous products having simple geometries, such as rods of square and circular cross section, short cylinders, spheres, etc., charts[2,3] have been produced from which the temperature at any time and position within these bodies can be read directly, provided that the thermal diffusivity of the rubber is known.

In the event of problems falling outside the range covered by the available heat-transfer charts, as a result of the very low values of thermal diffusivity of rubbers in comparison with most other materials, the temperature distribution in a rectangular or square solid can be determined by combining solutions for one-dimensional systems. The "infinite" rectangular bar in Figure 6.1 (sufficiently long that the influence of heat flow in the y direction is negligible) constitutes a two-dimensional problem described by the differential equation

$$\frac{\partial^2 T}{\partial x^2} + \frac{\partial^2 T}{\partial z^2} = \frac{1}{\alpha} \frac{\partial T}{\partial t} \tag{6.5}$$

It can be shown[4] that the dimensionless temperature distribution may be expressed as a product of the solutions of two plate problems of thickness $2L_1$ and $2L_2$, respectively,

$$\left(\frac{T_s - T_{x,z}}{T_s - T_0} \right)_{bar} = \left(\frac{T_s - T_x}{T_s - T_0} \right)_{2L_1 \text{plate}} \left(\frac{T_s - T_z}{T_s - T_0} \right)_{2L_2 \text{plate}} \tag{6.6}$$

where the coordinates x and z define a position within the bar.

A similar approach can be adopted for three-dimensional problems,

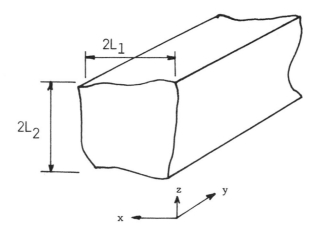

FIGURE 6.1. Rectangular bar of effectively infinite length (negligible heat transfer in y direction).

such as cubes or rectangular prisms, where the solution may be expressed as a product of the dimensionless temperatures for three plates having the thickness of the three sides of the block. Heat transfer in products with curved surfaces, such as cylinders and spheres, is either two- or three-dimensional and solutions cannot be determined from the product of one-dimensional cases. With a few exceptions it is necessary to solve the two- or three-dimensional problem directly. These exceptions occur when a solution for a body, such as a rod of infinite length, is available in graphical form and when the temperature distribution in a short rod of similar cross-sectional shape is required. The product of the dimensionless temperature for the infinite rod and a slab or plate having a thickness equal to the desired length of the short rod will give the solution of this three-dimensional problem.

Many practical problems of heat transfer in vulcanization cannot be solved using the techniques discussed previously. In addition to the problems of dealing with curved surfaces, a wide range of products include components made of metal and a variety of other materials, which must be bonded to the rubber during vulcanization. Heat may be conducted through these components far more rapidly than through the rubber, significantly influencing the temperature distribution in the rubber and resulting in the temperature boundary conditions of the rubber changing with time. A similar situation exists in the molding of complexly shaped components, when mold parts having large surface areas cool significantly between vulcanization cycles. To determine the temperature distributions for those more complicated products, it is necessary to use numerical techniques, such as the finite difference and finite element methods, unless an opportunity for direct measurement of temperatures by thermocouple probes presents itself.

The finite difference method is suitable for evaluating two-dimensional problems and is capable of being extended to some three-dimensional cases through careful selection of the coordinate system. The finite element[5,6] method is inherently capable of being applied to both two- and three-dimensional problems, but requires a computer program of considerable complexity for its use. Both methods are capable of dealing with products of complex shape and temperature boundary conditions which change with time. The finite element method is beyond the scope of this book. Detailed descriptions of this method are given in the texts referred to, and commercial programs are available.

The finite difference method approximates the governing differential equation for heat transfer, and the resulting calculations may be performed on an advanced calculator or a microcomputer. In fact, in its simplest form, a finite difference approximation can be evaluated without any computing aids, although the process is tedious.

The first step in the finite difference method[4,7] is to superimpose a grid on the cross section of the product to be analyzed, as shown in Figure 6.2. Given that the subscript m denotes the x position and the subscript n denotes the z position, the second partial derivative of Eq. (6.5), which describes two-dimensional heat flow, can be approximated by[8]

$$\frac{\partial^2 T}{\partial x^2} \approx \frac{1}{(\Delta x)^2} (T_{m+1,n} + T_{m-1,n} - 2T_{m,n}) \tag{6.7}$$

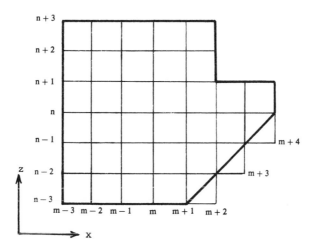

FIGURE 6.2. Finite difference mesh superimposed on a product cross section.

and

$$\frac{\partial^2 T}{\partial z^2} \approx \frac{1}{(\Delta z)^2} (T_{m,n+1} + T_{m,n-1} - 2T_{m,n}) \tag{6.8}$$

The time derivative in Eq. (6.5) is approximated by

$$\frac{\partial T}{\partial t} \approx \frac{T_{m,n}^{P+1} - T_{m,n}^{P}}{\Delta t} \tag{6.9}$$

where the superscripts designate the time interval. Combining Eqs. (6.7)–(6.9) gives the finite difference approximation of Eq. (6.5):

$$\frac{T_{m+1,n}^{P} + T_{m-1,n}^{P} - 2T_{m,n}^{P}}{(\Delta x)^2} + \frac{T_{m,n+1}^{P} + T_{m,n-1}^{P} - 2T_{m,n}^{P}}{(\Delta z)^2}$$
$$= \frac{1}{\alpha} \frac{T_{m,n}^{P+1} - T_{m,n}^{P}}{\Delta t} \tag{6.10}$$

Thus, if the temperatures of the various nodes (junctions of the finite difference grid or mesh in Figure 6.2) are known at any given time, then the temperatures after a time interval Δt may be calculated by writing an equation like Eq. (6.10) for each node and obtaining the values of $T_{m,n}^{P+1}$. The procedure may be repeated to obtain the temperature distribution after any desired number of time increments.

This method can also be used to determine the temperature distribution in a mold or in a component to which the rubber is to be bonded, provided that appropriate values of thermal diffusivity are used. Hence, the variation in temperature at the boundary of the rubber, which occurs with both time and position, can be determined.

If the intervals of the space coordinates are chosen such that $\Delta x = \Delta z$, the resulting equation for $T_{m,n}^{P+1}$ becomes

$$T_{m,n}^{P+1} = \frac{\alpha \Delta t}{(\Delta x)^2} (T_{m+1,n}^{P} + T_{m-1,n}^{P} + T_{m,n+1}^{P} + T_{m,n-1}^{P})$$
$$+ \left(1 - \frac{4\alpha \Delta t}{(\Delta x)^2}\right) T_{m,n}^{P} \tag{6.11}$$

If the time and distance intervals are conveniently chosen such that $M = 4$ in the equation

$$M = \frac{(\Delta x)^2}{\alpha \Delta t} \tag{6.12}$$

the temperature of node (m, n) after a time increment is simply the arithmetic average of the four surrounding nodal temperatures at the beginning of the time increment.

When a one-dimensional system is involved, the equation becomes

$$T_m^{P+1} = \frac{\alpha \Delta t}{(\Delta x)^2} (T_{m+1}^P + T_{m-1}^P) + \left(1 - \frac{2\alpha \Delta t}{(\Delta x)^2} \right) T_m^P \qquad (6.13)$$

and if the time and distance intervals are chosen such that $M = 2$ in Eq. (6.12), the temperature of node m after the time increment is given as the arithmetic average of the two adjacent nodal temperatures at the beginning of the time increment.

Once the distance increments and the value of M are established, the time increment is fixed and may not be changed unless the value of Δx or M is changed. Clearly, the larger the values of Δx and Δt, the more rapidly will the solution proceed. However, the smaller the value of these increments, the greater will be the accuracy of the solution. Also in one-dimensional systems $M \geqslant 2$ and in two-dimensional systems $M \geqslant 4$; otherwise a condition violating the second law of thermodynamics is generated.

If the mesh size or time intervals required for effective evaluation of a problem do not give $M = 2$ for the one-dimensional systems or $m = 4$ for two-dimensional systems, then the simple method of averaging the temperatures of adjacent nodes cannot be used. In addition, if the number of nodes is large, it is sensible to seek methods of reducing the time for a solution. The alternative methods of computation are well documented[8] and programs are available which are feasible with some of the more advanced calculators.[7] The amount of computational work can be reduced by taking advantage of any symmetry which may exist in the product being analyzed. If a product is symmetrical about one axis, the finite difference mesh need only be applied to one half of it. In more fortunate cases even less of the product may be evaluated.

6.2.3. Convection Boundary Conditions

The preceding analyses for conduction assume that the surface temperature of the rubber and the temperature of the heating medium are the same at all times during the heating operation. While this is a reasonable assumption for most molding operations, it does not apply to either heating or cooling by immersion in a gas or liquid, or to open steam vulcanization. Hence, air cooling after vulcanization is generally a slow process and elevated temperatures may be maintained in the center of the product for useful periods of time, unless the cross-sectional thickness of the product is small. This provides some compensation for the slow temperature rise in the center of a thick product and enables the residence time in the vulcanization process to be minimized.

Problems in convection are difficult to analyze and can vary widely in their accuracy depending on the precision with which the convective heat-transfer coefficient may be estimated. This depends on the flow of the fluid over the rubber. The value of the heat-transfer coefficient will be low when the flow is only due to thermal currents, but will achieve a higher value if a positive flow is generated. Keeping these facts in mind, practical measurements using thermocouple probes are preferable to calculation, where possible. For cases where direct measurement is impractical the following finite difference method for one-dimensional conduction with convective boundary conditions is given. Adams and Rogers[7] have published a computer program capable of solving this problem.

When convection occurs at a boundary, one can write

$$k \left(\frac{\partial T}{\partial x} \right)_s = h(T - T_\infty)_s \qquad (6.14)$$

This equation states that the energy conducted to the surface is equal to the energy leaving the surface, or vice versa. A finite difference approximation of Eq. (6.14) solved for T_0^p gives

$$T_0^p = \frac{T_1^p + (h\Delta x/k) T_\infty}{1 + h\Delta x/k} \qquad (6.15)$$

referring to Figure 6.3 for the nomenclature. The transient temperature distribution in the plate may be solved using Eq. (6.13) for $1 \leqslant n \leqslant N - 1$ along with Eq. (6.15) for $n = 0$, assuming that the problem is symmetrical. If the problem involves a nonsymmetrical initial temperature distribution or if the heat-transfer coefficients for the two surfaces are different, an equation similar to Eq. (6.15) will be needed for T_N^p. The surface temperature of the plate is given by

$$\theta_0^{p+1} = \left(1 - \frac{2F_0}{(\Delta x)^2} - \frac{2F_0 B_0}{\Delta x} \right) \theta_0^p + \frac{2F_0}{(\Delta x)^2} \theta_1^p \qquad (6.16)$$

where $\theta = (T - T_\infty)/(T_1 - T_\infty)$ and the Biot number $(Bi = hl/k)$ and the Fourier number $(F_0 = \alpha \Delta t/l^2)$ are used to simplify the equation. Equation (6.13) still applies for all other nodal points of θ_n^{p+1}. This set of equations can be solved individually to obtain the values of θ_n^{p+1}, once the initial

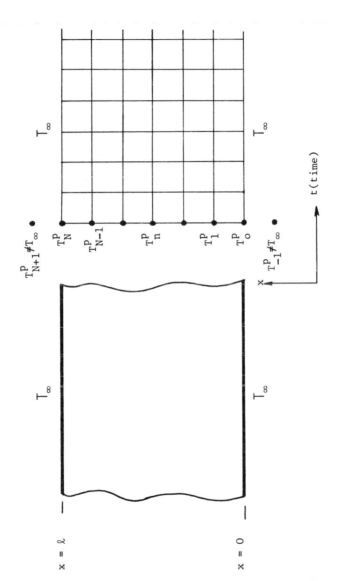

FIGURE 6.3. One-dimensional unsteady-state conduction with convective boundaries.

values of θ_n^p are known. The values for subsequent time intervals may be solved in a similar manner by using the previously calculated values as a starting point.

6.3. PREDICTION OF STATE OF CURE

6.3.1. Cure Simulation Instruments

Having determined the way in which temperature changes with both time and position within a product during the vulcanization process, the next step is to use this information to determine the state of cure or cross-link density distribution. The simplest and most direct method requires a cure simulator of the type manufactured by Göttfert, enabling the temperature–time profile for a particular position within a product to be reproduced in the test cavity of a curemeter. Alternatively, the cross-link density distribution may be determined graphically or from expressions describing the dependence of state of cure on time and temperature.

While the cure simulator method eliminates any errors arising from the approximations and assumptions of the graphical and mathematical methods, it does not allow the effect of changing the vulcanization conditions to be investigated, unless further cure tests are carried out to simulate the new conditions. The graphical method is relatively simple; but it is tedious and lengthy if a detailed analysis of a number of conditions is needed. The mathematical method is suitable for programming into a calculator or computer, permitting detailed analyses of a range of conditions to be carried out very rapidly; but it is first necessary to develop the program.

6.3.2. Time–Temperature–Fractional Conversion Charts

The purpose of the time–temperature–fractional conversion (TTC) chart is to present the information obtained from a series of curemeter traces at different temperatures in a form suitable for the superimposition of the temperature–time profiles derived from the heat-transfer analysis.[9] The fractional cure conversion (multiply by 100 for % cure) is defined by

$$\text{Fractional Conversion} = \frac{M_t - M_{\min}}{M_{\max} - M_{\min}} \qquad (6.17)$$

where M_t is torque at time t, $M_{\max}$ is maximum torque, and $M_{\min}$ is minimum torque, measured from the curemeter trace. Figure 6.4 shows that the TTC chart is a contour diagram. Each contour is constructed by

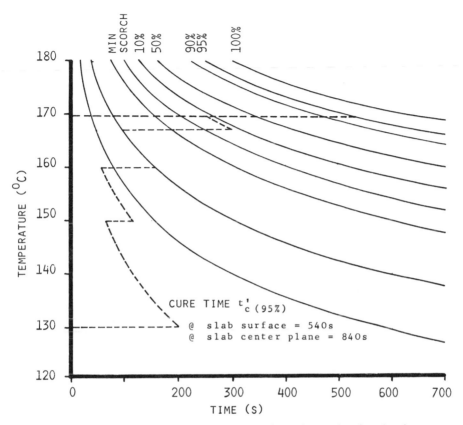

FIGURE 6.4. Time–temperature–percent conversion chart, showing heating-curve data transferred from Figure 6.5.

transferring the temperature and time coordinates for a given value of fractional conversion from each cure trace to the TTC chart. The "minimum" contour identifies the conditions for the transition from the minimum torque plateau, if it exists, to the cross-link insertion slope. Contours to the left of this are drawn by dividing the times to the minimum contour, at different temperatures, into an equal number of increments.

The procedure for the transfer of temperature–time profiles to the TTC chart first requires that the former is approximated by a series of steps, as shown in Figure 6.5 for a slab of 10-mm thickness. Taking the first step of this approximation, the time represented by this step can be plotted on to the TTC chart parallel to the time axis, starting from zero time and at the appropriate temperature. The increment to the next temperature takes zero time and therefore no progress is made toward vulcanization. For this reason

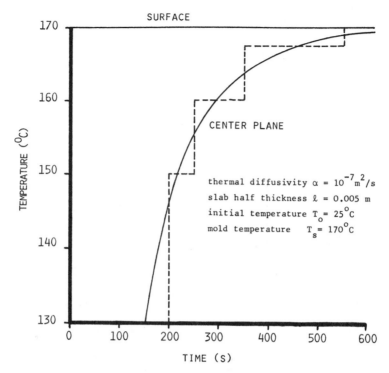

FIGURE 6.5. Heating curves for the surface and center plane of a rubber slab of 10 mm thickness.

the increment is drawn parallel to the nearest contour, as shown in Figure 6.4. After this a further line is drawn parallel to the time axis, to represent the time spent at the next higher temperature, starting from the end of the contour line representing the increment from the lower temperature. This procedure is repeated until the required value of fractional conversion is reached or until the mold surface temperature is reached. The lines representing the slab surface temperature in Figures 6.4 and 6.5 show that it achieves the temperature of the heating medium instantly, indicating that the value of the heat-transfer coefficient is infinity.

Reducing the size of the "steps" and increasing their number will obviously improve the accuracy of the prediction, but will increase the time to a solution. A compromise must be sought, based on the accuracy needed for the process being investigated. Further savings in time can be made by ignoring low temperatures where the reaction rate is very slow and by starting with large steps and then decreasing their size as the temperature and reaction rate increase.

6.3.3. Cure modeling

The simplicity and accuracy of the graphical methods described in the previous section make them very attractive; but they become lengthy and tedious when the cross-link density distributions at a number of heating medium temperatures and cure times are required for comparison. It is desirable to have a mathematical analogue of the techniques which can be written into a computer program and used directly in conjunction with the heat transfer calculations, avoiding the manual transfer of data. To do this, expressions which describe the cure curve and its temperature dependence are required. A considerable amount of work has been done in this area but, until recently, gross simplifications have been made to render the resulting expressions tractable for available computational methods. The recent work of Hands and Horsfall,[10] which provides an accurate model of cure characteristics, depends on computer methods.

Cross-linking at constant temperature (isothermal) is a function of temperature T and time t, enabling the cure level c to be expressed as

$$c = f(T, t) \qquad (6.18)$$

This can be differentiated to give a cure rate q, which is a function of both temperature and state of cure,

$$\frac{dc}{dt} = q(T, c) \qquad (6.19)$$

The function q can be obtained from cure data, preferably obtained from a curemeter which gives sensibly isothermal conditions, indicating the value of the new Wallace isothermal curemeter.

Hands[10] represents the S-shaped cure-time traces by two exponentials joined together at a point on the cure curve defined by $c = c_m$, the rate of cure being continuous across the junction. For $c < c_m$ the level of cure is given by

$$c = c_0 + \left(\frac{G_0}{\alpha_1}\right) \exp(\alpha_1 t - 1) \qquad (6.20)$$

where

$$\alpha_1 = \left(\frac{G_m - G_0}{c_m - c_0}\right) \qquad (6.21)$$

In these equations c_0 is the apparent cure at zero cure level, being the inherent elasticity of the unvulcanized rubber compound which gives the minimum torque on the curemeter trace. G_0 is the initial gradient of the cure trace, G_m is the gradient at the junction point, and c_m is the cure level at the junction point. Similar expressions can be written for $c > c_m$,

$$c = c_\infty + (c_m - c_\infty) \exp(\alpha_2(t - t_m)) \qquad (6.22)$$

where

$$\alpha_2 = \frac{G_m}{c_m - c_\infty} \qquad (6.23)$$

and

$$t_m = \frac{\ln(G_m/G_0)}{\alpha_1} \qquad (6.24)$$

in which c_∞ is the maximum cure level. Both G_m and c_∞ are temperature dependent:

$$G_m = \exp(\mu_1 + \mu_2 T) \qquad (6.25)$$

and

$$c_\infty = \exp(\omega_1 + \omega_2 T) \qquad (6.26)$$

Values of the seven parameters $c_0, G_0, c_m, \mu_1, \mu_2, \omega_1,$ and ω_2 which minimize the differences between the simulated cure curves and the experimental curves now have to be found, for the whole set of isothermal cure-time curves covering the temperature range of interest for the rubber compound and products being studied. Hands[10] proposes a grid-search method for this, in which an upper and lower limit for each variable is chosen, together with a number of equispaced intermediate values. The predictions resulting from selected combinations of values, represented by junctions in the grid, are then systematically tested against the experimental data. If S_i is the sum of squares of the differences between calculated and measured cure levels at a particular temperature, then the root mean square (rms), which must be minimized for the best fit, is given by

$$\text{rms} = \left(\frac{\sum_{i=1}^{n} (S_i/N_i)}{n} \right)^{1/2} \qquad (6.27)$$

where N_i is the number of cure levels being examined at each cure temperature and n is the number of temperatures used.

The computer time needed for a full grid search in seven variables would be prohibitive. Fortunately, Hands[10] has identified groups of variables which influence behavior in specific parts of the cure curves, precluding the need for a full search. These groups are:

1. c_0 and G_0 which mainly influence the start of the cure curves.
2. c_m, μ_1, and μ_2 which mainly influence the middle region of the cure curves.
3. ω_1 and ω_2 which mainly influence the upper end of the cure curves.

A complete search now consists of applying a grid search to each set of variables in sequence; the variables not being examined are assigned their best values from a previous grid search.

The whole search is repeated several times until the rms drops by less than 0.001 between consecutive searches. The upper and lower limits of the parameters for the first complete search are either guesses or estimates obtained from graphs of $\ln G_m$ and $\ln c$ against temperature. For subsequent complete searches, the lower and upper limits are equally spaced on either side of the previous best value.

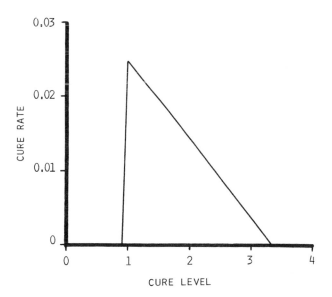

FIGURE 6.6. Cure rate plotted as a function of cure level, as measured by the swelling of isothermally cured samples in toluene (courtesy RAPRA).

Differentiation of Eqs. (6.20) and (6.22) enables Eq. (6.19), relating the cure rate to the level of cure, to be evaluated. For $c < c_m$,

$$\frac{dc}{dt} = \frac{G_0(c - c_m)}{c_0 - c_m} + \frac{G_m(c - c_0)}{c_m - c_0} \tag{6.28}$$

and for $c > c_m$,

$$\frac{dc}{dt} = \frac{G_m(c - c_\infty)}{c_m - c_\infty} \tag{6.29}$$

These equations represent the two straight lines shown in Figure 6.6. The first line connects the points (c_0, G_0) and (c_m, G_m); the second connects the points (c_m, G_m) and $(c_\infty, 0)$, completing the relationships needed for cure simulation.

6.4. MOLDING

6.4.1. Review of Molding Methods

Molding operations are concerned with the sequential shaping and cross-linking of a rubber mix. They must take into consideration the interaction between the requirements for viscous flow and vulcanization, and the dependence of both of these on temperature. The design or selection of a molding process is then a compromise imposed by the need to complete the shaping operation prior to the onset of cross-linking and to achieve a fairly uniform state of cure. As usual, technical and economic factors must be considered together. High-volume production will justify sophisticated techniques, while small numbers indicate that expenditure on equipment dedicated to specific products should be minimal. To fulfill these requirements three main techniques have been developed: compression molding, transfer molding, and injection molding, with many variations on each to suit individual requirements of dimensional accuracy, complex shape, thick sections, etc. In general the capital cost of machines and molds increases from a minimum for compression molding, through transfer molding, to a maximum for injection molding.

In compression molding a rubber charge or blank is made to flow to the shape of the mold cavity by the action of mold closure, the motive force for the closure being provided by a hydraulic press. Compression molding encompasses open flash and positive and semipositive methods. Each of these methods is illustrated in Figure 6.7. In open flash molding the shape,

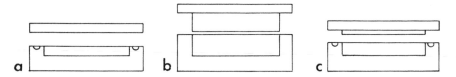

FIGURE 6.7. Schematic diagrams of the three basic types of compression mold. (a) Open flash mold. (b) Positive or plunger mold. (c) Semipositive mold (omitting guide pins and bushes).

size, and positioning of the charge is very important, since it must flow fully to the form of the mold cavity in preference to flowing out through the gaps between the mold components during closure. Also, excessive outflow or flash can cause the mold components to be separated by a layer of rubber which varies in thickness between molding cycles, resulting in poor control of dimensions. Positive- or plunger-type molds prevent escape of rubber due to the long and narrow flow path between the mold body and the plunger. For this reason high pressures can be applied effectively to the rubber charge, causing it to flow fully and consistently to the form of the mold cavity. However, the absence of flash or mould spew results in the dimensions of the molding, which are controlled by the position of the plunger, being entirely dependent on the accuracy with which the charge or blank can be produced. Semipositive molds are intended to embody the advantages of both open flash and positive molds. The very shallow plunger results in a flow path through which rubber may flow only when a high pressure is applied to the rubber in the mold cavity. Hence, the excess rubber should only escape after the molding has been fully formed.

In transfer molding rubber flows from an auxilliary cavity into the mold cavity through a narrow channel. Figure 6.8(a) shows a "loose-tool" (not attached to the press) transfer mold, while Figure 6.8(b) shows a transfer press or ram injection molding machine with a "fixed-tool" mold. The reasons for using transfer molding in preference to compression molding are threefold. First, the mold cavity is closed prior to the entry of rubber, giving considerably improved dimensional accuracy. Second, during transfer, new and clean surfaces are generated on the rubber, which is extremely important for achieving a strong and consistent rubber–metal bond. Third, with transfer presses, lower unit manufacturing costs are possible. This is due to reduced charge preparation; shortened cure times resulting from heating of the rubber during flow through sprue, runner, and gate systems; and the reduction in downtime due to only one charge being necessary, even when a multicavity mold is used.

Further reductions in cure times for bulky articles are possible in both

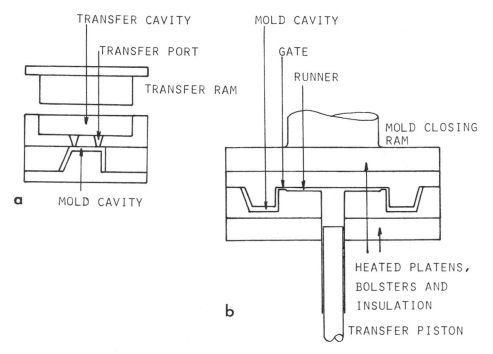

FIGURE 6.8. Transfer molding methods. (a) Loose-tool transfer mold. (b) Fixed-tool mold in transfer press (RAM injection molding machine).

compression and transfer molding by preheating the rubber charge. The most common and effective method for this is the microwave heater, which is capable of raising the temperature of a charge rapidly and uniformly throughout its whole volume. When used with transfer molding, preheating has the further advantage of reducing viscosity, permitting either a reduction in mold filling time or a decrease in the sprue and runner cross section, thus giving a reduction of the material "wasted" in each molding cycle.

Screw injection molding machines are designed with two configurations. The reciprocating screw machine of Figure 6.9(a) follows thermoplastics practice and is generally best suited for the production of thin-walled articles with very short cure times. This is because the shot volume is linked to the plasticization rate (the rate at which the screw conveys material forward from the feed pocket) by the diameter and length of stroke of the screw. This results in a high plasticization rate capability in comparison with shot volume. Since many injection molded rubber articles have comparatively long cure times, the plasticization capacity is generally not fully utilized. This situation is avoided by separating the plasticization and injection

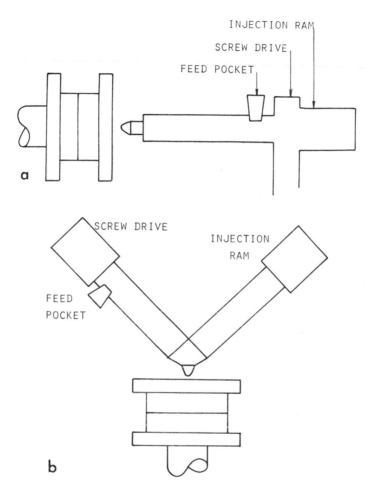

FIGURE 6.9. Injection molding machines. (a) Horizontal reciprocating screw machine. (b) Separated screw injection ram machine in vertical V-Shaped configuration.

functions as shown in Figure 6.9(b), enabling both functions to be given capacities appropriate to the range of moldings they are intended for.

The main advantages of screw injection molding machines over transfer presses arise from the much better control of heat history and the elimination of the thixotropic contribution to viscosity during screw plastication. Better control of heat history permits the temperature of the rubber prior to injection to be considerably higher than with transfer press molding, without incurring the danger of scorch, resulting in a further reduction in viscosity.

This enables the rubber to be injected into the mold at a higher rate and at a higher temperature than is possible with transfer molding, giving a reduction of both injection time and cure time. Also, the cross sections of the sprues and runners can again be made smaller to reduce the material "wasted" during each molding cycle.

6.4.2. Compression Presses

Compression presses generally have the configuration shown in Figure 6.10. The mold clamping force is invariably supplied by a hydraulic ram and the press of Figure 6.10 is termed "upstroking," since the platens are closed by an upward motion of the ram. The clamping force is reacted against the tie bars or side frames of the press, which also serve to maintain the alignment of the platens. Hydraulic rams can be either single or double acting. The former are only suitable for upstroking presses where opening by gravity is feasible, whereas for positive mold opening and for downstroking presses the latter type are required.

Opening and closing of modern presses is usually controlled by a solenoid-actuated two-position valve, which permits the press to be operated from a conveniently positioned instrument panel and gives the option of an automatically controlled curing cycle. In addition to opening the press when the cure time has elapsed, automatic control can include any "bumping" and "breathing" operations which are necessary to eliminate trapped air from a molding. Bumping consists of operating the hydraulic valve to open the press by a few millimeters and then closing it again. Breathing only involves the release of the mold clamping pressure and its subsequent reapplication. The timing of these operations from the start of the cure cycle and their duration is important for their success. Automatic control not only releases the press operator for other duties, but ensures that the operations are carried out consistently. Further controls for the hydraulic system include safety interlocks, to ensure that all the guards needed to protect the operator are in place and that actions needed to safeguard the integrity of the press are carried out prior to operation. Adjustable pressure relief valves, to control clamping pressure, and flow rate control valves, to set the press closing speed, are usually standard fittings.

Large presses, 2.5 MN clamping force and above, usually have individual hydraulic drive units, some of which are integral with the press, reducing the floor space it occupies. Smaller presses are often positioned in lines or groups and served from a common hydraulic drive unit. In both cases it is common practice to use an indirect hydraulic system of the type shown in Figure 6.10. The term indirect indicates that the press is supplied with hydraulic fluid via a hydropneumatic accumulator, allowing the oil

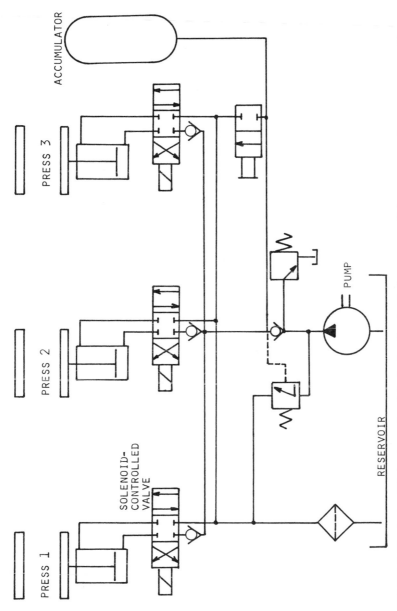

FIGURE 6.10. A simplified schematic of the hydraulic circuit for a group of three compression presses served by a single pump with an accumulator.

stored in the accumulator to be supplied at a higher rate than the hydraulic power unit is capable of delivering. For large presses this allows the high oil flow rate necessary for closing and opening the press at a viable speed to be decoupled from the delivery rate of the hydraulic drive unit, permitting the necessary volume of oil to be supplied to the accumulator over a considerable portion of the molding cycle, which results in a reduction in the size of the hydraulic power unit required. If a number of presses are supplied from one drive unit, the indirect system is necessary to maintain the press closing rate when two or more presses are operated simultaneously; otherwise, the volumetric flow of oil from the hydraulic power unit would be shared among the presses, with a proportional reduction in closing rate.

In compression presses heat is supplied to the molds by conduction from the press platens. The three main methods of platen heating are steam, electrical resistance heating, and induction heating. When steam heating is used the direct relationship between the temperature and pressure of saturated steam[11] makes temperature control via pressure regulation a relatively simple matter. However, the accuracy of this control depends on the efficient evacuation of condensate from the platens via a steam trap. Electrical resistance heaters are used widely but require careful arrangement in the platen to avoid temperature gradients. Also, the thermocouples used for feedback to the temperature controller must be positioned to measure a temperature which accurately reflects the mold temperature.

Development of compression press design has led to the Daniels Hydramould, which has a hinged upper platen which tilts backwards when the press is opened to permit easy access to the lower platen and mold cavity plate. This may preclude the need for a sliding mould base for loading and stripping, enabling it to remain directly above the heated platen and to be solidly fixed to it. The former reduces the mold cooling while the press is open and the latter can lead to more efficient platen to mold heat transfer.

Both Daniels and BIP have developed vacuum chambers which are attached to the press platen and enclose the mold both when the press is closed and during the latter stages of closing, before mold flow commences. Evacuating the mold cavities of air prior to the flow of rubber can practically eliminate trapped air and has led to substantial improvements in the precision of forming products with fine lips and undercuts, such as lip seals. The advent of vacuum presses has almost eliminated the need for machining of sealing lips after molding.

Vacuum chambers can also be beneficial in transfer and injection molding, if it is not possible to design the mold so that filling sweeps the air out of the mold via the split line.

6.4.3. Ram and Screw–Ram Injection Molding Machines

Ram injection molding machines or transfer presses are similar in construction to a downstroking compression press, with the exception of having a second upstroking hydraulic ram to provide the motive force for transfer of the rubber into the mold cavities. Control of this ram is important for the versatility of the process and for the quality of the moldings produced. Regulation of both the rate of injection and the pressure applied to the rubber enables injection to proceed rapidly without high pressures being produced in the mold cavity. The rate of injection is normally controlled indirectly via the hydraulic pressure applied to the ram, and ram displacement is used to meter the volume of rubber transferred into the mold cavities. Relay switches, activated by adjustable contacts attached to the ram, operate electrohydraulic valves to reduce the hydraulic pressure to a "hold" value when the mold cavities are filled. To avoid flashing of a mold, the relay switches are often activated just prior to mold filling. Thus the mold separating force due to the pressure in the cavities is significantly less than that indicated by multiplying the injection pressure by the projected area of the cavities and the runner system. Given the normal transfer pressure range of 60–90 MPa, Gardner[12] suggests that the locking force in meganewtons should be approximately equal to 30 times the projected area in square meters (Figure 6.11).

Injection molding machines having separated plasticization and injection units can be considered as ram injection molding machines with extruder feed. Valving systems direct flow from the plasticizing unit to the injection unit and provide a shutoff during injection to prevent backflow into the plasticizing unit under the high pressures. Reciprocating screw machines are often fitted with nonreturn valves at the screw tip for the same purpose. Plasticizing screws are normally driven by hydraulic motors for compatibility with the other press functions and versatility of speed control. Temperature control is generally by circulating fluid. For reduction of total cycle time, the rapid opening and closing of injection presses is desirable. Two-stage hydraulic systems and toggle locks have been developed to give this capability and also to avoid the expense of very large-diameter, long-stroke hydraulic actuators.

Control systems for rubber injection molding machines are usually of the "set-point" type, which maintain press variables, such as injection pressure, screw speed, and injection unit barrel temperature at preset values. Some machines are fitted with facilities for mold cavity pressure control, to detect the filling of the mold, to reduce injection pressure, and to minimize or eliminate flash; but adaptive control, through a microprocessor system, is largely restricted to the thermoplastics sector. However, it can be argued that

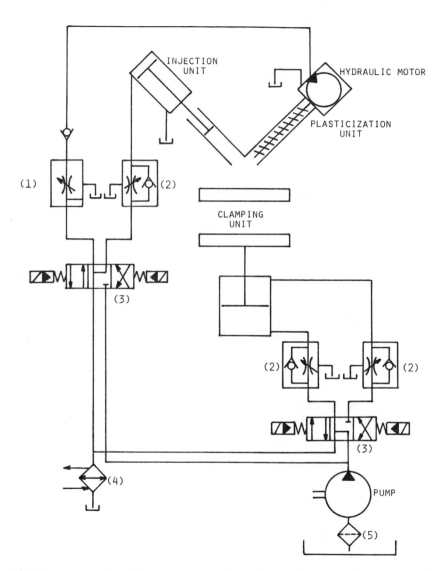

FIGURE 6.11. A simplified schematic of the hydraulic circuit for an injection molding machine having separate plasticization and injection units: (1) flow rate control for screw speed; (2) flow rate control valves for injection and clamping unit opening and closing; (3) solenoid and pilot operated valves; (4) an oil cooler; and (5) an oil filter.

a system which adapts to changes in the feed material is far more important for rubber injection molding, where batch internal mixing is used to prepare the feedstock, than for thermoplastics, where more uniform materials are encountered.

A number of steps in the progression from set-point control to fully adaptive control can be identified, together with the benefits accruing from them. The first step is digital setting of process variables, replacing handwheel, cam, and limit-switch controls, and the placement of the digital controls in a mimic-type control panel. This is not an improvement in the control system but a measure to reduce the time and possibility of errors in setting up a press. From here progress can be made to closed-loop control, such as the cavity pressure control mentioned earlier, where the measurement of a material variable is used for control rather than a machine variable. However, cavity pressure can only be used for the control of the injection ram in the final stages of its total travel. An additional method is needed to ensure that material is fed into the mold in a consistent and repeatable manner.

Starting from the conventional "open-loop" pressure and flow rate controls, two steps are available. The first is the use of an electrohydraulic valve, such as the Vickers KG valve, to provide a precisely metered oil flow rate in response to an external control signal and to correct automatically errors resulting from oil pressure and temperature variations. This system depends on the integrity of the total hydraulic system of the injection molding machine, since leakage at any point in the flow path from the valve to the injection ram, including flow past the piston of the hydraulic actuator, will result in a loss of precision and repeatability. This problem is overcome by the more expensive servohydraulic systems, where the feedback for correction of deviations from programmed conditions is obtained directly from the ram movement, via a linear potentiometer, which gives a voltage proportional to movement, or from a nozzle pressure transducer. Within reasonable limits servohydraulic systems render the precision of control independent of machine condition. Control of injection by nozzle pressure is particularly attractive for rubber injection molding, since the pressure drop through the nozzle, sprue, runners, and gates is proportional to the temperature rise in the rubber.

6.4.4. Compression Molds

The material selected for a compression mold will depend on the pressure to be applied to the clamping surfaces and the service life required of the mold. Hard steels give a good service life, provided that they are of a type which retain their hardness and dimensional stability at elevated temperatures, but are difficult to machine. In this category are nickel–chromium–molybdenum steels of hardnesses in the range 37–64 Rockwell C.

The general attributes of the three main types of compression mold shown in Figure 6.7 have already been discussed. The choice of mold split-line position on the molded product is often dictated by the function of the product. Other points to consider in choosing the split-line position are the flow of material within the cavity, the ease with which the mold may be loaded and stripped, and the complexity of the machining involved in manufacturing alternative mold designs.

Most compression moldings are produced to give "tear-off" type removal of the "flash" or "mold spew." This requires that a channel or "spew groove," sufficiently large to accept the excess material flowing from the mold cavity, is machined adjacent to it, with 2 or 3 mm separating the two. The "land" between the spew groove and the cavity then produces a thin membrane connecting the molding and the excess material, which can easily be removed by tearing, hence the name. The small mark or "witness" left on the molding is acceptable for most applications.

Loose-tool molds are prone to damage since they are often opened manually using levers. One alternative is to provide fixed jigs for this purpose and to avoid the use of metal tools for extracting components from the mold; a compressed air line is usually suitable. Fixed-tool molds attached to the press platens are preferable, if justified by production volume. These are mostly multicavity molds, where the mold cavity plates or units are attached to backing plates or bolsters. Cavity units are preferable to cavity plates since a damaged or worn unit can be replaced, without removing the mold assembly from the press in many cases. The bottom bolster is usually mounted in slides, enabling it to be moved forward by a pneumatic actuator to a convenient position for stripping and loading.

In determining the dimensions of a mold cavity, an allowance must be made for thermal contraction of the molding due to cooling from the mold temperature.[13] This allowance is required as a result of the considerable difference in the coefficient of thermal expansion between steel and rubber. Linear mold dimensions must be increased by

$$S = \Delta T (C_1 - C_2) R \tag{6.30}$$

where ΔT is the difference between the molding temperature and room temperature; C_1 is the coefficient of thermal expansion of a gum compound of the rubber to be molded; C_2 is the coefficient of thermal expansion of the mold material; and R is the proportion of rubber, by volume, in the compound. Typical values (per °C) of the coefficient of linear thermal expansion for gum rubbers are 195×10^{-6} for NBR and IIR; 206×10^{-6} for CR; and 220×10^{-6} for NR and SBR. If flow in the mold is essentially unidirectional, anisotropy of shrinkage (nonuniform shrinkage) can occur.

Thermal expansion due to the temperature rise in the rubber after the start of a curing cycle can result in large pressures being generated in a mold. If these pressures are relieved by local, rather than volumetric, deformation of the rubber after vulcanization has started, large stresses can be cured into the molding. Such stresses will result in distortion and, in extreme cases, local failure. The latter fault is known as "backrinding."

6.4.5. Transfer Molds

This section deals with molds which are used in conjunction with a compression-type press and, as for compression molds, they can be either loose or fixed tools.

The first aspect of these molds to consider is that the closing of the press provides both the motive force for flow of rubber from the transfer cavity and the clamping force to hold together the mold components. To prevent the mold cavity from opening as a result of the hydrostatic pressure of the rubber, the transfer cavity must have a greater projected area than the mold cavity. The closing force which must be generated by the press is then determined by the pressure necessary to transfer the rubber in a time which is sufficiently short to ensure complete filling of the mold cavity prior to the onset of vulcanization. The size of the transfer ports will obviously influence the pressure necessary for a desired transfer rate; and the relation between the two can be derived directly from the results of a capillary rheometer experiment, carried out at the temperature at which the blank is introduced into the transfer cavity.

The closure of the mold cavity prior to the entry of the rubber offers the opportunity of minimizing or eliminating flash. This requires that the mold components fit together precisely, even under the high forces and temperatures involved in molding. This precision of fit can be achieved for single-cavity moulds and, provided that the mold surfaces remain in good condition, consistently flashless moldings may be produced. But this cannot generally be achieved with conventional multicavity molds due to the deflection of the press platens and bolster plates being greater, as a result of forces being exerted further from the vertical axis of the press. Increasing the thickness of the bolsters gives little improvement. A solution to this problem, pioneered by the U. S. Rubber Company, is the use of a mold cavity plate which is locally deformable, enabling it to conform to the surface of the bottom bolster.[14] The remaining problem of conforming to the upper bolster has been overcome by the use of a bottomless transfer cylinder, as shown in Figure 6.12.

In the flashless molding system waste is also minimized by the use of a "cold" reservoir of rubber in the transfer cylinder, insulated from the mold cavities and maintained at ~80°C by oil circulation in the upper platen of

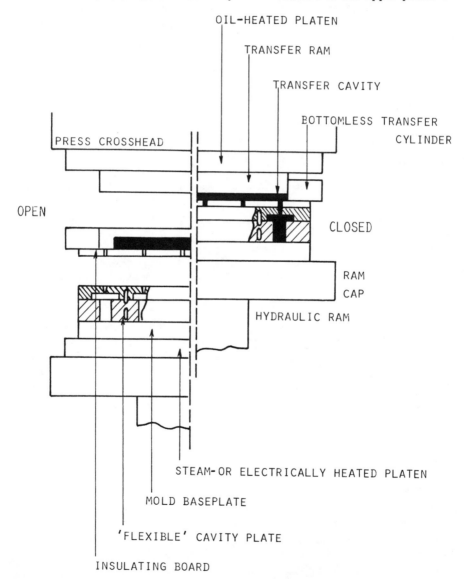

FIGURE 6.12. Schematic cross section through a "flashless" transfer mold.[14]

the press. The mold temperature is achieved by electrical or steam heating in the lower platen. Sufficient rubber can normally be contained in the transfer cylinder for a number of curing cycles. The only waste material produced by this system is small cured sprues from the transfer ports. However, it does require a high-quality press of extremely rigid construction to prevent lateral movement of the mold components.

6.4.6. Injection Molds

Molds for transfer presses (ram injection machines) and screw–ram injection molding machines differ only in detail, having similar constructional features. The main difference usually arises between vertically oriented machines, which include both ram and screw–ram presses, and horizontally oriented machines, which are of the reciprocating screw type. In the former the mold temperature is maintained by conduction from the heated platens of the press, whereas the heaters are integral with the mold in the latter case. The configuration of molds for horizontal machines often results in large surface areas for potential convective heat loss and long paths for conductive heat flow from the press platens. The usual solution to this problem is to place cartridge or flat electrical resistance heaters adjacent to the mold cavities and to isolate the heated portions of the mold by inserting layers of insulating material between the mold components, as shown in Figure 6.13. Molds for vertically oriented machines are usually larger in plan area, giving improved conductive heating, and also expose smaller areas for convective heat loss. Hence, conduction from the press platens, which are insulated from the rest of the machine, is adequate to give efficient and uniform heating, unless the product geometry results in a "tall" mold configuration, requiring that ancillary heaters are used to maintain a uniform temperature distribution. Alternatively, mold insulation can be used. In fact, mould insulation, to reduce both convective losses to atmosphere and conductive losses from the press platens, has been the subject of considerable study, particularly by the RAPRA, in order to reduce energy costs.

The factors which dictate the dimensions of sprue, runner, and gate systems for injection molds, namely, pressure drop, temperature rise, and injection rate (time), are interrelated. The calculation of these dimensions is difficult, due to the need to solve coupled flow and heat-transfer equations, and most mold manufacturers adopt a "past-experience" approach. However, valuable guidance can be obtained from simplified flow analyses.

In addition to the pressure drops and temperature rises in channels of constant or smoothly changing cross section, the transitions between them must also be taken into consideration, particularly when the change in cross section is considerable, as is the case for runners and gates. The "entrance

CARTRIDGE
HEATERS

CYLINDER CAVITY

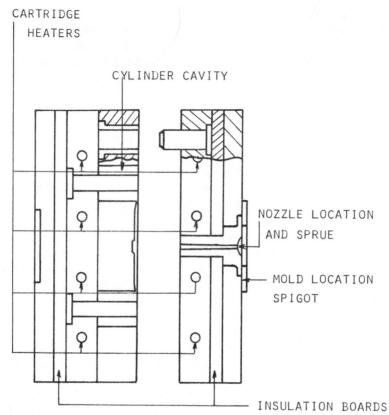

NOZZLE LOCATION
AND SPRUE

MOLD LOCATION
SPIGOT

INSULATION BOARDS

FIGURE 6.13. Conventional injection mold for a horizontal machine, showing insulation boards and electrical cartridge heaters.

effects" for capillaries and dies can provide some guidance here (see Sections 2.3.2 and 4.5.3).

In the scheme of mold design which follows,[15] it is first necessary to make an initial estimate of the number of cavities which may be included in the mold, the layout of the sprue and runner system required to feed them, and the volumetric flow rate for a desired injection time. The length of the flow path to each cavity should be the same wherever possible, to give an equal and simultaneous fill in each cavity. After setting the preceding parameters an initial selection of the sprue cross sections and runner cross sections is required; the dimensions of the nozzle and gates should also be included at this stage, since they are important parts of the total flow path during injection. The gate type and cross section is often fixed by the requirements of the product.

Having arbitrarily fixed the dimensions of the rubber flow path and the desired volumetric flow rate to give a preferred mold filling time, the next step is to determine the pressure drops and temperature rises in each flow path element, to arrive at a total pressure drop and a total temperature rise which can be compared with the available injection pressure and the scorch characteristics of the rubber, respectively. This requires flow property data for the rubber, taken over a range of shear rates and temperatures, for which the capillary rheometer is the most appropriate instrument.

For isothermal flow conditions the pressure drop to maintain a given flow rate in a channel of constant cross section and known length is given by

$$\Delta P = 2\eta_0 L \left(\frac{Q}{\pi R^{(3n+1)/n}} \frac{3n+1}{n} \right)^n \tag{6.31}$$

which is a rearrangement of Eq. (2.35), showing clearly the very strong influence of channel cross section in the $R^{(3n+1)/n}$ term. The temperature rise, assuming conductive heating to be negligible, is then given by

$$\Delta T = \frac{\Delta P}{\rho c_p} \tag{6.32}$$

In channels which are long in comparison with their cross section, the flow properties of the rubber at inlet and outlet will be different due to the temperature rise. Channels of changing cross section, such as the sprue and nozzle, must also be taken into consideration. These difficulties can be overcome by a simple application of the "lumped parameter method"[16] for which a channel is divided into a number of elements, as shown in Figure 6.14, within which uniform flow properties and channel dimensions are assumed, enabling Eqs. (6.31) and (6.32) to be applied to each element. Starting from a known rubber temperature prior to injection, the temperature increments, due to each element, can be added together to give an approximately appropriate temperature for each element. Hence the procedure is to use the temperature increment from the previous element plus all the other previous temperature increments to establish the flow properties for the element being considered. From these the pressure drop and temperature increment for the element being considered can be calculated and the procedure repeated for the next element. For nonround channels an alternative expression for pressure drop can be used:

$$\Delta P = 2 \left(\frac{3n+1}{n} \right) \eta_{wa} \frac{QL}{Ar_v^2} \tag{6.33}$$

where η_{wa} is the apparent viscosity at the channel wall, A is the cross

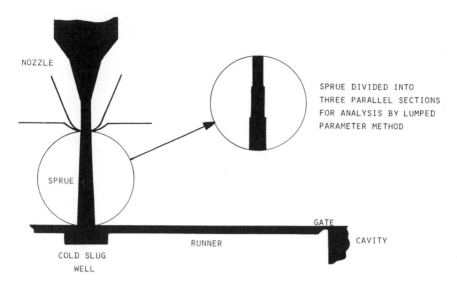

NOZZLE

SPRUE DIVIDED INTO
THREE PARALLEL SECTIONS
FOR ANALYSIS BY LUMPED
PARAMETER METHOD

SPRUE

GATE

CAVITY

RUNNER

COLD SLUG
WELL

FIGURE 6.14. Rubber flow path in a conventional hot-runner injection mold.

sectional area of the channel, and r_v is the channel's effective radius (see Section 4.5.6). Hence, for power-law fluids, η_{wa} is proportional to $R^{(3-3n)}$ and the pressure drop is inversely proportional to $R^{(1+3n)}$.

The lumped parameter method, as described here, has two inherent sources of error. The first of these arises from the inlet temperature being used to determine the flow properties for the element, rather than the average temperature in the element. The second arises from the assumption of uniform flow properties across the channel cross section. However, as temperature increases, the slope of the viscosity–temperature curve decreases, resulting in these sources of errors being less important at elevated temperatures. In fact, for most general-purpose rubbers, at temperatures above 130°C, it is reasonable to assume constant flow properties for the whole length of a sprue or runner. At bulk rubber temperatures below 130°C the assumption of uniform flow properties across the channel cross section and the use of Eqs. (6.31) and (6.32) can result in the pressure drop being overestimated. This overestimation arises both from wall slip and from conductive heat transfer from the mold creating a surface skin of rubber at a higher temperature, which has a lower viscosity than the bulk of the rubber in the channel, providing a lubricating boundary layer.

Having established pressure drops and temperature rises for nozzle, sprues, runners, and gates, including estimates for "entrance effects" caused by abrupt changes in channel cross section, the total design can be examined.

First, the total pressure drop can be checked against the available injection pressure and the heat history of the rubber can be checked against the scorch characteristics, with due allowance for temperatures and residence times prior to injection. If the arbitrarily chosen first design has faults, as it most likely will have, then its characteristics can be used for guidance in a second design, which should then only need minor modifications. In correcting undesirable features of an initial mold design, the influence of changing the volumetric flow rate (fill time) and the cavity layout may need to be examined, in addition to the obvious modifications to sprues, runners, and gates. The effect of the rubber temperature, prior to injection, on pressure drop and temperature rise may also require further consideration.

In the conventional type of injection mold described previously the runner system is an integral part of the mold and is held at the vulcanization temperature, so that the volume of rubber contained in it is vulcanized along with the moldings and then "scrapped" on removal from the mold. For multicavity molds used to produce large quantities of small or thin-walled moldings, the volume of rubber so wasted can form a large proportion of the total injected material. To reduce wastage and to gain other advantages, "cold"-runner systems have been developed.[17] The flow channel manifold necessary to convey rubber from the injection unit nozzle to the gates of the individual cavities is insulated from the cavity plates, enabling it to be maintained at a suitably low temperature. The concept is similar to that used for the flashless transfer molding method described earlier.

To maintain the total pressure drop within the capabilities of available injection pressure and to avoid undesirably high rubber temperatures in the "cold" part of the mold, the sprue and runner cross sections are much larger than for the equivalent conventional mold. Most of the pressure drop and temperature rise is concentrated in the secondary sprues and gates. This has the advantage of minimizing the heat history of the rubber, enabling "scorchy" mixes to be molded with a greater degree of safety than is feasible with conventional molds. The temperature at which rubber enters the mold cavity is generally lower using a cold-runner mold; but it is usually possible to include more cavities in the mold plates to offset the loss of productivity caused by the necessity of slightly lower mold temperatures and longer cure times, due to the pressure drop per cavity for mold filling being less (Figure 6.15).

The complexities of mold construction and temperature control associated with cold-runner molds are obviously greater than for the equivalent conventional mold. However, the REP Company points out in their literature on the subject[17] that, with reasonable planning, a common cold-runner block can be used for a family of molds, and the increase in complexity is normally offset by reduction in the number of opening levels in

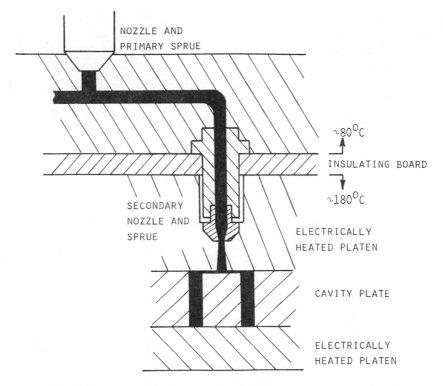

NOZZLE AND
PRIMARY SPRUE

∿80°C

INSULATING BOARD

∿180°C

SECONDARY
NOZZLE AND
SPRUE

ELECTRICALLY
HEATED PLATEN

CAVITY PLATE

ELECTRICALLY
HEATED PLATEN

FIGURE 6.15. Rubber flow path in a cold-runner injection mold.[17]

the mold. Further advantages may be gained when injection is required at points which are difficult of access for stripping the cured rubber contained in the runner and gate system of a conventional mold.

6.4.7. Materials Handling and Mold Stripping

The ease with which rubber deforms, both in the unvulcanized and vulcanized states, makes automatic handling of feedstock and finished components a difficult objective to achieve.

In the progression from compression molding, through transfer and ram injection molding to screw–ram injection moulding, feedstock handling is improved. Compression molding normally requires that a separate charge or blank of a particular shape be placed in each mold cavity to commence each molding cycle. With transfer or ram injection processes a single charge can be used to fill a number of cavities via a sprue and runner system, with the possibility of a single charge serving for two or three molding cycles in the case of a mold with an insulated transfer cylinder.[14] While charge weight is important for the former case, the only requirement of charge shape is that it should fit the transfer cylinder. Most screw–ram injection machines accept a continuous strip feed and are self-feeding; the rotation of the plasticating screw causes the strip to be pulled into the feed pocket. A large roll or coil of feed strip can last for many molding cycles.

The use of integral ejector pins to remove components from a mold is generally only feasible for the injection molding of components for which the forces exerted by the ejector pins result in body movement rather than local deformation. Such moldings include bonded components, where the ejector pins can bear against a rigid insert and "stiff" moldings, having a bulky section or made from a high modulus compound. When moldings need to be deformed during extraction, to remove them from undercuts in the mold cavity form, the manipulation required is usually complex and simple ejectors are ineffective.

A number of aids to improve press utilization are available for each molding method. For the loading of compression molds which produce large numbers of small components, a single loading tray consisting of two plates with holes corresponding to the cavity positions can be used. Aligning the holes allows the rubber charges to fall into the cavities. This reduces loading time and press open time, which improves press utilization and reduces the drop in mold temperature caused by convective heat loss; but the operator time is not reduced. Laying an extruded strip or cord across the cavities achieves this, and is feasible if the flow during mold closing is adequate to fill the cavities consistently and if the waste material is not excessive.

The basic aids for opening molds and extracting components from the cavities are hand tools. It is desirable to use fixed mold opening jigs (for loose-tool molds) and compressed air lines for component extraction, to avoid damage to either. The shape of the nozzle on the compressed air line is important and can influence strongly the ease with which a component may be extracted. Unloading jigs also aid mold stripping. These vary from a simple clamp, to hold the mold plate in a suitable position for manual work, to mechanisms which provide automatic unloading. Such devices are usually specific to a product and should be designed side by side with the mold.

The use of a rotating brush, traversing across the mold plate, has proved to be successful for extracting small moldings of simple shape and avoids the necessity for product-specific devices; but the greatest potential for general-purpose automatic unloading arises from the development of programmable manipulators (industrial robots). These enable the amount of specialized equipment associated with a specific product to be minimized and provide a unique opportunity for automating short-run and jobbing manufacture.

The potential benefits of robots for rubber molding are substantially greater than for plastics moldings, where a properly designed ejector system can often render a robot totally unnecessary. The complex manipulation required to remove many rubber products from their molds will generally require robots of greater sophistication than the simple devices being marketed in the plastics industry; but it can be argued that the need to introduce them is substantially greater, due to the concern being expressed about the effect of fumes from high-temperature vulcanization on the health of operatives.

Apart from reducing the number of operators and removing them from the immediate vicinity of the presses, robots confer a number of other economic and technical advantages. Current measures taken to render the atmosphere and temperature immediately adjacent to the press suitable for operatives involve the moving of large volumes of air. The equipment needed to do this is costly to purchase, install, maintain, and run. It also has a serious cooling effect on mold temperatures, and results in the energy taken to replace this loss being a substantial proportion of the total press energy requirements. Even so, the cooling which occurs can exert a significant influence on the state of cure of products. Placing the presses in a hostile environment enclosure, which simply isolates them from people, practically eliminates the continuing environmental control costs and results in much improved temperature control.

The successful introduction of robots into a molding shop starts with plant layout and mold design considerations. The former is dealt with in Chapter 8 and the latter involves considering the design of the manipulator on the end of the robotic arm in conjunction with the mold design. Consideration must be given to the manipulative actions which can be achieved, assisted by a compressed-air blast if necessary. The presentation of the products in the opened mold should also be arranged so that access is adequate and the actions required for product removal are within the movement repertoire of the manipulator. It is necessary that each press in a group served by a single robot should have molds which can be unloaded with the same manipulator. However, the actions required for product removal can be different in each case, being simply a matter of programming.

6.4.8. Mold Lubricants, Surface Treatments, and Cleaning

Lubricants and surface treatments for molds serve two purposes. They aid the filling of the mold by promoting wall slip and minimize the forces which are needed for mold stripping. The term "lubricants" can be applied to both internal lubricants and surface treatments. The former are compounding ingredients, such as fatty acids and their derivatives, microcrystalline waxes, or low-molecular-weight polyethylenes. Surface treatments are surface-active materials, such as detergents and silicone emulsions,[18] or "dry"-lubricant types, based on PTFE or polyethylene, usually carried in solution and applied from an aerosol. Internal lubricants have the advantage of being present in controlled amounts, giving minimal mold fouling and, with careful selection, not interfering with bonding to nonrubber components.

Although careful compound design and avoiding the use of surface treatment can inhibit the onset of mold fouling, there will eventually be a need for mold cleaning. This can be signaled either by the loss of an important surface finish or by an increase in the number of defects due to poor mold flow.

The aim of any mold cleaning technique is to remove the fouling layer of decomposition products while minimizing the wearing away or degrading of the mold surfaces resulting from the treatment. These requirements are met by the Vaqua blast system,[19] in which the particles used to remove the fouling layer are carried in a high-velocity water jet, and by electrolytic oxidation in salt baths. The latter method has the advantage that clearing of complex undercut mold surfaces presents no problems of access; but it is hazardous in use, requiring that the stringent safety precautions associated with caustic chemicals be used. It is also possible to use "cleaning compounds,"[20] containing ingredients which remove the mold fouling layer, giving the very positive advantage of in-press mold cleaning.

6.4.9. Deflashing and Finishing of Moldings

This section is concerned with operations used to compensate for the inadequacies of molding processes for producing completely finished components. Assembly, which is extremely product specific, is omitted.

The selection of methods of deflashing moldings depends on the size of molding, the mold design, production rate, and the characteristics of the rubber compound. For small components produced in large quantities, cryogenic trimming is widely practiced. This generally involves lowering the temperature of moldings placed in a vibratory or rotary insulated vessel, to take the flash below the glass transition temperature of the rubber compound. The motion of the vessel then causes collisions between the moldings, resulting in the flash being removed. Precise control of the freezing medium, usually liquid nitrogen, is necessary to ensure that only the thin flash is reduced below T_g; the molding is then protected from impact damage. Sometimes it is necessary to use media, which are small objects of varying shapes and materials, to ensure that the flash removal is effective.

This is generally undesirable since the media and the moldings then have to be separated from the flash.

Cryogenic trimming is unsuitable for large moldings and generally requires a tear-off groove in the mold, to ensure that the line of fracture does not extend into the body of the molding. Also, the glass transition temperature of materials such as silicone rubber cannot be achieved, rendering the method ineffective. In this latter case the incentive to use a flashless molding method is very strong, to avoid the need for trimming.

Finishing is concerned with meeting requirements for dimensions and surface finishes which cannot be achieved, or which are uneconomical to achieve, in the molding process. For example, it has been common practice to give the sealing surfaces of lip seals their precise dimensions and geometry by a postmolding lathe-cutting operation; but improvements in molding techniques, particularly the use of vacuum, are eliminating the requirement for this operation.

Although many surface finishes can be achieved by texturing the mold surface, it is sometimes necessary to carry out a grinding operation when the frictional characteristics of the molding surface are important. This generally applies to rollers in feed mechanisms.

6.4.10. Determination of Molding Process Operating Conditions

This section will be mainly concerned with injection molding. Conditions for compression and transfer molding can be determined using the viscous flow, heat-transfer, and vulcanization analyses decribed in the previous sections. While these analyses are adequate for guiding injection mold design, they are insufficiently precise to provide a viable method of selecting machine settings for maximum productivity. However, once a newly designed mold is made available, a practical approach can be adopted, by running trials with it on an injection molding machine of the type used in production.

The injection pressures and temperatures around which the mold was designed should guide the selection of the ranges of machine settings to be investigated. The process variables will include, in progression from the material feed to the finished molding:

1. Screw speed.
2. Screw-back pressure.
3. Preplasticization unit temperature.
4. Injection unit temperature.
5. Head temperature.
6. Nozzle temperature.
7. Injection pressure or speed.

8. Point of switchover for injection to hold-on pressure.
9. Hold-on pressure.
10. Mold temperature.
11. Cure time.

The preceding list assumes that the machine is fitted with a conventional hot-runner mold and has separate preplasticization and injection units; in horizontal screw–ram machines variables 3 and 4 will be combined.

Considerable work has been carried out on the influence of machine settings on process performance, but this work has concentrated on their individual effects[21] and has largely ignored the strong interactions between variables which can be expected. These interactions will produce an operating window, within which the process can be operated successfully, of complex shape, making it unlikely that optimum productivity will be achieved using an intuitive approach to machine setting. Since injection molding is primarily for high-productivity, long-run jobs, a more rigorous approach to setting is likely to produce sound economic benefits.

The injection molding machine can be treated in two stages. The first stage includes all the operations in which viscous flow occurs and the second is concerned with the curing operation. The objective of the first stage is to fill the mold with rubber at the highest possible temperature which does not incur the danger of scorch before the mold has been completely filled. The second stage then involves setting a mold temperature and a cure time appropriate to the initial temperature of the rubber in the mold and the progress it has made toward vulcanization due to the heat history accumulated in the first stage, using the techniques described in previous sections.

Starting with the mold temperature and cure time set at values indicated by prior experience, a factorial experiment design (FED), described in Section 7.3, can be devised to investigate variables (1) to (7). The point of switchover from injection pressure to hold-on pressure and the value of the hold-on pressure need not be included in the FED, and can be adjusted independently to ensure consistent filling of the mold without excessive flashing.

Whether the rubber has scorched prior to mold filling or not will be immediately apparent from the molding. The combinations of variables in the FED which seem likely to incur a danger of scorch should be explored first, to determine the boundaries of the operating window. Adjustment of these values may then be necessary prior to starting the main experiment.

To determine the influence of each of the combinations of machine settings on the temperature at which the rubber enters the mold, it is necessary to measure the injection pressure and the rubber temperature at the nozzle. A nozzle designed to take a pressure transducer and temperature sensor can be used for this purpose; but it should be noted that the temperature sensor must have a very fast response, rendering thermocouples unsuitable. An infrared sensor with a fiber-optic measuring head, such as the one manufactured by Vanzetti, is recommended for this application. Using the relationships between pressure drop and temperature rise in the sprue, runners, and gates of the mold developed for design purposes (Section 6.4.6), the temperature at which the rubber enters the mold cavities can now be estimated for each set of molding conditions.

Using multivariable regression analysis (MVRA), also described in Section 7.3, the influence of the machine settings on mold filling temperature can be quantified by a polynomial equation, and the settings giving the maximum temperature located. The viability of the value for practical production conditions must then be determined, in terms of the consequences of the normal variations in rubber compound properties and machine conditions. It may be preferable to choose conditions which give a measure of processing safety, rather than those which give the maximum temperature.

The influence of the heat history experienced by the material on its progress toward the onset of cross-linking can be estimated, for the machine settings selected, using the graphical methods described in Section 2.5.2 or the procedures described in Section 6.3.3, defining both the material condition and the initial temperature. The procedures decribed in Sections 6.2 and 6.3.3 can then be used to find a mold temperature which gives an adequate product in a minimum cure time.

In the preceding procedures it has been assumed that the plasticization time is substantially less than the cure time plus the mold open time. In this case an additional measure of processing safety and/or productivity can be gained by delaying the start of plasticization, so that it is completed only when the material is required for injection, minimizing the residence times at high temperature. However, if the plasticization times are in the same region as the cure plus mold open times, machine settings which result in the former being larger than the latter should obviously be avoided.

All the procedures described in this section depend on computer methods. While the practical trials are probably no more expensive, in terms of resources and time, than the intuitive methods of machine setting, the time needed for the analysis of the results would be prohibitive without substantial computer aid. Commercial computer programs are available for multivariate regression analysis and for heat-transfer analysis. The program for relating pressure drop to temperature rise should be available from the mold design stage, pointing to one of the benefits of in-house mold design. This leaves the cure modeling program, which will provide obvious benefits in return for the time spent in its development.

6.4.11. Molding Faults and Their Correction

Molding faults may arise during process development, as a result of the initial selection of inappropriate equipment or processing conditions, and during manufacture, as a result of deviations from the specified operating procedure or excursions outside the tolerance bands of the material properties and processing conditions. When these occur it is desirable that the causes be swiftly identified, for effective remedial action. The following paragraphs describe the common molding faults and their causes and suggest remedies.[21-24]

Distortion. This is elastic recovery of the molding, which takes place on release from the mold. It is caused by the flow in the mold occurring after the rubber compound has started to cross-link. The obvious remedy is to modify the processing conditions, the cure system of the compound, or the mold geometry (nozzle and gates). These remedies are listed in ascending order of undesirability, and the best methods to use to establish which remedy is necessary are those described in the previous section. In production, a check of machine conditions and compound scorch characteristics should be sufficient to identify the source of the problem. Particular attention should be paid to the press closing rate for compression and transfer molding and to the injection time for both screw and ram injection molding.

Delamination (onion skin, orange peeling). This fault is due to material continuing to enter and fill the mold after the initial layers of rubber laid down on the mold surface have vulcanized. In addition to having the causes described for the previous molding fault, delamination is exacerbated by long flow paths, thin sections, and high mold temperatures. Since high mold temperatures are desirable with thin sections, to give short cure times, the main remedy for this problem is to lengthen the scorch time of the compound, provided that the procedures described for the previous fault do not reveal another remedy.

Backrinding. This is a severe local deformation at the split lines of compression molds and at the gates of transfer and injection molds. It is caused by thermal expansion of the rubber during vulcanization and can result, on demolding, in local failure of the rubber. In all cases, raising the temperature of the rubber entering the mold will reduce or eliminate the problem. Compression mold design can influence the tendency to backrinding and care should be taken to choose a split line which does not concentrate the thermal expansion of the rubber into a small region of the mold, otherwise large local deformations will occur. In transfer and injection molding a substantial pressure drop occurring between the mold cavity and the runner system after cure has started will result in a large elastic deformation around the gate. This can be substantially reduced by avoiding a sustained high pressure in the mold, from either injection or hold-on stages. These measures will also result in a reduction of the tendency of the mold to produce flash.

Porosity. This is the expansion of the molding, on removal from the mold, to give a cellular or sponge structure. It is caused by undercure and by the presence of volatile materials, mainly water, in the rubber compound. While it is necessary to maintain the level of volatile materials in a compound below limits dictated by their influence on vulcanizate properties, porosity is invariably a sign of undercure. For bulky moldings it may be possible to raise the preheat temperature and thus avoid extending the cure time. For thin-section moldings, raising the cure temperature is desirable.

Blisters. These are mainly due to air trapped or entrained in the rubber. This can occur either in preparation or, in the case of screw injection machines, in the preplasticization unit. In the latter case it is possible to reduce or eliminate the problem by adopting one or more of the following—increasing the screw-back pressure, injecting at a slower rate, and venting the mold adequately (vacuum may be desirable).

Air traps. These are faults due to air being trapped between the rubber and the surface of the mold. In all methods of molding air traps are due to the mold filling pattern, and molds should be designed so that air is swept from the cavity by the advancing rubber. This is a result of the correct choice of split line and, for compression molding, blank shape and placement and, for transfer and injection molding, choice of the gating position(s). For some complex moldings these objectives cannot be achieved and vacuum should be used. In fact, where very precise corners and edges are required, as is the case for lip seals, the use of vacuum can save an off-line machining operation.

6.5. BATCH VULCANIZATION

6.5.1. Hot-Air Ovens

Hot-air ovens are used to postcure components molded from rubber compounds with a high resistance to oxidation. The hot-air postcure can serve two purposes. The first is to continue the cross-linking process to improve the physical properties of the rubber compound, particularly, its time-dependent behavior, as measured by creep, stress relaxation, and compression set. Secondly, the postcure is used to drive off volatile materials from components intended for service at high temperatures, by raising them above their service temperature. Potentially toxic fumes are disposed of in this way and the product undergoes a small change in dimensions, reducing it to its correct size. Postcuring for this latter reason is mainly practised with silicone and fluorocarbon rubber compounds.

The type of oven chosen depends on the precision of temperature control required. Ovens which depend on simple convection can have temperatures which vary by ±5°C, with position in the oven. This is improved to ± 3°C by using fan-assisted convection. Jacketed air-flow ovens, in which the air passes over heating elements between the inner and outer walls of the oven, driven by a fan, typically give ± 1.5°C. A ducted air fan oven, in which the air flow is further improved, give ± 1°C. Ovens of the latter two types are generally only required for materials testing laboratories, where high precision is important.

Where toxic fumes are likely to be given off by the components being postcured, it is important to vent the oven directly to the atmosphere and to ensure that the fumes do not reenter any of the work areas.

6.5.2. Autoclaves and Steam Pans

Autoclaves are used for a variety of purposes, including vulcanization of large products unsuitable for molding, vulcanization of extrudates and forming of extrudates (e.g., car radiator hoses), and secondary vulcanization of large moldings, to improve press and mold utilization. Two types of autoclave are in common use—jacketed and unjacketed. The jacketed type has a double wall, so that steam may be circulated in the jacket, to provide heating without direct contact with the products. An inert gas can then be introduced into the autoclave, to eliminate oxidation and permit brightly colored articles with a good surface finish to be produced. In the unjacketed type the steam is introduced directly into the autoclave and may cause condensation marking on the product. Preventive measures usually result in a dull surface finish.

Large articles usually require a "stepped cure," which involves raising the autoclave temperature to the maximum vulcanization temperature through a series of steps,[1] to give a relatively uniform temperature distribution through the product during heating. This technique is used when bulky moldings are given a second cure in the autoclave, after being vulcanized to the point of dimensional stability in the mold.

The mode of heat transfer in an autoclave, from the heating medium to the product, is convection. Therefore the convective heat-transfer coefficient will control the rate of temperature rise in the products. This must be taken into consideration when cure times and temperatures are being determined.

Autoclaves are pressure vessels and need to comply with the general regulations for pressure vessels, which require regular testing and foolproof interlock mechanisms to prevent the door from being opened while there is pressure in the autoclave. Even very low pressures can exert considerable forces over the door area.

6.6. CONTINUOUS VULCANIZATION

6.6.1. Drum Vulcanizers

Continuous drum vulcanization is used for sheet products and is an off-line process, due to the throughput rate being much lower than that of calendering or spreading. The configuration of a drum vulcanizer is shown in Figure 6.16; the flexible steel belt provides the consolidation pressures needed for composite products and imparts a good surface finish.

The rate of rotation of the drum can be determined from simple one-dimensional heat-transfer calculations and reference to a TTC chart. This speed can be increased if a microwave preheating unit is used to raise the initial temperature of the product.

The thickness of product which may be vulcanized is limited mainly by the curvature of the vulcanizing drum. A thick product straightened out after vulcanization will be subject to permanent surface stresses. Thus, determination of the permissible thickness of product must take into consideration the radius of the drum and the end use of the product.

6.6.2. Fluidized Beds

Two types of fluidized beds are available, one of which operates at atmospheric pressure[25] and one which operates at pressures up to approx-

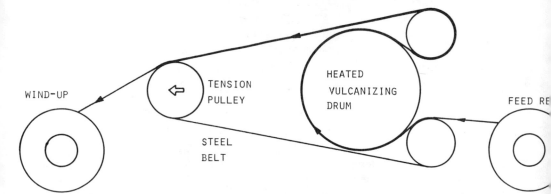

FIGURE 6.16. Continuous drum vulcanizer for sheet products.

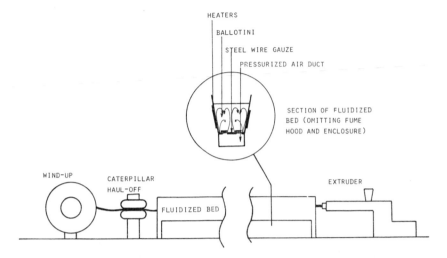

FIGURE 6.17. Extrusion and fluidized bed continuous vulcanization line.

imately 0.7 MPa.[26] The former type is used for homogeneous or unreinforced extrudates, whereas the latter has been developed specifically for cable and hose.

Fluidized beds are on-line vulcanization processes[27] designed to accept a product direct from the extruder, requiring that the throughput rate is synchronized with that of the extruder. Figure 6.17 shows a typical extrusion and atmospheric-pressure continuous vulcanization line. A vented extruder is required to remove volatile materials from the rubber compound prior to vulcanization, to minimize porosity in the extrudate. Desiccants are usually included in the compound for the same purpose.

The principles of fluidization are relatively simple: an air stream passing through a powder bed at an adequate velocity will cause the particles to become suspended on the air stream and separated from each other. In this state the bed takes on all the attributes of a liquid; objects can float on the surface of the bed or sink depending on their density and convection currents can be set up in the bed, driven by temperature differences. Thus externally applied heating will be transferred to objects in the bed by convective heat transfer. In fluidized-bed vulcanizers the design is required to give uniform fluidization, uniform temperature, and rapid heat transfer, by means of forced convection from external heaters. The powder bed consists of glass beads or ballotini. Thus the bed provides both support and heat to the extrudate, ensuring that delicate and complex cross sections undergo minimal distortion.

Unlike steam vulcanization, the temperature of a fluidized bed does not depend on pressure. Therefore, in the pressurized fluidized bed, the pressure is only required to be sufficient to eliminate porosity, which is essential for cable and hose. This pressure is usually in the region of 0.5 MPa, which can be achieved with much lighter and more compact equipment than is necessary for an equivalent steam system.

The cure time of a fluidized-bed vulcanized extrudate is a function of bed length and haul-through speed. Determining appropriate cure times and temperatures involves a two-dimensional heat-transfer problem, which can be solved by the techniques described in Section 6.2. The cure time and temperature then depend on the cross-sectional shape and size of the extrudate, the thermal diffusivity of the rubber compound, the convective heat-transfer coefficient for the fluidized bed to rubber, and the required uniformity of cure in the extrudate.

As for any vulcanization system, venting of fumes is essential and most fluidized beds are "enclosed," ensuring that all the fumes generated can be drawn off and vented to the atmosphere. Fluidized beds which operate at atmospheric pressure are only required to be opened during startup and pressurized beds, due to the nature of the products they are designed for, do not require opening, except for maintenance.

6.6.3. Salt Baths

Salt baths use a mixture of salts, usually potassium nitrate and sodium nitrate, which are molten at typical curing temperatures. Salt baths are very similar in external form and heating arrangements to the fluidized bed. However, the salts are much denser than rubber compounds, requiring the use of rollers or a submerged conveyor belt to hold the extrudate under the surface of the molten salt.

The methods used to determine haul-through speed and bath temperature are similar to those outlined for the fluidized bed. Also, given the nature of the salts used, the requirements for effective fume extraction are more stringent than those for the fluidized bed.

6.6.4. Microwave Units

Microwave vulcanization[28] has the advantage of providing uniform and rapid heating throughout a product, eliminating the pronounced temperature distributions resulting from convective and conductive heating. The majority of microwave units are designed to accept extrudates; but they are also used to preheat conveyor belting, prior to semicontinuous press vulcanization and drum vulcanization.

Microwave vulcanization units usually consist of a microwave section followed by a hot-air section; the function of the latter section is to maintain the temperature established by microwave heating. The whole system is usually about 7 m in length and a conveyor belt, made from materials not susceptible to microwave heating, is used to transport the extrudate through it.

The temperature rise ΔT in a material subjected to microwave radiation is proportional to the absorbed power, to the length of time t the power P is applied, and is inversely proportional to the mass ρV of material being heated:

$$P = \frac{\rho V c_p \Delta T}{t} \qquad (6.34)$$

However, P in Eq. (6.34) is the power absorbed by the rubber, which depends on both the characteristics of the rubber and the additives used. Nonpolar rubbers, such as NR, SBR, and EPDM, are poor absorbers of microwaves and only become viable propositions for microwave vulcanization when mixed with carbon black, or with a polar polymer, such as NBR or PVC. Figure 6.18 shows the percentage of applied energy which is absorbed by a number of rubbers, with a range indicating the influence of carbon black as a filler.

Two further problems encountered with continuous microwave vulcanization are dimensional stability and oxidation. During vulcanization the extrudate is unsupported and can distort after heat softening, simply by the effect of its own weight, and it will then be vulcanized in the distorted shape. This problem is particularly severe for complex extrudates with thin sections. Surface oxidation may occur with NR and SBR, due to the high temperatures achieved and the exposure of the extrudate to the atmosphere during vulcanization, although modern antioxidants can provide sufficient protection in most instances.

In addition to requiring an efficient fume extraction system, microwave emission must be below the limits specified by legislation, and emission levels are required to be monitored at regular intervals.

6.6.5. High-Pressure Steam Tube

In steam-tube vulcanization[29] (often called the CV process) the extrudate is unsupported and is subjected to tensions which could cause total failure of a homogeneous extrudate. For this reason the technique is only suitable for composite products having a continuous nonrubber reinfor-

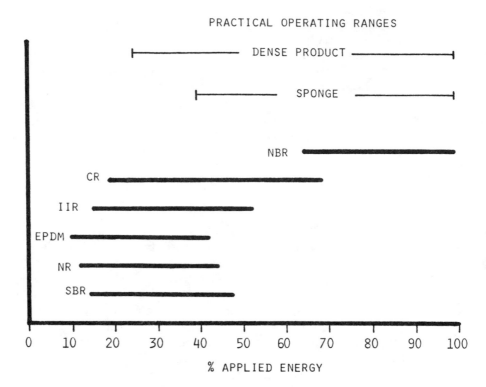

PRACTICAL OPERATING RANGES

FIGURE 6.18. Typical microwave power absorption ranges for rubber compounds.

cement which remains relatively inextensible at the vulcanization temperature. This gives two major applications: cables and hose.

The steam tube is attached to the extruder via a splice box, as shown in Figure 6.19, giving positive sealing. A step-down system is used at the exit, passing from high-pressure water, to low-pressure water, and then to the atmosphere. Dual seals are used to separate these zones. The water level in the steam tube can be adjusted to give an optimum steam/water ratio for uniform cross-link density in the product being vulcanized.

Three types of steam tube are used: inclined, catenary, and vertical. The choice between them is determined by the tension needed to hold the product away from the tube wall. For very light cables the tension needed to produce a very shallow catenary is small, and an inclined (straight) tube can be used. For heavier cables and hoses the tube must conform to the catenary "droop," as shown in Figure 6.19, to avoid excessive tension in the product. When the sheaths of very large cables are to be vulcanized, or products manufactured which have a low tension limit, a vertical tube must be used, where the

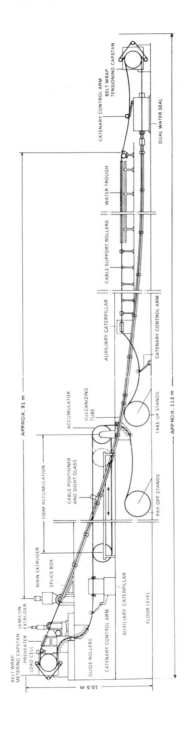

FIGURE 6.19. Catenary continuous vulcanization (CCV) line. (Illustration Courtesy of Shaw Davis–Standard.)

product is subjected to little more than the tension produced by its own weight.

The average length of a vertical tube is 50 m, compared with 60 m for the inclined tube and 100 m for the catenary tube. The larger products and shorter tube dictate a slower transit speed. Also, a high tower or a substantial shaft in the ground is needed for the vertical tube.

6.6.6. Shear Heads

Shear heads are devices for raising the temperature of a rubber compound to a suitable vulcanization temperature, prior to it being forced out of the die of an extruder to form a desired extrudate cross section.[30] Therefore the shear head forms an integral part of an on-line vulcanization system, although it replaces the conventional extruder head. A cross section of a shear head is shown in Figure 6.20; a narrow annular gap is formed between the screw extension or mandrel and the rotary sleeve of the shear head, which forms the flow path from the screw to the die. In this annular gap high shear rates and stresses are generated, providing an efficient conversion of mechanical energy to heat energy via viscous dissipation. The result is a rapid rise in the rubber compound temperature, which is controlled by the speed at which the rotary sleeve is run. The setting of this speed is dependent on the temperature rise required, the screw speed (since the relative speeds of the screw and sleeve determine the shear rate), and the

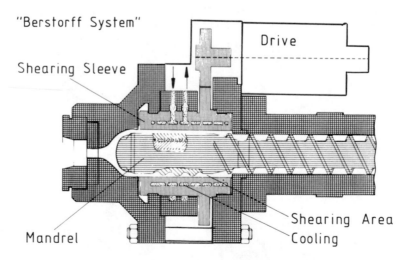

FIGURE 6.20. Extruder fitted with a shear head. (Illustration Courtesy of Berstorff)

throughput of the extruder, the temperature rise being dependent on energy input to the rubber compound per unit volume.

In addition to the smooth-surfaced mandrel shown in Figure 6.20, ones with mixing elements are available, to give improved temperature uniformity in the rubber compound at the die. The shear head shares the advantages of uniform heating with the continuous microwave vulcanization method; but it is not limited to polar and carbon-black-filled rubbers.

The shear-head method is viable because of the nature of the temperature and time dependence of the onset of scorch. The very short residence time between the shearing gap and the die allows the final dimensions of the extrudate to be achieved before effective vulcanization starts, provided that the process and compound variables are suitably adjusted.

After leaving the die the extrudate must be held at the curing temperature sufficiently long for the vulcanization to be completed. A hot-air tunnel with a belt conveyor may be used or, for thin-walled or dimensionally demanding extrudates, a fluidized bed may be substituted, to provide support as well as heating. The length of the bed required as an adjunct to the shear head is much less than if the fluidized bed were used alone.

REFERENCES

1. Hills, D. A., *Heat Transfer and Vulcanisation of Rubber*, Elsevier, London (1971).
2. Heisler, M. P., *Trans. ASME* **69**, 227 (1947).
3. McAdams, W. H., *Heat Transmission*, 3rd ed., McGraw-Hill, New York (1954).
4. Holman, J. P., *Heat Transfer*, 4th ed., McGraw-Hill, New York (1976).
5. Fenner, R. T., *Finite Element Methods for Engineers*, MacMillan, London (1975).
6. Zeinkiewicz, O. C., *The Finite Element Method in Engineering Science*, McGraw-Hill, London (1971).
7. Adams, J. A., and D. F. Rogers, *Computer Aided Heat Transfer Analysis*, McGraw-Hill, New York (1973).
8. Smith, G. D., *Numerical Solution of Partial Differential Equations: Finite Difference Methods*, 2nd ed., Oxford Univ. Press, New York (1978).
9. Weber, J. R., and H. R. Espinol, *Rub. Age* **100** (3), 55 (1968).
10. Hands, D., and F. Horsfall, *Kaut. und Gummi Kunst.* **33** (6), 440 (1980).
11. Mayhew, Y. R., and G. F. C. Rogers, *Thermodynamic and Transport Properties of Fluids, SI Units*, Blackwell, Oxford (1968).
12. Gardner, H. M., Chapter 2 in *Injection Moulding of Elastomers*, ed. by W. S. Penn, Maclaren, London (1969).
13. Juve, A. E., and J. R. Beatty, *Rub. Chem. Technol.* **28**, 1141 (1955).
14. Jurgeleit, H. F., *Rub. Age* **90** (5), 763 (1962).
15. Jansen, R. B., M. Sc. Dissertation, Loughborough University, U.K. (1979).
16. Parnaby, J., "The Application of Polymer Rheological Behavior in Relation to Machine Design and Control," Proceedings of the Joint PRl/BSR Conference on Polymer Rheology and Plastics Processing, Loughborough University, U.K. (1975).

17. Cottancin, G., "Cold Runner Injection Moulds," Plastics and Rubber Institute Conference 'Mouldmaking '82', Solihull (January 1982).
18. Thimineur, R., *Elastomerics* **110** (5), 36 (1978).
19. Pullen, P. M., *Vaqua Process of Mould Cleaning*, Spadone Machine Co. Literature (1972).
20. Murtland, W. O., *Elastomerics* **110** (4), 46 (1978).
21. Wheelans, M. A., *Injection Moulding of Rubber*, Newnes–Butterworths, London (1974).
22. Bament, J. C., Contribution to *Injection Mouldings of Elastomers*, ed. by W. S. Penn, Applied Science Pub., London (1969).
23. Rienzner, H., *SATRA Translation No. 19* (1967).
24. Wagner, D. F. S., *Rub. World* **166** (1), 60 (1972).
25. Davey, A. B., *Chem. Proc.* **12** (4), 7 (1966).
26. Anon, *Plast. Rub. Wkly.* **474**, 22 (1973).
27. Anthony, J. G., Paper given at Meeting of ACS Rubber Division, Chicago (May 1977).
28. Gardiner, R. A., Paper given at Meeting of ACS Rubber Division, San Francisco (October 1976).
29. Santer, S. R., *Rub. J.* **151** (1), 57 (1969).
30. Berstorff Technical Information Booklet, Berstorff Company, West Germany, 1982.

7

Process Control and Quality Control

7.1. THE INTERACTION OF PROCESS CONTROL
AND QUALITY CONTROL

Quality control has traditionally been concerned with the setting of standards which must be maintained at each stage of manufacture, followed by the manual monitoring of process supplies, operations, and products, to check if the specified standards are being maintained. In contrast, process control has been almost entirely concerned with the design and performance of systems for maintaining *machine* conditions and controlling *machine* operations. However, both quality control and process control have the common objective of enabling products which are acceptable to the customer to be produced at a cost which ensures both a competitive price and a viable profit margin.

The dichotomy between process control and quality control arose as a natural consequence of the character of manufacturing technology and equipment available prior to the introduction of the microprocessor. In situations where the opportunities for automatic process monitoring, data analysis, and control are limited and where the major contribution to the success of an operation is made by an operator, quality control is naturally concerned with operator performance and is also influenced by the limitations of manual monitoring methods. Consequently, with automatic systems exerting a relatively small influence on quality in comparison with the operator, except in cases of improper setting of controllers or malfunctions, the whole responsibility for automatic systems has tended to be placed with the engineering and maintenance groups, as a natural extension of their responsibility for the proper functioning of manufacturing equipment. As a result, the monitoring of output quality has generally involved off-line inspection, often after a delay of many hours from the time of manufacture, coupled with a limited amount of on-line sampling inspection. Neither of these inspection methods are conducive to the rapid detection and correction of defective work.

The introduction of microprocessor technology into manufacturing gives the potential for radical improvements in production monitoring and control capabilities but it must be accompanied by an integration of process-control and quality-control objectives and responsibilities for successful exploitation. With progressive improvements in process-control methods and the subsequent automation of operations, the contribution of control systems to product quality increases, while the contribution of the operator diminishes. In many sectors of the rubber industry this trend can be sharply accelerated by the effective application of microprocessor technology, which itself is being rapidly improved and reduced in cost. In addition to leading to greater consistency of operation, computer methods can also be used to replace manual inspection in many cases. They also have the advantage of monitoring *current* production and can provide a continuously updated analysis of the quality of manufacture, creating a powerful tool for the rapid detection and correction of defective work.

Unlike conventional control systems, which perform fixed tasks and are usually concerned with the regulation of machine conditions, microprocessor systems are programmable and can perform sophisticated analyses of the data read from measuring instruments, thus providing an assessment of product quality which is essential for advanced control and the replacement of manual inspection. This demands an insight into the contribution of manufacturing operation variables to product quality which is beyond the scope of most engineering groups, requiring that the quality-control and technical-support groups be involved in the specification and selection of system hardware. These latter groups should also make a major contribution to the programming of new systems and the reprogramming of existing systems in order to improve their performance.

As reliance on microprocessor monitoring and control systems grows it becomes increasingly important to ensure that they are functioning properly and accurately at all times. Validation and calibration can be identified as an important and growing responsibility of the quality-control group.

7.2. SPECIFICATIONS

7.2.1. The Writing of Specifications

The writing of specifications[1,2] generally starts with a product performance specification, which describes in quantitative terms what the product should do, the conditions under which it is required to perform, dimensional constraints on size and shape, and, sometimes, the expected service life. After product design and prototype or sample manufacture have been completed and customer approval has been gained, a product specification can be written. This is a detailed description of the materials to be used and the properties required of them, the dimensions of the product and of subcomponents (if there are any), and the attributes which a product must have in order to be acceptable to the customer. The required attributes can range from force-deformation characteristics to cosmetic appearance.

To enable a manufacturing specification to be devised from a product specification, tolerance bands or limit values need to be defined for each of the items in the product specification. Only when the required precision of manufacture is defined can effective selection or development of manufacturing methods proceed. The setting of tolerances or limit values requires a clear understanding of the influence of deviations from the mean or target values in the product specification on product performance. Some of these values are easy to establish, such as dimensional tolerances, but many, particularly those associated with physical behavior and service life, are more difficult, due to the problem of establishing measurements by which they may be judged, other than direct service trials. Despite the high cost which is often associated with service trials, they often provide the only route by which realistic tolerances can be established.

The product specification is not always originated by the manufacturer of the product. Jobbing companies are often given a product specification by a customer. Also, product specifications are often guided or constrained by national standards and customers' own standards.

Once the product specification, together with appropriate tolerances on dimensions, properties, and so on, has been established and, if necessary, agreed upon with the customer, work can proceed on the manufacturing specifications. This work is normally carried out in conjunction with the process development group and the production management of the areas concerned. Figure 7.1 shows a simplified flow diagram of the procedures involved.

The manufacturing specifications are instructions to machine setters and operatives, defining their contribution to the product, whereas quality audit specifications define the monitoring techniques to be used and the standards to be applied to the product, at each stage of manufacture. Taking the typical example of rubber mixing, three separate manufacturing specifications are needed:

1. The materials to be used and their weights (with tolerance bands).
2. The internal mixer settings and operating procedure.
3. The two-roll-mill settings and operating procedure.

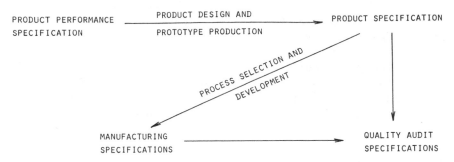

FIGURE 7.1. A simple sequence for specification development during the introduction of a new product.

There then needs to be a quality audit specification to complement the manufacturing specifications. This specification includes the following:

1. The methods and frequency for checking the weighing systems, with tolerance bands for accuracy and a directive on the number of checks on manual weighing accuracy.
2. The methods and frequency of comparing the record of mix temperature and power (if available) with standards having tolerance bands at key points. Also, the method and frequency for checking machine functions and instrument calibrations.

Item (2) is applicable to both the internal mixer and the two-roll mill. In addition to specifications for the monitoring of manufacturing operations, standards need to be established against which the results of the operations can be judged. Continuing with the example of mixing, a specification of the test methods and testing conditions for the mixed batches is required, with tolerances or limits on the results which should be achieved. Of course, measurements are only worth taking if the testing equipment is operating consistently and with the precision demanded by the use of the measurement. Normally, the precision of an item of testing equipment is considered adequate if it is 10 or more times that of the operation which is being monitored. In quantitative terms, the standard deviation of the results of replicated measurements on the testing equipment using standard samples should not be more than one-tenth of the standard deviation of a number of results from a manufacturing operation consistently producing items which are acceptable to the customer. This concept will be explored further in Section 7.3, dealing with process-capability studies. In addition, periodic calibration of testing equipment should be specified, to ensure that any movement of recorded values from the correct values (defined by the movement of the mean of replicated values, and not by a single measurement) is corrected.

As indicated by the preceding examples, manufacturing and quality audit specifications generally form a three-tier system:

Manufacturing specifications

↓

Manufacturing quality audit specifications

↓

Specifications for the validation and calibration
of instruments and testing equipment

To communicate specifications to the manufacturing and quality audit personnel responsible for implementing them on a routine basis, a clear and consistent format is required. This should be standardized throughout the company, although some differences will inevitably arise due to the diversity of processes used by a company. However, these should be differences of detail rather than basic format. Use of computer storage of specifications usually dictates a common format.

7.2.2. Materials

Specifications for materials are required at three stages:

1. Raw material.
2. Mixed unvulcanized compound.
3. Vulcanised compound.

The manufacturing specifications for materials are generally very simple, initially serving to identify the type and quantity of each constituent of a rubber mix and then providing an identifier, usually a code number, for the mix during its progress through the manufacturing procedures. The quality audit specifications are more complex, being required to define the variations in material properties which can be tolerated without problems in manufacturing or customer acceptance of the finished products. Two major difficulties in achieving this objective are the selection of testing methods capable of producing results which correlate with manufacturing and product performance and setting tolerance bands for these measurements.

The problem of selecting measurements by which raw materials can be characterized is particularly acute, due to the wide range of materials used in rubber compounds and their often complex interactions with other ingredients. The guidance provided by the standards organizations is not very helpful in this area, and the response of many rubber companies has been to rely on raw-material manufacturers to supply consistent materials. The total lack of safeguards resulting from this policy can lead to serious problems. It also implies that the product manufacturer does not know what makes a particular polymer suitable for processing operations. This lack of knowledge can make a changeover to an alternative supplier's material very problematic. Only when a low level of precision is required in the product and processes are operated in a region where they are relatively insensitive to changes in material behavior, can material testing be safely ignored. However, process insensitivity is usually located in regions of low process productivity. Clearly, the problems of specifying raw-material tests need to be tackled.

A typical rubber mix consists of bulk particulate fillers, a cross-linking system, oils or plasticizers, and antidegradents, as well as one or more types of rubber. Each of these groups of materials have their own testing requirements for characterization of their influence on manufacturing operations and product performance. They will also depend on the product. For example, a rubber mix which is to be used in a thin film to retain a gas pressure cannot tolerate grit inclusions in the filler.

The SPRI Ltd TMS Rheometer, described in Section 2.3.1, is well suited to the goods inwards testing of raw elastomers; but tests which can be routinely used for testing the nonrubber constituents of a rubber mix are less well developed. An exception is the DBP test[3] for carbon black, which gives a measure of both particle size and structure. Also the moisture content of fillers is often important and can be readily determined, as can the presence of grit in fillers, by a sieve test. The accelerators and antidegradents, which exert a large influence on processing and product behavior, are very rarely checked, although it is known that some accelerators undergo storage deterioration. Characterization methods, such as differential thermal analysis (DTA)[4,5] and differential scanning calorimetry (DSC), can be of considerable assistance here. Oils can be checked by their specific gravities and viscosities.[6]

7.2.3. Subcomponents

Subcomponents for rubber products include metal, ceramic and plastic items for molding and bonding operations, textile and metallic fabrics for flexible reinforced composite products, and a variety of items which are assembled with rubber components to give complete products. Most of these subcomponents will be bought in from external suppliers and have to conform to standards laid out in specifications. These standards should ensure that a subcomponent passed as being acceptable is suitable for the manufacturing operations it will pass through and will enable the finished product to perform its function. It is desirable that the precision or standard of subcomponent manufacture necessary for the product performance should also be adequate for the manufacturing operations, since raising the precision or standard of subcomponent manufacture above the level required by the customer will undoubtably raise direct costs.

Subcomponent tests generally fall into two categories: the first one is concerned with dimensions and appearance and the second with strength and stiffness. The measurement of metal, plastic, and ceramic subcomponents is carried out using standard engineering measurement practices. These are well documented[2] and can be readily located. Strength and stiffness tests are usually confined to the reinforcements of flexible composites. Rigid components, such as fixing plates for rubber–metal bonded mountings, rollers, and seal housings usually exceed their minimum strength and stiffness requirements by a substantial factor of safety, often as a result of the need for a particular shape and size. Plastic subcomponents, metal fabricated and welded subcomponents, and some ceramic subcomponents can vary in their strength and may justify the specification of testing methods. Standard gradual loading tests, such as ones carried out with a universal tensile testing machine, are often adequate; but ceramics and plastics are susceptible to shock loading and, depending on the application, impact tests[7] may be more appropriate.

It is worth noting that some customers supply jobbing companies with subcomponents, to which a rubber element must be added, which has the advantage, for the rubber company, of avoiding a large capital investment in stock. This is particularly important when the subcomponent cost vastly exceeds the value added by the rubber element. Dimensional specifications may still be necessary in these circumstances, to ensure that the subcomponent is suitable for the manufacturing operations.

7.2.4. Machine Settings

Specifications of machine settings, such as times, temperatures, positions, and pressures, are fairly straightforward to deal with, provided certain basic rules are observed. The primary rule is that all the machine settings which influence productivity or quality should be specified. These will initially be the settings determined by the design and development group, and agreed upon by the production management at the time of the transfer of the particular product/machine combination from development to production. There then needs to be a procedure for determining, authorizing, and recording changes to the specification, to attempt to avoid unauthorized and (worse) unrecorded changes to the machine settings.

Determining changes in machine settings, in response to trends apparent in the product performance data generated from the production operation, should be the responsibility of the technical problem-solving group. After authorization for a change has been given by the production manager, the specification can be altered and the nature of the alteration, together with the reason for it, recorded.

It must be accepted that machine performance drifts over a period of time, as a result of wear and general aging. Also changes in material suppliers may occur. Both can be expected to require changes in machine settings. A history of such changes, accompanied by the reasons for them, is a major asset for understanding and controlling each product/machine combination and continually reviewing process capabilities.

The format of machine-setting specifications is required to be easily understood by all those who have to use them and to be as consistent as possible across the whole range of machines in use in the manufacturing operations. The storage of specifications and their accompanying records can range from the traditional folder and filing cabinet system to computer disk storage. The three essentials which any system must provide are accuracy, updating, and accessibility. It is also desirable for the storage system to be similar to that used for other important records, such as standard costs, since these records are referred to by people from a number of different areas of the company. The introduction of computer data processing systems generally results in this standardization, by making it absolutely necessary.

7.2.5. Manual Operations

The term manual operations is used to describe tasks ranging from craftsperson-type jobs, where the only machine assistance comes from hand tools, to jobs where the operator's contribution is largely indirect, through the controls of a machine. In the former case the specification will give the operations to be performed on the product, using a fairly standard kit of tools. In the latter case the specification is usually unique to a particular product/machine combination. If a task can be performed on more than one type of machine, a specification will be required for each machine.

A specification is required to provide the information an operative needs, in addition to the basic job training, to complete successfully the particular operation on the product. Avoiding the repetition of procedures which are common to all tasks at a work station enables specifications to be written in a concise format, which is then likely to be sufficiently brief to be understood and complied with. Actions and procedures should be specified in a quantitative manner wherever possible, together with the expected results of actions, to enable an operator to assess the progress or outcome of the task.

The training-within-industry (TWI) system[8] provides a sound support for semiskilled jobs. Each job is broken down into a sequence of elements, enabling the trainee to assimilate each element separately and then combine them. It is then helpful to the operator if those elements are reflected in the format of the specification, to provide a direct link with the training.

7.2.6. Products

A product specification can usually be divided into three parts:

1. Dimensions.
2. Appearance.
3. Function.

Dimensions are normally specified with tolerance bands, using standard engineering nomenclature; but it is important that the specified tolerances recognize that rubber is a flexible material. All too often tolerances more appropriate to metals are specified by a customer. These are usually unnecessary and only serve to raise the direct costs, which are then passed on to the customer. They also cause endless problems with customers, since changes in temperature are quite capable of causing a dimension to go out of tolerance. Customer liaison and education is essential in this area.

Product quality audit specifications for measuring methods should also recognize that rubber is a flexible material. Noncontacting methods are preferable, but when they are not suitable minimal contact pressure or force should be used. This contact force should be operator independent; but, again, it is not always possible, in which case inspector training should emphasize minimal contact force.

Visual inspection is the only mode of inspection in many sectors of the rubber industry. Its purpose is usually twofold. First, it is intended to locate physical defects which would compromise the functions of the product. Second, it is necessary when the appearance of the product is a factor governing the customer's acceptance of it. The main difficulty of specifying visual quality standards is one of imprecision and ambiguity, which can leave the inspector with a broad range over which a product could be judged to be either good or defective. It is useful at this point to break down the inspection procedure into two stages:

1. Identification of a defect.
2. Determining a course of action based on the severity of the defect.

Setting an inspection specification for the identification of defects is usually quite simple. It is far more difficult to quantify the severity of a defect, which is necessary for item (2). It is possible, through the use of physical examples on standards boards, to show defects on each side of a decision line. Just-pass and just-scrap examples would be given, with pass/repair and repair/scrap borderlines for high-cost repairable products.

Functional tests are usually derived from the product performance specification and are generally concerned with either physical or electrical performance. They are very product specific, but the main problem lies in specifying testing methods and expected results which correlate with service performance.

7.2.7. The Master Specification

A master specification contains all the information which is necessary to manufacture a product to the customer's requirements. Its purpose is to provide a data base from which all the subsidiary specifications dealt with in the preceding sections can be derived. The master specification is essential to the purchasing, production control, standard costs, and manufacturing groups, as well as the quality-control group. A master specification may include identification codes, storage times, and other organizational linking items, in addition to the materials, operations, and standards.

7.3. PROCESS-CAPABILITY STUDIES

7.3.1. Utility and General Considerations

Quantitative information on process or work station capabilities provides an essential foundation for efficient manufacture; its main uses can be identified as:

1. Evaluating alternative proposals for the purchase or modification of equipment for cost/performance effectiveness.
2. Ensuring that new products are assigned to manufacturing routes and work stations which are inherently capable of achieving the quality required by the customer, without constant attention and adjustment.
3. Enabling machine settings and operating procedures which yield optimum productivity to be determined.
4. Assessing the consequences of wear and general aging of equipment on the quality and cost of manufacture.
5. Providing a data base from which realistic estimates and quotations on new products can be composed.

From the preceding list it can be seen that the results of process-capability studies will be used by a number of different groups or departments. Also, the expertise needed to conduct a study may be distributed through a number of departments. While the quality-control department is often given the responsibility for coordinating process-capability studies and maintaining the records, it is a good policy to assemble study groups for specific cases.

Before process-capability study methods can be selected, it is necessary to define and categorize the aspects of process capability which require study. For equipment purchase or modification proposal evaluations, these aspects can be summarized as:

1. The range of inputs and outputs which can be successfully achieved.
2. The output rates, as functions of capital and operating costs.
3. The precision and quality of the outputs.
4. The simplicity and reliability of setting-up, startup, shutdown, and changeover procedures.
5. Reliability and maintenance requirements.

Purely manual operations are dealt with comprehensively in numerous texts on work study[9,10] and are not included here for that reason.

One of the prime requirements of many items of rubber processing equipment is that it should be undedicated. An internal mixer, an extruder, or an injection molding machine is usually required to operate with rubber compounds exhibiting a wide range of processing behavior and, in the latter two cases, to produce a wide range of products. A capability study designed to aid the selection of equipment must therefore establish if a machine is capable of operating with the combinations of inputs and outputs dictated by the company's products, both present and forecast. This depends on well-conceived practical trials, which in turn depend on a clear understanding of the process. Figure 7.2 shows a generalized block diagram of a typical

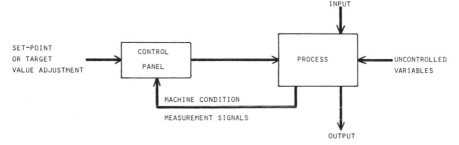

FIGURE 7.2. Layout of a conventional control system for a rubber processing operation.

rubber processing operation, such as extrusion or injection molding. It is somewhat different from the block diagrams to be found in most process-control books, but it does serve to show that in the majority of cases process control is concerned with maintaining machine conditions and is not directly controlling the product or output. As a result variations in the input can be expected to exert a direct influence on the output. Similarly, uncontrolled variables, such as the changes in machine temperature which occur during startup, will also influence the output. The quality or precision of a product will be defined by a distribution of values, bringing a statistical element into process capability.

For each product there will be set points which result in optimum productivity. To define process capability adequately it is necessary to measure performance at these optimum conditions, for a number of products which represent the required operating range. The experiment design and analysis methods necessary to achieve this are dealt with in the next section.

Although process performance can only be truly assessed if the optimum productivity is defined, it is not necessary to operate a process at optimum conditions to manufacture a product successfully. In fact, most processes have a fairly broad band of settings within which it is possible to manufacture successfully. This raises an interesting point,—with a short-run product the resources needed to optimize processing conditions may be greater than the benefit to be derived from optimization. In this case it is only necessary to find successful operating conditions, which is a much simpler task. Hence, the time and resources committed to a process-capability study or the determination of processing conditions for a new product should be closely related to the expected benefits.

7.3.2. Design of Processing Trials

Although a considerable amount of information on process performance can be obtained from equipment manufacturers, from independent experts in technical institutes, and from the technical literature, there is no real substitute for practical trials for making a detailed assessment of process performance.

Most polymer processes are multivariable in nature, having a number of adjustable functions or controllers which influence performance. Using conventional experimental methods, where one variable is investigated while an attempt is made to hold all others at fixed levels, does not allow the effect of different combinations of variables to be defined. The combined effect of changing two variables simultaneously is often greater than the effect of changing each separately by a similar amount. Because of the inability of the conventional experimental method (often called the screening variables method) to deal with interactions between variables, it cannot fully define the region within which a process can operate successfully. Also it will not enable optimum conditions to be defined unless they happen to lie within the limited portion of the operating region characterized by this method.

Factorial experiment designs[11-13] and evolutionary operation methods[14,15] both overcome the limitations of conventional experiment design. Factorial experiment design is particularly suitable for a process-capability study, in which an experimental region is defined by selection of upper and lower values of the process variables, to give a range within which optimum conditions are expected to be found. Combinations of variables at different levels are then defined according to the experiment design used, to give the process settings for the practical trials. Analysis of the results from these trials will enable the boundary of the operating region or "window" of the process to be defined and the optimum conditions to be located. Evolutionary operation simplex methods are well suited to establishing

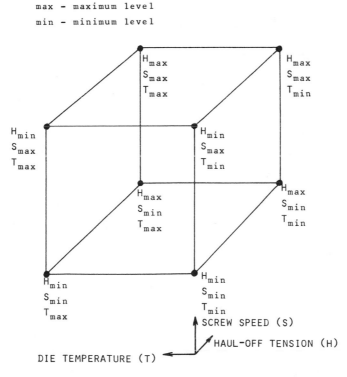

FIGURE 7.3. Two-level factorial experiment design for determining the influence of screw speed, die temperature, and haul-off tension on extruder performance.

optimum conditions for a new product on an existing item of processing equipment. They do not allow the whole operating region to be defined, but they can be used during the normal production operation of a process. Once successful operating conditions have been established and production has started, small adjustments can be made to settings and the effect of these on performance determined. Further adjustments, planned on the basis of the effect of previous adjustments, can then be made, if necessary. As a result the operating conditions are moved, in a series of small steps, to their optimum values.

Factorial experiment designs can be built up from very simple elements. Taking extrusion as an example, the influence of die temperature, screw speed, and haul-off tension on extrudate dimensions can be established initially by using two levels of each of the three factors in all of their possible combinations. This is shown in Figure 7.3, from which it can be seen that a two-level factorial design will have 2^n experimental points, where n is the number of independent variables.

After performing the extrusion trials specified by the experiment design, measurements of extrudate dimensions will have been obtained for each of the experimental points. Because there are only two values of each variable, it is only possible to obtain linear relationships; but these are often sufficient to give a useful insight into process performance. Simple factorial experiments of the type being used here can be interpreted by inspection and plotting of results, as shown in Figure 7.4, where the effect of die temperature on extrudate dimensions is plotted at each of the four permutations of the other two variables.

The usual method of treating the results of factorial experiments is by computer, using a multiple regression analysis program, of which there are a number commercially available.[16] These computer programs define the constants of a polynomial equation which relates the dependent variable, extrudate dimension in this case, to the independent variables. For the two-level factorial extrusion experiment, the equation would have the form:

$$\text{extrudate dimensions} = B_0 + B_1 S + B_2 T + B_3 H \tag{7.1}$$
$$+ B_{12} ST + B_{13} SH + B_{23} TH$$

The interpretation of the equation can be simplified by coding the values of S, T, and M so that they each have a value of zero at the center of the experiment design. The constant B_0 then represents the extrudate dimension at the center of the design, representing the average value of the measurements obtained at each experimental condition; B_1, B_2, and B_3 are the main effect constants, and B_{12}, B_{13}, and B_{23} are the interaction constants.

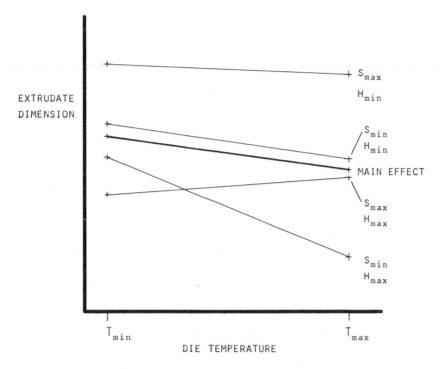

FIGURE 7.4. A plot showing the influence of interactions with screw speed S and haul-off tension H on the relationship between an extrudate dimension and die temperature.

The main effect constants are obtained by determining the separate effects of each independent variable *averaged over the levels of the other two*. Referring to Figure 7.4, the main effect of die temperature is obtained by averaging the four values of extrudate dimension at T_{min} and similarly at T_{max}. The slope of the graph between T_{min} (average) and T_{max} (average) then gives B_2. The differences between the slope of the main effect and the slope of the other lines shows that there are interactions between the independent variables. Only if all of the slopes are similar are there no interactions. Interactions cannot be evaluated as simply as the main effects; but it can now be seen that the constants B_{12}, B_{13}, and B_{23} will be needed to enable Eq. (7.1) to represent adequately the relationship between extrudate dimensions and the three extruder variables.

Simple two-level factorial experiment designs have two main faults: they do not allow any curvature in the relationship between the dependent and independent variables to be characterized and they do not allow an assessment of the influence of experimental error on the relationship obtained to be made. The former can be overcome by inserting more points into the experiment design, providing sufficient information for the true form of the relationship to be determined. The latter can be overcome by replicating experiments at some points in the design, enabling a statistical analysis to be performed. Both of these measures introduce additional experimental points, so designs which give the additional information without resulting in an unreasonable increase in the number of practical trials which have to be performed are clearly desirable.

There are two basic methods of ensuring that the resources committed to practical trials are modest in comparison with the expected benefits. One involves adding more levels of each independent variable to the experiment design but only using a portion of the total number of experimental points thus generated. Partially complete designs of this type are known as fractional replicates. Alternatively, by making some assumptions about the shape of the relationship between the dependent and independent variables (the response surface) in the case being studied, a highly efficient design can be chosen. This is purely a case of deciding if the form of the equation associated with the design will generate sufficient curvature to model adequately the process being studied. For polymer processes a polynomial equation of quadratic form usually gives an acceptable fit to experimental results, enabling the central composite rotatable design of Box et al.[17,18] to be used.

Central composite rotatable designs are based on a complete two-level factorial design, which is then supplemented by additional points to enable the curvature of the response surface and the experimental error to be estimated. They are called rotatable designs because the variance at any point in the design is a function only of the distance of that point from the center of the design, which simplifies the interpretation of the statistical analysis of the experiment. These designs have been widely used for the design of rubber compounds, under the general name of computer compounding.[19] A two-factor design is shown in Figure 7.5, to illustrate the relative positions of the experimental points. These points are now identified by code values, which clarify the design and are necessary for the computer statistical analysis of an experiment. The central composite rotatable design assigns five levels to each variable or factor:

$$-a \quad -1 \quad 0 \quad +1 \quad +a$$

where

$$a = 2^{k/4} \tag{7.2}$$

and k is the number of independent variables in the experiment. Table 7.1

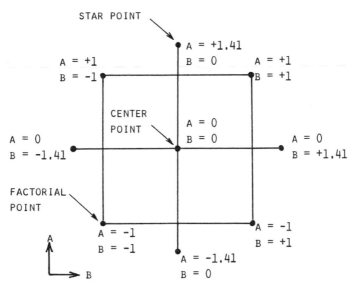

FIGURE 7.5. A central composite rotatable experiment design in two independent variables.

shows how real variables are assigned to the coded experimental points and Table 7.2 gives the full experimental design, for a three-factor internal mixer experiment. The number of center-point replicates is determined by the magnitude of the experimental error expected, in relation to the effect of changing the values of the independent variables on the measured response. In a batch process, such as internal mixing, there are numerous sources of experimental error, and a large (6–12) number of center-point replicates is necessary to obtain a good estimate of process performance at the center point. In a continuous process, such as extrusion, where the experimental errors should be small (given a uniform feedstock and feed rate), the number

TABLE 7.1

Code value	Fill factor	Unit work (MJ/m^2)	Rotor speed (rpm)
−1.68	0.53	266	13
−1	0.6	322	20
0	0.7	404	30
+1	0.8	484	40
+1.68	0.87	542	47

TABLE 7.2

	Mix number	Fill factor F	Unit work W	Rotor speed S
Factorial points	1	+1	+1	+1
	2	+1	+1	−1
	3	+1	−1	+1
	4	+1	−1	−1
	5	−1	+1	+1
	6	−1	+1	−1
	7	−1	−1	+1
	8	−1	−1	−1
Star points	9	+1.68	0	0
	10	−1.68	0	0
	11	0	+1.68	0
	12	0	−1.68	0
	13	0	0	+1.68
	14	0	0	−1.68
Center-Point replicates	15–20	0	0	0

of center-point replicates can be small (3 to 4), and in processes known to be very consistent, can be reduced to one. However, where there is any doubt about the consistency of process performance, it is advisable to opt for a larger number of center-point replicates.

The central composite rotatable design gives a good return of information for the number of experiments which have to be performed when three or four variables are investigated, but becomes less efficient as the number of experiments is increased. A redundancy factor R_f has been developed[20] to give an indication of experiment efficiency and is defined by

$$R_f = N/b \tag{7.3}$$

where N is the number of experimental points in the design and b is the number of terms in the polynomial response equation, which, for the three-factor internal mixing experiment detailed in Table 7.2, has the form

$$Y = B_0 + B_1 F + B_2 W + B_3 S + B_{11} F^2 + B_{22} W^2 + B_{33} S^2$$
$$+ B_{12} FW + B_{13} FS + B_{23} WS \tag{7.4}$$

The redundancy factor in this case is 1.4 because only the number of experimental points is counted and not the number of runs, which is greater if replicates are performed. Table 7.3 gives the redundancy factors for four-, five-, six-, and seven-factor experiments. Ideally, R_f should be as near to unity as possible (it cannot be less than unity); but an experiment can be

TABLE 7.3

Number of independent variables	Number of terms in response equation[a]	Number of experimental point in design	Redundancy factor
4	15	25	1.67
5	21	43	2.05
6	28	77	2.75
7	36	143	3.97

[a]For response equation of form given by Eq. (7.4).

considered to be acceptably efficient if $R_f \leqslant 2$. In effect this means that some of the experimental points can be removed from the five-, six-, and seven-factor designs without a catastrophic loss of information, although it must be emphasized that the choice of these points is critical.

As the number of independent variables is increased, the proportion of the total number of experiments contained in the two-level factorial element of a composite design also increases. Coupled with the fact that star points or center-point replicates cannot be removed without a serious loss of information, this points toward a reduction of the number of factorial points. Fractional replicates of two-level factorial experiments can easily be generated. Half-replicates, which use half the available experimental points, can be selected by taking a full factorial design and choosing those points for which

$$ABCDE = I \tag{7.5}$$

for a five-factor experiment, and

$$ABCDEF = I \tag{7.6}$$

for a six-factor experiment, where A, B, C, D, E, and F are the coded values of the independent variables and I is an identity which can take the value of $+1$ or -1. Table 7.4 shows portions of five- and six-factor designs, where the fractional replicate is selected by taking points identified by $ABCDE = +1$ and $ABCDEF = +1$; these points are indicated by asterisks. The points without asterisks can be selected by setting $ABCDE = -1$ and $ABCDEF = -1$, and they form an equally viable alternative half-replicate. As a result of using half-replicates of the two-level factorial elements of the composite designs, the redundancy factor is reduced to 1.29 for the five-factor case and 1.61 for the six-factor case. For the seven-factor design the redundancy factor would only be reduced to 2.19 and a further reduction, such as a quarter-replicate of the two-level factorial points, is indicated.

CHAPTER 7

TABLE 7.4

| Six factor | A | B | C | D | E | F | |
Five factor		A	B	C	D	E	
1	+1	+1	+1	+1	+1	+1	*
2	+1	+1	+1	+1	+1	−1	
3	+1	+1	+1	+1	−1	+1	
4	+1	+1	+1	+1	−1	−1	*
5	+1	+1	+1	−1	+1	+1	
6	+1	+1	+1	−1	+1	−1	
7	+1	+1	+1	−1	−1	+1	*
8	+1	+1	+1	−1	−1	−1	
9	+1	+1	−1	+1	+1	+1	
10	+1	+1	−1	+1	+1	−1	*
11	+1	+1	−1	+1	−1	+1	*
12	+1	+1	−1	+1	−1	−1	

However, this results in coupling between some of the interaction terms of the response equation, so that their effects cannot be separated.[11] Unless the fractional replicate can be chosen so that coupling occurs between interactions known to be insignificant, there is a definite loss of information.

If it is known that one or more of the independent variables have a particular form of relationship with the response, which may demand more curvature in the response surface than can be produced by a quadratic polynomial equation, a transformation of either the independent or the dependent variables can be used. A transformation is simply a method of changing the distribution of the values of a variable over its range. For example, the rate of change of some properties with respect to work input is very high in the early stages of internal mixing but decreases rapidly as the work input increases, giving an approximately logarithmic relationship. It is useful to transform the work input so that

$$x = \log \text{ (work input)} \tag{7.7}$$

The values of the transformed variable x are then assigned to the coded values for the experimental design. In fact any logarithmic, exponential, or reciprocal relationship should be considered a potential candidate for transformation.

It is preferable to transform the independent variable(s), rather than the response, for two practical reasons. First, the relationship between each of the independent variables and the response can be quite different, requiring that a different type of transform be used in each case. Alternatively, only one or two of the independent variables may need to be transformed. Second, the response is much easier to interpret if it is not expressed in transformed quantities.

After deciding on an experiment design appropriate to the process being investigated and tentatively assigning real values to the coded variables, some preliminary trials are usually necessary to check if experimental points at the extremes of the design are feasible or, alternatively, if the points need to be moved further out. In a process with strong interactions between variables, the most extreme conditions are often found at the wholly negative or wholly positive combinations of the coded factorial points; and from a knowledge of the dynamics of a process, it is usually possible to identify the star points which will exert the greatest influence on performance. During the preliminary trials it is also necessary to fix the levels of those process variables which are not included in the experiment design.

While all the details of a series of practical trials are being worked out, some thought has to be given to the effect of the logistics of the exercise on the results which will be obtained. If trials are spread over a period of days, there is a chance of uncontrolled variables exerting a significantly different effect on the process on each day. While this problem cannot be eliminated, measures can be taken to avoid its interfering with the interpretation of the results. If the trials are carried out in the order given in the experiment design, there is the possibility that the uncontrolled variables will produce a systematic bias in the results. However, if the run order is randomized and some center-point replicates are included in each day's trials, the error will be random and will appear, in electronics terms, as noise or pure error.

7.3.3. Empirical Modeling and Statistical Analysis of Process Performance

The mechanics of fitting a response equation to the results of a factorial experiment design, then analyzing goodness of fit and experimental error, are competently dealt with by a number of commercial statistical computer program packages. However, although this allows the technologist to adopt a "black box" approach to multivariate regression analysis and to statistical analysis, it does not absolve the technologist from the need to propose the response equation to which the data are to be fitted or from interpreting the statistical analysis of the model-fitting attempts. This section deals with the understanding of model fitting and analysis which is necessary in order to use an interactive computer program competently.

Starting with the input, it is usually necessary to create a file which lists the experimental runs, identifying them by the code values of the independent variables and the responses associated with them. At this stage it may also be necessary to identify the terms which could be used in the response equation,[16] although all the terms identified need not be used. Continuing from the previous section, central composite rotatable designs provide data for the fitting of a quadratic polynomial response equation which, in its general form, can be written as

$$
\begin{aligned}
Y = B_0 &+ B_1 X_1 + B_2 X_2 + B_3 X_3 + \cdots + B_N X_N \\
&+ B_{11} X_1^2 + B_{22} X_2^2 + B_{33} X_3^2 + \cdots + B_{NN} X_N^2 \\
&+ B_{12} X_1 X_2 + B_{13} X_1 X_3 + B_{14} X_1 X_4 + \cdots + B_{N-1N} X_{N-1} X_N
\end{aligned}
\tag{7.8}
$$

in which the interactions previously identified in Eq. (7.1) can be more precisely defined as second-order interactions, which differentiates them from the third- and higher-order interactions. Third-order interaction terms are not often statistically or practically significant and can usually be ignored. Table 7.5 gives the second-order interaction constants for experiments of up to nine factors. The constants required for a particular response equation can be identified as all those to the left of the diagonal line labeled with the number of independent variables used in the experiment. With all the linear, quadratic, and second-order interaction terms needed for the response equation, the computer analysis of the experimental results can be carried out. This is the simplest part of the whole operation.

The multiple regression analysis produces values for each of the constants in the response equation, but *it should not be assumed that this equation adequately models process performance until it is confirmed by a rigorous statistical analysis.* Response equations can be obtained from results where the combined effects of experimental error and lack of fit lead, in the absence of any check on validity, to totally erroneous recommendations being made, with disastrous consequences. The statistical analysis can usually be carried out by the computer in conjunction with the response equation fitting; but the technologist or engineer using the system must have some knowledge of the techniques used in the statistical analysis in order to interpret its meaning.

The primary method is analysis of variance (ANOVA),[19,21] which takes the total variation in the response and first determines how much of this variation is accounted for by the response equation. This is defined by the regression mean square (RGMS), which is the sum of squares of the differences between the fitted values Y_F at each experimental point and the mean of the observations $(\bar{Y})$, divided by the appropriate number of degrees of freedom. The degrees of freedom associated with each sum of squares in Table 7.6 are a function of the total number of experiments in the design, including replicates N, the number of terms in the response equation R, and the number of center-point replicates C.

TABLE 7.5
Second Order Interactions

Number of independent variables

2	3	4	5	6	7	8	9
B_{12}	B_{13}	B_{14}	B_{15}	B_{16}	B_{17}	B_{18}	B_{19}
B_{23}	B_{24}	B_{25}	B_{26}	B_{27}	B_{28}	B_{29}	
B_{34}	B_{35}	B_{36}	B_{37}	B_{38}	B_{39}		
B_{45}	B_{46}	B_{47}	B_{48}	B_{49}			
B_{56}	B_{57}	B_{58}	B_{59}				
B_{67}	B_{68}	B_{69}					
B_{78}	B_{79}						
B_{89}							

The difference between an experimental value Y_E and a fitted value Y_F is termed a residual, which identifies variation not accounted for by the response equation. The overall measure of this is the residual mean square (RSMS). An F ratio for the response equation, which can be tested for significance by reference to standard statistical tables, is given by

$$F_{EQ} = \frac{\text{RGMS}}{\text{RSMS}} \qquad (7.9)$$

With the degrees of freedom for both the numerator V_1 and the denominator V_2, the F ratio can be compared with the values in the F distribution

TABLE 7.6

Analysis of Variance Table

Source	Sum of squares (SS)	Degrees of freedom (d.f.)	Mean squares (SS/d.f.)
Response equation	$\sum (Y_F - \bar{Y})^2$	$V_1 = R - 1$	RGMS
Residuals	$\sum (Y_E - Y_F)^2$	$V_2 = N^a - R^b$	RSMS
Lack of fit	$\sum (Y_E - Y_F)^2 - \sum (Y_{EC} - \bar{Y}_{EC})^2$	$V_1 = N - R - C^c + 1$	LFMS
Pure error	$\sum (Y_{EC} - \bar{Y}_{EC})^2$	$V_2 = C - 1$	PEMS

$^a N$ = number of runs in the experiment design.
$^b R$ = number of terms in the response equation.
$^c C$ = number of center-point replicates.

tables.[22] If the ratio exceeds the upper 5% point of the F distribution, it is almost certain that the response equation has some explanatory power. If, on the other hand, the ratio lies well below the upper 5% point, there is a good chance that the equation is merely modeling random variation, *and it would be unwise to use it as a basis for decision making.*

To test the lack of fit of the experimental results to the response surface, it is first necessary to subtract the sum of squares of the variation due to pure error, which is estimated from the center-point replicates, from the residual sum of squares, giving the lack-of-fit sum of squares. Dividing this by the appropriate number of degrees of freedom gives the lack-of-fit mean square (LFMS). The pure-error mean square (PEMS) is obtained by dividing the sum of squares of the differences between the experimental results for the center-point replicates (Y_{EC}) and the mean of the center-point replicate results $(\bar{Y}_{EC})$ by the appropriate number of degrees of freedom. The F ratio for lack of fit can now be defined as

$$F_{LF} = \frac{LFMS}{PEMS} \tag{7.10}$$

If the response surface fits the results adequately then this calculated F ratio will be *less* than the numerical value given in the table for the upper 5% point of the F distribution. In consequence, the response equation must be rejected if this F ratio is significant at the 5% level (i.e., it exceeds the upper 5% point).

The calculation of the quantities in the variance table and the F ratios can obviously be carried out using the computer, either as an integral part of the regression analysis program package or as a subsequent operation using a program written in-house. However, if the F ratios indicate that the response equation does not adequately model the relationship between the results and the independent variables, it is the experimenter's task to decide upon a suitable course of action.

If the regression F ratio is not significant at the 5% level and appropriate transformations have already been employed in the experiment design and regression analysis, the main hope is to try a different form of equation. The simplest measure is to add more terms to the response equation, such as third-order interactions, and try again. If this does not work it is reasonable to conclude that the quadratic polynomial equation cannot attain the response surface curvature demanded by the experimental results and to consider other forms of the equation giving more acute curvature, such as ones with cubic terms. However, with an efficient experiment design of the control composite rotatable type, intended to be used specifically with quadratic polynomial equations, more experimental points will be needed to model adequately the coefficients of the higher-order equation. This goes beyond the scope of this section, and reference should be made to specialist texts on experiment design.

If the regression F ratio is significant at the 5% level, or only marginally fails to be significant at this level, but is accompanied by a significant lack-of-fit F ratio (indicating an unacceptable lack of fit), the first step is to plot the residuals *vs.* the fitted values at the experimental points on a scatter plot of the type shown in Figure 7.6. This will indicate if the problem is being caused by rogue results, known as outliers. The effect of an unrepresentative result, particularly at one of the star points, can exert a serious distorting effect on the response surface. If, from a knowledge of the process, the unrepresentative result(s) are considered to be due to gross experimental error two courses of action are open. The best one is to repeat the trial which produced the rogue result, but in many cases this may not be possible and the alternative is to manipulate the result. It can sometimes be removed completely (not set to zero) or adjusted to coincide with the fitted value, followed by a rerun of the regression and statistical analyses. Obviously, a good deal of common sense and understanding of the process is required here; it is very easy to obtain acceptable F ratios from fictitious results.

The lack-of-fit F ratio can be significant even though the response equation provides a satisfactory representation of the data. This anomalous situation arises when the denominator of the F ratio is small, as a result of a very narrow distribution of errors in the center-point replicates. Here again judgment is called for. If it is apparent that the lack-of-fit F ratio would not be significant if the distribution of errors in the center-point replicates was

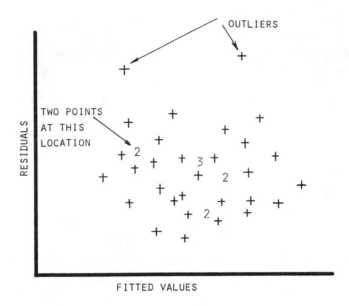

FIGURE 7.6. Scatter plot of residuals (fitted values — experimental values) *vs.* fitted values.

broader but still within the normal range of variations expected from the process, then the response equation can be accepted.

The predictions that are made using a polynomial model of the kind described in this chapter are only valid within the experimental region, which for a central composite rotatable design, is defined by

$$X_1^2 + X_2^2 + X_3^2 + \cdots + X_N^2 \leqslant N \tag{7.11}$$

which, for a five-variable design, can be written as

$$X_1^2 + X_2^2 + X_3^2 + X_4^2 + X_5^2 \leqslant 5 \tag{7.12}$$

Since the response equation is an empirical mathematical model, there is no guarantee that the response surface will continue to define the process performance outside the experimental region. In fact, it is very probable that it will diverge very strongly from it. In consequence, it cannot be overemphasized that *regression equations cannot be used for extrapolation outside the region defined by the experiment design.*

7.3.4. Representation and Optimization of Process Performance

Inspection of the coefficients of response equations gives an overall indication of process performance; but for a detailed examination of performance, a graphical representation of the response surface is required. For hard-copy graphs the two techniques normally employed are contour plotting (Figure 7.7) and isometric plotting (Figure 7.8). Both give the response as a function of two independent variables, while holding any additional independent variables at fixed values. The isometric plot probably gives a more immediate impression of the shape of the response surface than the contour plot; but it is difficult to locate values intermediate between the maximum and minimum. This is easily achieved on a contour plot and a further advantage is that a number of responses can be superimposed on the same axes, allowing a comparison to be made between different performance criteria. Contour or isometric plots can be produced with the additional independent variables, called nonaxial variables, at a number of different levels, to enable the influence of these variables on the responses to be observed. It is sensible to choose the two independent variables which exhibit the strongest interaction for plotting, by reference to the response equation.

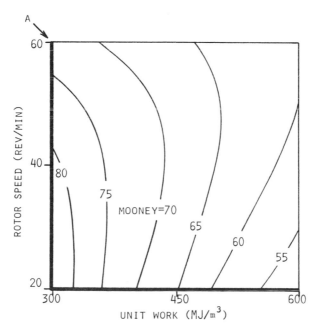

FIGURE 7.7. Contour plot showing the influence of internal mixer rotor speed and unit work on Mooney viscosity.

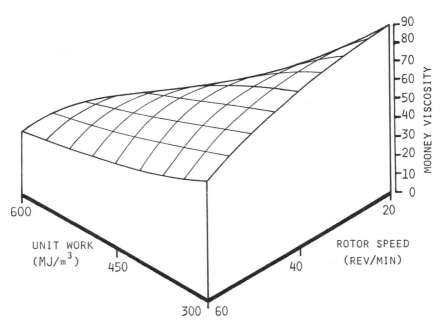

FIGURE 7.8. Isometric plot showing the influence of internal mixer rotor speed and unit work on Mooney viscosity (viewed from the direction of arrow A in Figure 7.7).

The values for the nonaxial variables can then be determined by again examining the response equation or by using the interactive methods which follow.

Contour and isometric plots are useful for inclusion in written reports; but they have some disadvantages for interactive examination of response surfaces if the full benefits of using a color graphics terminal, such as the type also described in Sections 11.1.5 and 11.1.6 for use with interactive work-station scheduling and loading, are to be obtained. A major drawback is the amount of computation needed to produce a plot, causing delays which are unacceptable for interactive work, even when a powerful computer is used. Also, the plots become difficult to interpret if several are displayed simultaneously on the screen. Most of these problems can be overcome by using shading diagrams of the type shown in Figure 7.9,[23] which require substantially less computation and time to produce. On the graphics terminal screen the numbered areas would be displayed in different intensities or shades of a single color, enabling a number of responses to be identified by color and plotted in adjacent or superimposed shading diagrams. By comparing Figure 7.9 with Figure 7.7 it can be seen that a shading diagram is essentially a coarse contour plot.

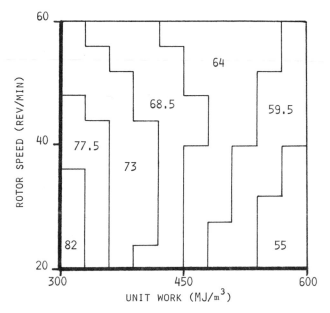

FIGURE 7.9. Shading diagram showing the influence of internal mixer rotor speed and unit work on Mooney viscosity.

The advantages of using an interactive technique for examining response surfaces increase as the number of independent variables is increased. With an experiment of four or more factors it is possible either to present a coarse representation of the whole response surface or a detailed representation of a small portion of it. Initially, a matrix of shading diagrams can be set up to represent the whole response surface. Four of the five variables in an internal mixer experiment are shown in Figure 7.10, with circulating water temperature at a preselected value. The influence of both the main effects and the interaction resulting from ram pressure and fill factor can be discerned from the differences between the shading diagrams. Three levels of circulating-water temperature can be examined in conjunction with the three levels of the other variables by foregoing some detail in each shading diagram. Three matrices of the type shown in Figure 7.10 can be set up side by side on the screen, each at a different level of water temperature. It has been found that sufficient resolution and discrimination are maintained by reducing the number of blocks within each shading diagram to 9 (a 3 × 3 matrix), as compared to 100 in Figure 7.9. For the representation of response surfaces with more than five independent variables, it is necessary to display the effects of some variables by setting up matrices of shading diagrams at different levels of the fixed variables sequentially.

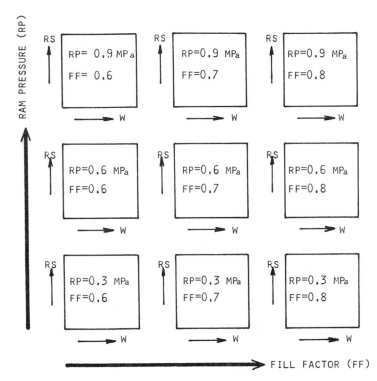

FIGURE 7.10. A matrix of shading diagrams showing the influence of rotor speed (RS), fill factor (FF), unit work (W), and ram pressure (RP) on internal mixer performance.

Graphics terminals which can store a diagram, enabling the user to "flip" from one diagram to another very rapidly, are desirable for this type of operation.[24]

During the scanning of the whole response surface, different combinations of independent variables can be tried on the axes of the shading diagrams, to gain the additional insight into process performance to be derived from a different viewpoint. Also, the nonaxial variables can be tried at any combination of values, within the experimental region, which seem likely to yield interesting information. There are then a number of options available for proceeding from this point which emphasize the value of an interactive method:

1. Expansion of regions of interest.
2. Imposition of constraints.
3. Simultaneous display of responses.

From a scan of the whole response surface it is usually possible to identify regions which yield response values which are of practical interest for technical or commercial reasons. These can be expanded by setting up shading diagrams which only cover the regions of interest, allowing a more critical examination of them to be made.

The identification of regions of interest can be aided by the imposition of constraints. If it is known that a response should not be greater than or less than a certain value, or should be between certain limits, then the regions of the response surface which do not conform to these constraints can simply be deleted. In addition, the regions deleted by the application of constraints to one response surface can be transferred to other response surfaces; additional constraints can be imposed in conjunction with the other responses to, hopefully, arrive at a "processing window" which satisfies all the constraints. If the whole experimental region is deleted, then it is obviously necessary to seek an alternative processing machine, or to reexamine the constraints, with a view to relaxing some of them.

Constraints can only be set on the values of responses if a tolerance band or target value can be precisely defined. This is more likely to be possible when a new product is being introduced onto existing equipment than when the purchase of new equipment is being considered. In the latter case a more flexible approach is desirable, since the setting of rigid constraints or expectations can reduce the range of choices in an unnecessary and counterproductive manner. This problem can be avoided by looking at a number of responses simultaneously.

If a number of separate and distinct responses define the performance of a machine, their relative values in regions of interest can be compared by setting them up on the screen side by side, as shown in Figure 7.11 for an internal mixer. Again the values of the nonaxial variables can be tried at any combination of values which lie within the experimental region; but the region for each response is changed simultaneously, to enable a direct comparison between them to be maintained. Using this technique, the "best" combination of the performance indicators (responses) for any particular machine can be discussed and decided upon. These can vary widely between alternative machines, reflecting the differing philosophies of design and operation of the machine manufacturing companies.

7.3.5. An Integrated Approach to Experiment Design, Model Fitting, Analysis, and Representation

Using the procedures described in Sections 7.3.2 to 7.3.4, it is possible to design an experiment, fit a mathematical model to the results, check the validity of both the results and the model, and then use it for assessing

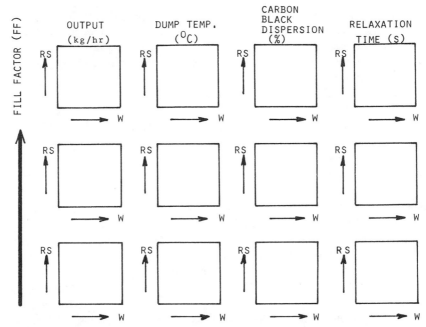

FIGURE 7.11. Simultaneous display of a number of measures of internal mixer performance and their dependence on fill factor (FF), rotor speed (RS), and unit work (W).

process capabilities or for other purposes, provided that the standard experiment designs described are appropriate to the system being investigated. Unfortunately, situations often arise in which standard designs are either inappropriate or inapplicable. For example:

1. The shape of the operability region may be incompatible with the symmetry required by a standard design.
2. It may be desirable or necessary to use a model for which no standard design exists. This is particularly likely if the form of the model has been chosen to reflect the fundamental principles and dynamics of a particular process.
3. Even using a standard model, such as a quadratic polynomial, we may know in advance that some of the terms—perhaps some of the interactions—are likely to be negligible. Ideally, this information should be used in selecting a design.
4. It may not be possible to carry out the number of runs or trials required by a standard design. This problem is likely to be particularly serious in an industrial context, where the cost of experimentation may be high or access to equipment may be limited.

The development of satisfactory experiment designs involves a substantial statistical effort, which is often beyond the resources of the engineer or technologist. To overcome these problems an approach known as optimal experiment design has been developed[25] and work is now going forward[24] to enable the experimenter to use optimal design methods interactively, through a suitable computer program.

The interactive optimal experiment design represents the first element of an interactive computer-based approach to experiment design, model fitting, statistical analysis, and representation. The examination of response surfaces using shading diagrams described in Section 7.3.4 is an integral part of the computer program package, and methods of examining the validity of results and fitted models using computer graphics are also being developed. This work is intended to make statistical experiment design and analysis a much more accessible tool for use by the engineer and technologist than it has been hitherto.

7.3.6. Tolerance Bands and Process Precision

The boundary of the operating region or processing window defined by the imposition of constraints is not absolutely fixed. The normal uncontrolled variations which occur in any process can result in an unacceptable product being produced when machine settings and procedures falling within the processing window are used, particularly when they are near the boundary. Determining the probability of this occurring is an important part of assessing machine performance.

The probability of a process achieving an acceptable result can be estimated from the distribution of results for the center-point replicate trials of a central composite factorial experiment design, provided that the preferred operating conditions lie close to the design center-point. It cannot be assumed that the amount of process variability will be constant over the whole operating region; and when the preferred operating conditions lie close to the boundary of the region, it is necessary to carry out replicated runs at those conditions. A measure of process variability can then be obtained using standard statistical methods.

The results of replicated runs from most processes form a normal distribution for which the population standard deviation is defined by

$$\sigma_x^2 = \frac{1}{n} \sum_{i=1}^{n} (x_i - \bar{x})^2 \tag{7.13}$$

where $\bar{x}$ is the population mean, n is the number of results in the population, and x_i are the individual results. A frequently used measure of process capability is $\pm 3\sigma_x$,[2] shown in Figure 7.12, which accounts for 99.73% of the process output. If all the values of a property defined by $\pm 3\sigma_x$ are within the range defined by the tolerance band, it can be concluded that the process is capable of achieving an adequate precision of output for most purposes. For a full assessment of process capability, it is usually necessary to check the $\pm 3\sigma_x$ range for a number of properties against the product or process specification. Alternatively, when the purchase of a new machine for manufacturing a range of products is proposed, selection can be made by comparison of ranges defined by the $\pm 3\sigma_x$ statistic for different machines, or by a comparison with previously defined expectations of the $\pm 3\sigma_x$ ranges.

When a product property, such as a dimension or a deformational stiffness, is specified with a tolerance band, the capability of a process for manufacturing to that precision can be simply determined by comparing the range defined by the tolerance band with the range defined by the $\pm 3\sigma_x$ statistic. If the latter is less than or equal to the former, the process can be deemed capable of achieving consistently that particular product requirement. A different situation arises when a property must not be greater or less than a certain value. In these cases the mean value $\bar{x}$ should be $3\sigma_x$ or more above the minimum permissible value or more than $3\sigma_x$ below the maximum permissible value.

The σ_x value defined by Eq. 7.13 identifies the population standard deviation, whereas in a process-capability study it is only possible to take a relatively small sample of the population, from which the sample mean is

$$\bar{x} = \frac{1}{n} \sum_{i=1}^{n} x_i \qquad (7.14)$$

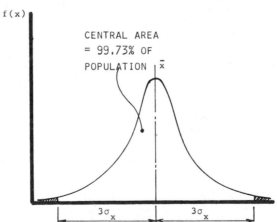

FIGURE 7.12. The normal distribution showing the area defined by ± 3 standard deviations ($\pm 3\sigma_x$).

An unbiased estimate of the population standard deviation is now given by

$$s^2 = \frac{1}{n-1} \sum_{i=1}^{n} (x_i - \bar{x})^2 \qquad (7.15)$$

The sample mean can be different from the population mean and, similarly, the sample standard deviation may not coincide with population standard deviation. The probability of this resulting in a greater percentage of the values in the population distribution being outside the tolerance band is examined in Section 7.6.2, in the context of sampling inspection by variables.

7.4. PROCESS MONITORING

7.4.1. General Considerations

Process monitoring consists of measuring process variables which give useful indications of process performance and of providing analyses of the data collected in three main categories:

1. The quantity of process output with respect to time.
2. The quality of process output with respect to time.
3. Process (equipment) condition.

This section will deal primarily with computer process monitoring systems which use instruments permanently installed on processing equipment and which form an integral part of the production supervision and communication scheme. The analyses in the preceding list would be used by the production control, quality-control, technical-support, and maintenance groups, in addition to the production management, as a basis for both work planning and immediate action.

Process monitoring does not involve systems for taking corrective action at the process. This comes under the category of process control, to be dealt with in Section 7.5. Unless the measurement of process variables is being used for input to an automatic controller in addition to being recorded by the monitoring system, any adjustments to the process have to be made manually, acting on the information provided by the monitoring system. The main advantages to be gained from using a computer process monitoring network are:

1. Rapid detection of the onset of production problems, often avoiding defective work or damage to equipment.
2. Detailed diagnosis of process performance, enabling causes of low productivity to be identified and corrected.
3. Reduction of manual checking and inspection.

At this point it is worth considering the technical difficulties of setting up monitoring systems. Those which only give a record of output quantity for production management and planning purpose are fairly simple, whereas taking and analyzing measurements which accurately reflect the quality of a product is much more difficult. The former is dealt with in Section 11.3.3, in conjunction with manual work recording, while the latter will be dealt with in this section. The main justification for monitoring output quality and making manual corrections to a process, instead of installing an automatic control system, is again technical difficulty. Devising control systems which will adjust process set points in response to analyses of measurements which reflect product quality is usually a complex and lengthy task.

To provide a useful service, process monitoring must be continuous, with on-line analysis of data and warnings to operators, supervisors, and other personnel concerned with the process when corrective action is required. The elements of a typical computer process monitoring system are:

1. Transducers or sensors.
2. Power supply (for the transducer).
3. Signal conditioning unit.
4. Signal transmission lines.
5. Computer(s) for data collection, analysis, storage, and communication.

Transducers or sensors (the terms are interchangeable) are measuring devices which convert the measured quantity into a form which can be readily manipulated and transmitted to a remote point. In the majority of cases this involves the production of an electrical signal which is related to the measured quantity. There are two basic types of transducer: analogue and digital.[26-28]

Analogue transducers produce an electrical signal which changes in value in some proportion with changes in the measured quantity, in a continuous manner. For the case of linear proportionality, which is desirable for simplicity, the measured quantity y is related to the transducer output x by

$$y = mx + c \qquad (7.16)$$

where m is a constant of proportionality and the constant c defines zero offset (if any), which is the transducer output when the measured quantity value is zero.

Digital transducers produce either a train of pulses or a binary signal, both of which can be handled by a computer in a simpler manner than an analogue signal. In pulse train transducers the pulse-to-pulse separation is calibrated against a known change in the measured quantity. These transducers are most often used for monitoring rate of change (e.g., velocity), using the internal clock of the computer to measure the time between pulses and thereby determining the amount by which a variable changes. They can only be used for absolute measurements if a record of the number of pulses from a fixed datum or reference value is maintained. The output from a transducer producing a binary signal provides a direct reading of the measured quantity, without reference to a record of previous counts, making this type of transducer particularly suitable for absolute measurements.

The direct proportionality between the output from an analogue transducer and the measured quantity must be maintained during transmission to the computer, otherwise the analyses will be based on inaccurate data. Sources of variation can be summarized as:

1. Transducer power supply.
2. Signal conditioning.
3. Noise

The output of most analogue transducers which require a power supply is directly proportional to the energizing voltage. As an approximate guide, the maximum permissible variation due to the power supply should be less than 10% of the minimum change which could be considered significant for the process. Signal conditioning for a transducer which gives linear proportionality is simply amplification of the signal to a level at which it can be transmitted to and accepted by the computer. Many transducers produce signals in the mV range, which are very susceptible to distortion by "noise"; and computers typically accept signals in the ranges 0–5 and 0–10 V. It is a good policy to amplify as close to the transducer as possible, to minimize the influence of the electric fields generated by motors, transformers, and power cables. Most companies manufacturing transducers also produce combined power supplies and amplifiers for their instruments.

To further minimize noise, careful earthing and screening of the monitoring equipment is necessary, coupled with the use of screened cables for transmission. In summary, noise and unstable power supplies cause short-term fluctuations of the signals from analogue transducers, while amplifiers and the transducers themselves can be subject to zero-offset drift and changes in the proportionality relationship. Calibration at regular intervals is essential to ensure that the data being recorded are accurate.

In contrast digital transducers are far less sensitive to both noise and drift. This is because the pulse or "bit" is the information and its voltage can vary considerably without interfering with the measurement it represents. However, amplification close to the transducer is usually necessary, so that spurious pulses due to noise are avoided; but it does not need the high precision required for analogue signals and is much cheaper. Unfortunately, not many analogue transducers have digital equivalents.

7.4.2. Temperature Measurement

Temperature is one of the most important and frequently measured process variables; but great care must be taken with the selection and positioning of sensors to obtain readings which represent accurately process conditions and performance. To start, it is necessary to distinguish between machine temperatures and rubber temperatures. Machine temperatures, which include those of metal components, heat-exchange fluids, and hydraulic fluids, are generally maintained by local automatic control systems. In contrast, rubber temperatures depend on the physical properties of the rubber compound and, usually, the interactive influence of a number of process variables. This indirect control of rubber temperature emphasizes the need for monitoring in circumstances where it is an important process variable. It is also necessary to distinguish between the measuring of temperatures for monitoring and control. The positioning of the sensors can be quite different, particularly when heating or cooling by conduction is used.

Sensors giving an output for remote monitoring of temperature can be separated into two categories: thermoelectric and infrared, with the use of thermoelectric devices, such as wire-wound resistance thermometers and thermocouples, predominating over infrared devices.

Thermoelectric sensors are best suited to measuring machine temperatures and rubber temperatures in situations where the rate of change of temperature is slow and the frictional heat generated by the movement of the rubber over the sensor does not significantly affect the measurement. The accuracy with which a thermoelectric sensor can follow a changing temperature is quantified by a time constant Γ, which depends on the mass or thermal inertia of the sensor and is defined as the time for the sensor output to rise to 63.2% of full-scale output, in response to a step input of temperature. Assuming that the output of the sensor is a voltage V, which is related to temperature T by a constant of proportionality m under steady-state conditions, as in Eq. (17.16), the transient response to a step input of temperature is given by

$$V = \frac{1}{m} T(1 - e^{-t/\Gamma}) \qquad (7.17)$$

which is shown graphically in Figure 7.13. If temperature increases

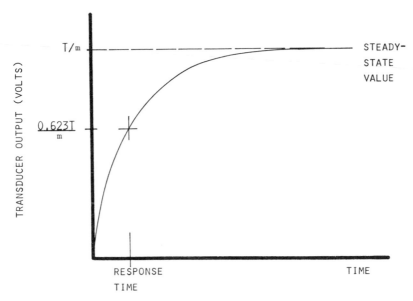

FIGURE 7.13. Response of a temperature-measuring transducer to a step change of temperature, showing the response time at 62.3% of the new steady-state value.

progressively, which is a more practical case than the step input, the transducer response will lag behind the temperature, as shown in Figure 7.14 and described by

$$V = \frac{1}{m} R[t - \Gamma(1 - e^{-t/\Gamma})] \qquad (7.18)$$

where R is the ramp rate $(°Cs^{-1})$. It is interesting that these equations are similar in form to those for a Maxwell liquid in Section 2.2.2. Figure 7.13 shows that the time constant should not be taken as the total rise time of the sensor output. Only at approximately $t = 5\Gamma$ can the output be considered to be practically equivalent to the steady-state output. Therefore a sensor with a time constant of 2 s will have a total rise time of 10 s.

The time constant of a thermoelectric sensor depends on the amount of heat it needs to absorb in order to achieve the temperature to be measured and the rate at which it can absorb it. It helps if the sensor is made from a material which has a high thermal conductivity and a low specific heat; but the most important factors are size and position. As the size of a sensor increases its heat capacity will increase and unless it has a flat, thin geometry, the time for heat conduction through it will also increase. This will be exacerbated by the thermal mass of the components between the sensor

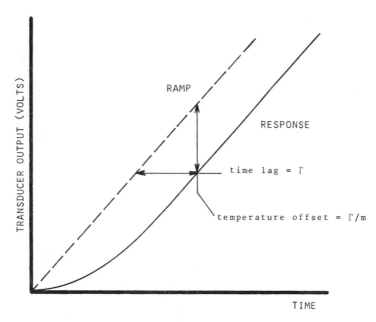

FIGURE 7.14. Change in output of a transducer in response to a ramp increase in temperature.

and the heat source being measured. Protective shields, location fittings, or simple poor siting can increase substantially the time constant. In fact, the important time constant is the one which is obtained from the installed sensor.

If a steady-state temperature gradient exists between the sensor and the desired measurement point there will be a constant error or offset. This is particularly important for measuring rubber temperature in continuous processes such as extrusion, where the sensor is sited in the barrel or in the head, both of which are used to transfer heat to or from the rubber. For a representative measurement of rubber temperature to be made, it is necessary to insulate the sensor from the surrounding metal, using an epoxy resin or an alternative insulating material.

Thermoelectric sensors may be divided into two classes:

1. Temperature-dependent resistance elements, of which "resistance thermometers" and semiconductor thermistors are the main examples.
2. Thermocouples, which fall into the "self-generating" category.

Wire-wound resistance thermometers usually employ platinum or nickel as the element material; they have a linear temperature dependence, with resistance/temperature coefficients of about 0.4% per °C for platinum and 0.6% per °C for nickel. The semiconductor material used for thermistors has a negative temperature coefficient generally lying between 3 and 6% per °C, but this tends to be nonlinear with temperature in most cases. Thermistors also tend to be less stable with time than metallic resistance thermometers, especially in the long term, and to have a much more restricted temperature range (−70°C to 260°C) than metallic resistance thermometers (−170°C to 810°C), although this is not important for the majority of rubber processes. In addition to their higher sensitivity, thermistors are capable of a much more rapid thermal response (shorter time constant) than resistance thermometers because of their low-heat-capacity small block or bead form of construction.

Thermocouples are also low-heat-capacity devices but require a reference cold junction, which is held at a constant temperature, or a temperature-tracking equivalent, which increases their complexity. They are based on the Seebeck effect, in which a junction of two dissimilar metals emits an electromotive force (emf) related to temperature. A representative junction pair nickel chrome/nickel aluminum (Chromel/Alumel) has an output constant of 0.0385 mV per °C.

Due to their thermal inertia, wire-wound resistance thermometers are best suited to measuring steady-state metal temperatures, such as mold temperatures. Thermisters and the far more widely used thermocouples are also best suited to measuring metal temperatures, although it is possible to obtain a reasonable indication of rubber temperature with an insulated sensor of low thermal mass. One of the most common rubber temperature monitoring applications is the batch thermocouple in an internal mixer.

Infrared sensors give a fast response to changing temperature and are not influenced by thermal inertia, making them particularly suitable for measuring rubber temperatures. When a free rubber surface is available, such as the band on a two-roll mill or calender and the extrudate emerging from a die, the infrared radiation can be detected from a distance and focused onto the sensor by lenses. The sensor can then be sited where it will not interfere with the operation of equipment or be prone to damage. When a free surface is not available for measurement, as in the head of an injection molding machine or in the chamber of an internal mixer, it is necessary to provide a route for the transmission of the infrared radiation from the measurement point to the sensor. Fiber optics provide this facility.

Typical rubber processing temperatures are close to the minimum measurable temperatures of some commercial instruments. This temperature is determined by the sensitivity of the infrared detector and the amount of energy which can be focused onto it from the IR source. This obviously depends on the temperature of the source, but it is also strongly influenced by the source's emissivity, size, and distance from the sensor. Emissivity is a measure of the efficiency with which a material surface radiates infrared energy. Low-emissivity surfaces radiate less infrared energy for a given temperature than high-emissivity ones. Figure 7.15 gives typical emissivities, which must be compensated for in measurement. Infrared sensors have a control calibrated in emissivity values which alters the gain or amplification of the sensor output, to provide a signal which gives an accurate indication of temperature.

If the source does not fill the field of view of the sensor, then the temperature reading will be averaged over the source plus some "background," giving a spurious result. The field of view is defined by an angle, so that the diameter of the measuring area depends on the distance of the sensor from the source. Kane-May,[29] Heimann,[30] and Barnes[31] all manufacture systems with fields of view in the range 2° to 3° and contactless remote sensing at maximum distances of 1 to 2 m, which operate in temperature ranges suitable for rubber processes. The Kane-May instrument has a very low cost (£400) in comparison with the others, which are in the range £700–£2000.

When fiber optics are used, the field of view is very much reduced and

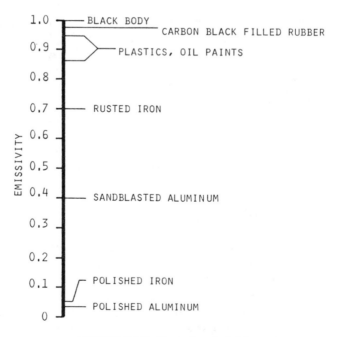

FIGURE 7.15. Typical emissivities.

there is also some loss in transmission through the fiber optic. On some equipment this seriously limits the minimum measuring temperature. The Vanzetti system,[32] which has been used in plastics extruders and injection molding machines and also in rubber mixing research, gives a minimum measuring temperature of 50°C with a solid fiber rod of 150-mm length and 60°C for a 300-mm fiber rod; but this increases to 130°C for a 300-mm flexible fiber bundle. As a result the infrared detector has to be placed in close proximity to the measuring point and siting can be difficult. This limitation has only recently been overcome by Luxtron,[33] which uses an extremely sensitive detector to make measurements in the range −30°C to +200°C, with a flexible fiber optic probe of up to 15 m in length. As could be expected this instrument is expensive, at £6000 in comparison with £1500 per channel (multisensor systems are possible) for the Vanzetti equipment. In addition to these general-purpose instruments, an infrared sensor has been developed for measuring the batch temperature in internal mixers, by Carter Bros. Ltd[34] in conjunction with Land Pyrometers Ltd.

7.4.3. Pressure Measurement

The manufacture of rubber products depends on numerous pressures being maintained within defined limits. These include steam, water, oil, air, and rubber pressures. As with temperature, pressures related to machine functions are usually subject to local automatic control; but rubber pressure is more amenable to direct control than rubber temperature, and is increasingly being subjected to closed-loop control in extrusion and injection molding.

The two main types of pressure transducer used in rubber processing equipment are based upon strain gauges and on the piezoelectric effect. In the first type the deformation of a metal diaphragm under pressure is converted to an electric signal by the change in electrical resistance of strain gauges bonded to it. This resistance is then converted to a voltage output proportional to the pressure sensed by a Wheatstone bridge arrangement within the transducer, which also allows full compensation for temperature changes. The output is usually in the range 0 to 20 mV for zero to maximum pressures, indicating the need for amplification and precautions against "noise." Maximum pressures range from less than 10 kPa (0.1 bar) for gases at low pressure to 200 MPa (2000 bar). Strain gauge transducers are produced by a number of companies and those suitable for measuring rubber pressures, which usually have maximum pressures in the range 20–200 MPa, are available from Dynisco,[35] Gentron[36], and others, including some combined pressure and temperature (thermocouple) transducers.

Piezoelectric materials produce an electric charge when they are deformed. This effect is used for commercial pressure transducers, in addition to a number of other devices, such as load cells and vibration sensors. A quartz crystal giving a charge proportional to the pressure applied to it is the sensing element, and this charge is then converted to a voltage by a charge amplifier. One of the major problems of the charge output from the transducer is that any attempt to measure it will tend to dissipate it, requiring that very high-impedance amplifiers be used. Even so, steady-state or slowly varying pressures cannot be measured, since stabilities are in the order of a few minutes. This generally limits their use to injection molding, but in this application they are extremely useful because of their small size. With sensing elements down to 4 mm in diameter, compared to 10-mm diameter for the strain gauge transducers, they can be fitted directly into mold cavities, where they are used to monitor mold filling.

The major manufacturer of piezoelectric transducers for polymer processes is Kistler,[37] with rated maximum pressures up to 500 MPa.

7.4.4. Position, Displacement, and Velocity Measurement

Monitoring of position, displacement, and velocity[28] provide useful indicators of the performance of many processes and are often essential to the operation of sequence control systems. In the context of process monitoring, they are often more useful for measuring the quantity of output rather than the quality.

Measurement of position implies a single-point detection system, such as a limit switch or a proximity sensor. This is adequate when it is only necessary to know when a moving element of a machine is at certain points, as in the cases of the drop door of an internal mixer or the clamping unit of an injection molding machine. Both limit switches and proximity sensors (which are noncontacting devices and much more precise than limit switches) make or break an electrical circuit when the moving machine element is in a predetermined position. Monitoring the time for which the circuit is open or closed can then give a useful insight into the operation of the process.

Measurement of linear displacement requires the distance of a moving machine element from a datum or reference point to be detectable over its entire range of movement. This is usually necessary for monitoring the quality of performance of a process and is being increasingly used in sophisticated control systems. Probably the most common application is to injection molding, where monitoring of injection ram movement in a normal injection pressure-controlled machine provides a very good measure of performance. Other applications which should yield useful information are the movement of the ram of an internal mixer during the mixing cycle and the last stages of the closing of a compression press, when rubber flow is occurring.

Direct measurement of linear displacement can be achieved with three devices: the linear potentiometer, the linear variable differential transducer (LVDT), and the inductive displacement transducer. Linear potentiometers utilize either a wire-wound or a conductive plastic track, on which a wiper rides. This wiper is attached to the moving machine element via a sliding rod, so that the change in resistance of the potentiometer and, in consequence, the output voltage, are directly proportional to displacement. This linearity, and a voltage output which does not require amplification, are the main advantages of the linear potentiometer. The disadvantages are wear at the wiper contact and a susceptibility to dirt on the track. However, a life of 80×10^6 cycles is quoted for the conductive-plastics type and effective sealing is possible.

The LVDT produces an alternating voltage output which is directly proportional to displacement, needing signal conditioning equipment to convert it to a dc voltage. Its main advantages over the linear potentiometer are the absence of physical contact between the moving parts and infinite resolution. Both of these derive from the principle on which it operates,

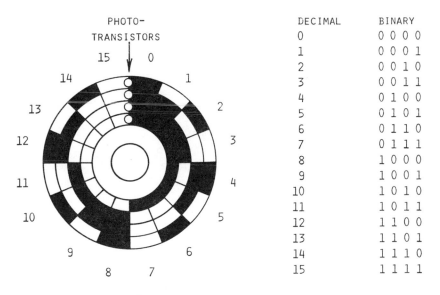

DECIMAL	BINARY
0	0 0 0 0
1	0 0 0 1
2	0 0 1 0
3	0 0 1 1
4	0 1 0 0
5	0 1 0 1
6	0 1 1 0
7	0 1 1 1
8	1 0 0 0
9	1 0 0 1
10	1 0 1 0
11	1 0 1 1
12	1 1 0 0
13	1 1 0 1
14	1 1 1 0
15	1 1 1 1

SHADED SEGMENT = 0
BLANK SEGMENT = 1

FIGURE 7.16. A four-track disk encoder for measuring angular displacement and position (resolution $360°/16 = 22.5°$).

which is that of a moving magnetic core altering the magnetic flux between primary and secondary coil windings. The major disadvantages are the complex power supply and signal conditioning and the limitation imposed on the rate of displacement by the coil excitation frequency. Displacement frequencies of only up to 10% of the excitation frequency can be measured, although this is not usually a problem for rubber processing machinery.

The inductive displacement transducer uses a change of coil inductance due to the displacement of a magnetic core inside the coil. This principle enables transducers with up to 2.5 m travel to be built, compared with 0.25 m for the linear potentiometer and 0.65 m for the LVDT. The advantages and disadvantages of this transducer are similar to those cited for the LVDT, but it does have poorer linearity ($\pm$0.5% of full scale, compared with 0.25%).

Measurement of angular displacement can be achieved with two main devices: the rotary potentiometer and the disk encoder. The rotary potentiometer operates on the same principle as the linear potentiometer and produces a dc voltage proportional to angular displacement, but the disk encoder is a digital transducer. It works on the principle shown in Figure 7.16, where light is transmitted through the disk to the phototransistors, or not, depending on the disk position. The resolution of the device depends on the number of tracks. In general, the maximum number which can be represented is given by 2^n for an n track disk. For eight tracks,

$$\text{maximum number} = 2^8 = 256$$

and

$$\text{resultant resolution} = 360°/256 = 1.4°$$

and for a nine-track disk the resolution would be 0.7°.

For disk encoders used in position control it is extremely important that no ambiguity in position measurement arises from the change from one segment to another. This problem is overcome by using Gray-coded disks, which require a Gray-code to binary-code converter.

Rotary displacement transducers are often used for linear position sensing by employing a rack and pinion arrangement to drive the input spindle. This arrangement gives practical advantages in environments where foreign matter falling on the rod of a linear transducer could interfere with its functioning. Rotary spindles are far easier to seal and the whole transducer can often be sited away from problem areas. There can also be a cost advantage in using rotary transducers in preference to linear ones.

All displacement transducers can be used to measure velocity, by recording movement against a time base. In microprocessor and computer systems this is a "software" function, taking the time count from the internal quartz "clock." However, there are a number of devices simpler than displacement transducers which are designed specifically for angular velocity measurement.

Tachogenerators are very widely used as angular velocity transducers in electrical motor-speed control systems. They are analogue devices, producing a voltage proportional to angular velocity, and can be based on either ac or dc generators. In the former case it is necessary to use rectifier and filter circuits for signal conditioning.

When angular velocity is measured using a computer system, it is often more convenient to use a digital pulse system, such as the electromagnetic pulse generator, or an optoelectric technique. In the former method a toothed wheel is attached to the rotating shaft and a coil wrapped around a permanent magnet placed in close proximity to it. The change in air gap between the wheel and sensor then induces an alternating voltage in the coil. This can be conditioned by a pulse shaper, such as the Schmitt trigger, and the resulting square waves or pulses fed to the computer, where they can be counted with respect to time. Alternatively, an optoelectric system can be used in which a light source and phototransistor are sited on each side of a disk, having holes or slots in its periphery, attached to the rotating shaft. The light beam then activates the phototransistor when any of the slots or holes are aligned, but is interrupted when they are not, causing a chain of pulses to be generated. These can be counted with respect to time to determine angular velocity.

Angular velocity transducers can also be used for measuring linear velocities by the method described for angular displacement transducers. They are perfectly adequate for this purpose unless position is needed as well as velocity.

7.4.5. Product Dimensions and Weight

On-line measurement of product dimensions and weight is largely restricted to continuous processes, such as calendering and extrusion. Direct measurement of the thickness of sheet products is relatively simple in concept, but extrudates of complex cross section present serious difficulties. Continuous monitoring of weight per unit length is often used as an indirect measure of dimensions.

Sheet-thickness monitoring methods are separated into two distinct groups: contacting and noncontacting. Contacting methods invariably involve a measuring roller running on the surface of the sheet where it passes over a datum surface, such as a rigidly located roller. The vertical movement of the measuring roller from the datum caused by the thickness of the sheet is then monitored by a displacement transducer. A modified technique is used to measure the thickness of the rubber film on a calender roll, prior to it being brought into contact with the fabric during topping or frictioning. Due to the movement of the calender roll resulting from nip adjustments and "roll float," it cannot be used as a datum for a measuring system mounted on the calender frame. To overcome this problem a differential system is used, where a circular rotating blade cuts through the rubber layer and runs on the roll surface.[38] The distance between this and a conventional roller running on the rubber surface is then monitored. All contacting methods of thickness measurement suffer from some inaccuracy due to deformation of the sheet under the roller contact pressure. This can be minimized by making an appropriate correction, and the importance of the error does depend on the degree of measuring precision required by the product. In general terms, the error band of the measuring system should be less than 10% of the tolerance band of the product being measured.

The main noncontacting method of thickness measurement uses transducers incorporating sources of ionizing radiation. There are two types of radiation thickness sensors. The most widely used type is based on the transmission principle, where the attenuation of the radiation passing through the rubber sheet is related to the sheet thickness. The amount of radiation falling on a detector sited opposite to the source then provides a measure of sheet thickness. When only one side of the rubber sheet is accessible, as in the case when the sheet thickness on the calender roll needs to be measured, it is necessary to use a backscatter or reflectance technique. The reflection of radiation from the calender roll is attenuated by absorption in the rubber sheet, enabling the reading from a detector mounted on the same side of the rubber sheet as the radiation source to be related to thickness.

Figure 7.17 shows a calender set up for the skim coating of a cord fabric, with two backscatter sensors for measuring the thickness of one skim coat and a transmission sensor for the total thickness of the product. The backscatter gauges are usually fixed in location but the transmission sensors are often of the scanning type, traversing across the sheet on a track. This enables the uniformity of sheet thickness in the transverse direction, as well as longitudinally, to be measured. Corrections can then be made for a "wedge-shaped" cross section, which is caused by a difference in the nip setting on each side of the calender, and for a convex or concave sheet, caused by inadequate camber correction, provided that roll crossing or bending facilities are available. Again, the precision required depends on the needs of the product and the economics of the process.

Most radiation thickness sensors use β sources, which are suitable for measuring sheets of up to 3.75 kg m^{-2} (3.75-mm thickness for materials with

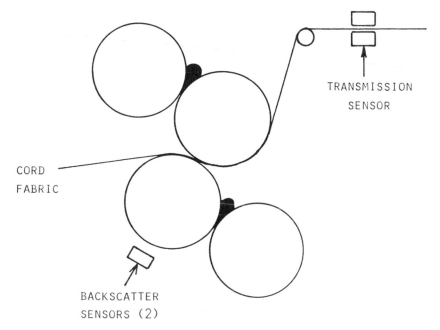

FIGURE 7.17. Radiation-type thickness sensors on a four-roll calender used for skim coating of cord fabric.

a density of $1000 \, \text{kg m}^{-3}$). For thicker sheets sensors incorporating γ sources are available.[39]

Direct on-line measurement of extrudate dimensions is possible using both contacting and noncontacting sensors but with both methods there are serious problems in obtaining a useful unambiguous measurement. One of the main problems arises from the tendency of extrudates to undergo some axial rotation on emerging from the die. Figure 7.18 shows the effect of axial rotation of a square-section extrudate on measuring accuracy. If a measurement can be made at specific points, as indicated by the arrows, the error is small; but this may be difficult to achieve and often only overall measurements, indicated by the heavy parallel lines, are possible. In this latter case, the measuring error caused by rotation is substantial. It is possible to prevent rotation by the use of guide rollers, particularly for wide, flat extrudates such as tire treads; but care must be taken that such guides do not cause distortion, thus invalidating the measurement.

For a complex extrudate the main problem is often one of not being able to gain access to any of the critical dimensions. It is then necessary to monitor dimensions indirectly by measurement of weight per unit length.

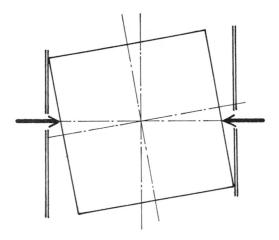

FIGURE 7.18. Sources of measurement error arising from changes in alignment of an extrudate.

This is a traditional method of checking an extrudate and a common procedure is for an extruder operator to cut a specified length of extrudate, using a gauge cutter, and weigh it. However, this technique is periodic, operator dependent, and impractical when continuous vulcanization is used. A better method is to conduct the extrudate over a weighing unit consisting of a roller or conveyor platform supported on electronic load cells. Most load cells incorporate strain gauges and have power supply and signal conditioning requirements similar to those of the pressure transducers described in Section 7.4.3.

7.4.6. Energy and Power Measurement

With energy costs forming an ever-increasing proportion of the total manufacturing overhead, energy measurement and management is a matter of prime importance. On individual processes the pattern of energy usage can also provide a substantial amount of information about the quality and consistency of performance. This approach is widely used in internal mixing, where motor current or power is displayed on a meter or recorded against time on a chart recorder. More recently, the power integrator has been used to monitor and control mixing by the energy input from the main drive motor (energy = power × time).

Transducers for electrical measurement which are computer compatible are widely available and well documented.[26,27] They can be usefully applied to the majority of rubber processes. In fact, many machines are supplied with ammeter and voltmeters as standard fittings; but as emphasized in the previous sections, a continuous record, with respect to time, is necessary for effective process monitoring.

Steam and compressed-air services are notorious sources of energy "loss," due both to inefficient use and leakage. It is generally well worth continuously monitoring the energy input to steam raising and air compression: first, to establish the pattern of usage under well-regulated and energy-efficient conditions and, second, to detect any changes in this pattern of usage. Electrical energy for air compressors is relatively simple to monitor, as previously mentioned. The fuel consumption of oil- and gas-fired boilers for process steam and space heating can also be simply measured with a flow transducer, giving the facility of continuous monitoring.

7.4.7. Data Acquisition Systems

Data acquisition systems utilize computers which read the output from transducers at closely spaced intervals, producing a record which is essentially continuous.

Signal conditioning for transmission to a computer has been dealt with in the previous sections, but in addition to that, interfacing devices are necessary to allow the computer to read the signals from a number of transducers. Figure 7.19 shows a typical layout. Starting with analogue inputs, which are usually in the majority, an analogue-to-digital converter (ADC) is necessary. These devices are usually capable of sampling and signal conversion rates far in excess of those needed for individual transducers, so it is a normal practice to use a single ADC to convert the signals from a number of transducers. For this purpose, a multiplexer is necesary. Modern solid-state multiplexers are programmable, enabling the order in which readings are taken and the intervals between readings to be specified. It is therefore possible to sample different transducers at different rates. The number of inputs which a single multiplexer can handle ranges from 16 to more than 1000, at rates in the region of 100,000 samples per second, if required.

The input from digital transducers is often in the form of a pulse train, which requires the use of a pulse counter to convert it into a digital quantity. Pulse-generating transducers usually measure velocity or rate, so in the typical operation of a pulse counter an internal register is loaded with the number of pulses to be counted. At the start of the routine a clock is initialized and as each pulse in the pulse train is received, the register subtracts one. When the register reaches zero, the clock time is recorded and the "rate" of the pulse chain is determined by dividing the number of pulses by the clock time.

When it is only necessary to know whether a limit switch, valve, or motor push-button control is open or closed, a contact input interface can be used. This interface consists of a relatively simple contact switch which can

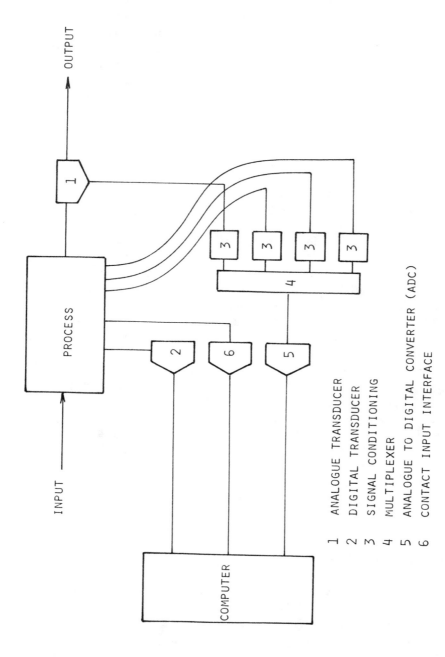

1 ANALOGUE TRANSDUCER
2 DIGITAL TRANSDUCER
3 SIGNAL CONDITIONING
4 MULTIPLEXER
5 ANALOGUE TO DIGITAL CONVERTER (ADC)
6 CONTACT INPUT INTERFACE

FIGURE 7.19. A dedicated data acquisition system.

be opened or closed by a signal from the process. The computer is programmed with the desired status of these contacts and periodically scans their actual status for comparison.

All the interfacing units mentioned in this discussion are either integral with the computer or available as simple plug-in units. The general trend in microprocessor usage is away from the early concept of "build your own dedicated unit," except when large numbers of similar units are needed for a specific job, making it economical.

One of the most important trends that has developed, both in data acquisition and in computer-aided manufacture, is the arrangement of a company's computers in an interconnected hierarchical system.[28] Figure 7.20 shows a complete system; but it is important to recognize the value of the concept for the planning and progressive building of a comprehensive data acquisition system from very small beginnings. Most companies start their data acquisition system by installing a dedicated unit on one of their more complex and critical processing operations. Often, injection molding is chosen because of the intensive development of this area in the thermoplastics field. This type of "stand-alone" system aids management and technical staff by providing a detailed information service, either at the machine or at a remote terminal. The main advantage of the progressive approach in building a data acquisition system is that experience can be gained with the expenditure of comparatively little capital, and mistakes can be more readily avoided or corrected. However, it is extremely important to have an overall plan or scheme before starting, so that the capabilities of the computers at each level in the hierarchy can be approximately defined and the problems of compatibility and communications between computers recognized. Careful selection of hardware can alleviate the latter problems. A further advantage of a hierarchical system is the ability of the first-level computer to revert to stand-alone operation in the event of a fault in the communications links or in a computer higher in the hierarchy.

There are three types of first-level units:

1. Integral with the processing machine.
2. Dedicated add-on monitors.
3. General purpose microcomputers.

The provision of process monitoring facilities as original equipment, integral with a processing machine, is a very recent innovation, mainly on thermoplastics injection molding machines. Although it relieves the product manufacturer of the responsibility for selection, installation, and commissioning, there are some disadvantages. A new machine must be purchased to obtain the monitoring system, which is then difficult to update

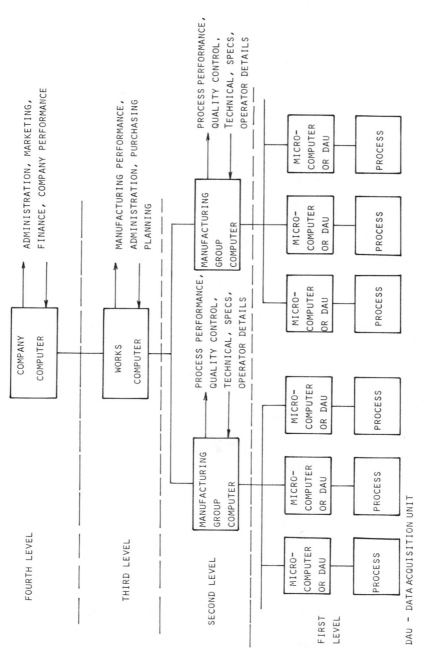

FIGURE 7.20. A hierarchy of computers for supporting company activities.

in line with new technology and ideas. Clearly, the capital investment in a new machine is large compared with the cost of monitoring equipment, and may delay the introduction of a monitoring system. Both the dedicated add-on monitors and general-purpose microcomputers provide a much greater degree of versatility and can be retrofitted to existing equipment. The former are generally preprogrammed to perform a limited number of functions, but can be introduced and commissioned fairly rapidly and do not require an in-depth knowledge of computer monitoring technology. Companies supplying systems for this market include Dextralog, SPL, and Lucas Logic. Using general-purpose microcomputers in the first level gives a high degree of flexibility but does require the development, installation, and commissioning to be carried out by personnel with a working knowledge of the company's operations and of computer monitoring technology. This points to the need for in-house expertise which, at the present time, is rare, but can be expected to become more common with the permeation of microprocessor technology into all aspects of industry and technical education.

In a system used purely for monitoring, the second-level computer is a "clerk." Its function is to organize and analyze the information transmitted from the first-level computers in formats which aid the manufacturing group management and technical staff in improving productivity and profitability. The third- and fourth-level computers perform similar collection, collation, and analysis tasks, for the executive management, including administrative and financial information. When a hierarchical computer system is installed, its uses extend far beyond the monitoring of machine function and product quality discussed here. The flow of information is two way—for process control, production planning, internal reports and for communicating instructions.

7.4.8. Data Analysis and Presentation

The clerk function of the second-level computer is crucial to the effectiveness of a process monitoring system. The ability to monitor current, rather than historical, machine and operator performance has been identified as one of the major attributes of a data acquisition system.[40] When a large amount of information is being processed a "management-by-exception" approach must be adopted, to avoid overwhelming line managers and thus reducing their effectiveness, instead of improving it.

The analysis of process performance must be summarized in exception reports. In general, a line manager will only want to be informed when a problem is likely to arise on a particular process and, if more than one problem arises simultaneously, which one to tackle first. This requires a computer program which compares the actual performance of a process with the desired performance (the latter is usually specified by targets or tolerance bands) and conducts a trend analysis or forecasting operation, to anticipate the onset of a problem. When a problem is forecast it needs to be communicated to the line manager as soon as possible. This can be achieved through a computer terminal, which is best sited in the manager's office. Once a problem has been notified, the manager will need to retrieve details of the changes in performance which lead up to the notification, in order to formulate a course of action.

One of the problems with a data acquisition system is that very large amounts of information often have to be stored. Mass-storage peripherals overcome the main problems, but they still have finite capacity, requiring a well-conceived data management plan. The trend analysis and the line manager often operate on a time scale of a few hours and have little use for detailed information older than this. It is the production group managers and the technical and quality-control staff who need to consider a broader spectrum and a longer time scale for planning and to monitor overall production group performance, including that of the line managers. This is the function of the third-level computer.

Once a "maximum" age has been defined for data stored in the second-level computer, all data older than this can be transmitted to the third-level computer at intervals and deleted from the second-level computer storage space. On arrival at the third-level computer, the information can be analyzed and condensed for the production group manager, then collated with data from other production groups and further analyzed for the executive management. After this it can be deleted from all records. For the production group manager and the works executive management, the isolation of the "critical few" problems from the acceptable majority of production is even more important than at the second level. The jobs which produce repetitive or chronic problems, losing the company a significant amount of revenue, need to be identified. Also, reports which analyze products with respect to individual operators and machines help to identify the sources of problems.

The continuous stream of data from a process can be analyzed using the concepts embodied in control charts.[2] These plot the movement of the mean and range (difference between maximum and minimum values) of successive samples of n units, with respect to time. While movement of the mean indicates a "drift" of the process operating point, an increase in the range indicates a deterioration of process precision.

If a sample size n, being a number of sequential data points from a process, is specified, the limits of the control charts can be determined, as shown in Figures 7.21 and 7.22. It is normal to use the $\pm 3\sigma$ (standard deviation) statistic referred to in Section 7.3.6 for the action limits and $\pm 2\sigma$

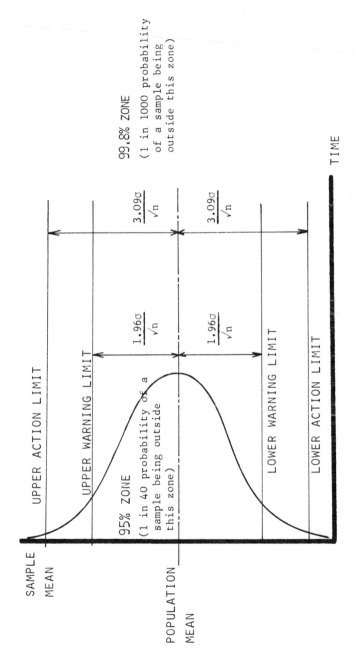

FIGURE 7.21. Control chart for sample means.

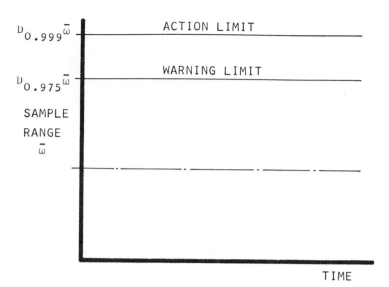

FIGURE 7.22. Control chart for sample ranges.

for the warning limits; but these limits can be altered to take into account the requirements of the products being manufactured. The D values in Figure 7.22, used to determine the warning and action limits, are obtained from tables.[2,41]

For conventional manual control chart usage, a value of sample range or mean falling outside the warning limits would require investigation and an intensification of sampling. The concept of intensifying sampling only when anomalies in process performance are detected can be valuable in computer monitoring, reducing the overall frequency of sampling, which also reduces the computer power and data storage capacity requirements. However, the main advantage of computer monitoring lies in the analysis and reporting of the process performance. In addition to sounding an alarm when a warning limit is exceeded, trend analysis can be used to determine the rate of drift or loss of precision of performance, enabling the urgency of a problem to be specified. The techniques which can be used for trend analysis are dealt with in Section 11.2.2.

For monitoring processes which have a manual element of operation, computer terminals designed specifically for work reporting (see Section 11.3.3) are valuable for correlating performance with individual operators. Without facilities such as these, the resources needed to track defective work to source are often prohibitively high and cannot be operated routinely, particularly when small products which do not have individual identification are being manufactured.

7.5. PROCESS CONTROL

7.5.1. General Considerations

Process control includes all actions taken at a process to ensure that the output is acceptable for downstream operations and to the customer. This sweeping definition is necessary because the concepts of process control must be extended to all processes, including those which are essentially "operator dependent." In the previous section, dealing with process monitoring, manual control was implicity involved, since corrections would be made to processes by the operator, line manager, or technical-support staff, in response to a report of the onset of a problem. In this section, the emphasis will be placed on automatic control systems; but these operate at many levels, from the very simple to the very complex, and automatic control methods are often used in conjunction with manual control. This apparently confusing situation can be clarified by considering the terminology of process control.

Open-loop control involves making an estimate of the necessary input for a desired output. Most rubber processes have this type of control. For example, the mixing time for an internal mixer is set to open the drop door and dump the batch when the mixture is predicted to have achieved certain desired properties. If factors which influence the properties of the mixture change, such as the timing of material additions or mixer temperature, the prediction will be in error and the desired mixture properties may not be achieved. There is no corrective action in open-loop control.

Processes which have open-loop control of the output or product will invariably have closed-loop control of some machine functions. A good example is the extruder, which will have closed-loop control of barrel, head, and die temperatures and screw speed, to maintain them at set points or target values which are predicted to be capable of giving an acceptable extrudate. In closed-loop control the output from a transducer is compared with a signal representing the set point and the resulting error signal is used to initiate an action, through a process controller, which will *return the transducer* to the set point. This is often referred to as local automatic control, to differentiate it from closed-loop control of the product or output.

Closed-loop control of the output from a process involves similar principles to local automatic control but the transducer measures some attribute of the output. This can be seen in internal mixing, where batch temperature or mixing energy are used to determine the time at which the batch is dumped.

Manual control is often used in conjunction with all the control methods described, through the adjustment of set points to overcome variations for which the control systems cannot provide adequate adjustments. Some processes may need frequent manual overrides of this type in order to maintain product uniformity, resulting in costly and inefficient operation. In such cases the time and cost of setting up an adaptive control system may be justified. Essentially, adaptive control systems make the adjustments to the settings of closed-loop controllers which would otherwise require the time of the operator, line managers, or technical-support staff.

With all control systems, the precision of control is entirely dependent on the degree to which transducer readings represent real changes in the process variables being controlled. As with process monitoring, this depends on transducer selection and positioning.

7.5.2. Local Automatic and Closed-Loop Control

Probably the most widely used controller familiar to the rubber industry is the temperature controller, making it a useful example with which to develop the general principles of control. For precise control of temperature it is now a common practice to use a three-term controller. The three terms—proportional, integral, and derivative action—give the alternative name of PID controller. Each term can be dealt with separately.[26–28]

The error signal, shown in Figure 7.23, is obtained by subtracting the feedback signal from the reference or set point signal. This is then used to activate the controller to take corrective action. In proportional control, the controller output x_p is proportional to the error signal ε, which can be expressed as

$$x_p = K_p \varepsilon \qquad (7.19)$$

where the constant of proportionality K_p is termed the gain or sensitivity. The gain also defines the *proportional band*, which is the error range required to change the power supplied to the heater from zero to maximum. Increasing the gain or sensitivity decreases the width of the proportional band.

On–off control is a special case of proportional control. If the gain is made very high, the power supplied to the heater will change from zero to maximum and vice versa in response to a very small deviation of the feedback signal from the set point. Because of this extreme response the machine element temperature will cycle above and below the set point, giving poor precision of control in comparison with the proportional controller.

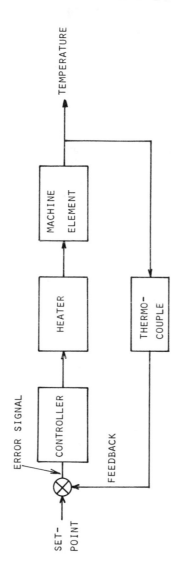

FIGURE 7.23. Block diagram for a local automatic temperature controller.

In the case of integral action, the power supplied to the heater is proportional to the time integral of error signal ε, giving

$$x_p = \frac{K_p}{\Gamma_i} - \int_0^t \varepsilon \, dt \qquad (7.20)$$

where Γ_i is the integral time, which can be alternatively defined by the reset rate, the reciprocal of the integral time. Integral action serves the purpose of correcting the offset error which arises when proportional control only is used. For this reason it is sometimes called "automatic reset."

In derivative control the power supplied to the heater is proportional to the rate of change of the error, which can be expressed by

$$x_p = K_p \Gamma_d \frac{d\varepsilon}{dt} \qquad (7.21)$$

where Γ_d is the derivative time. The effect of the derivative term is to reduce the power to the heater progressively as the set point is approached, avoiding a significant overshoot and quickly eliminating oscillations about the set point. This is particularly important in conjunction with integral action, which tends to cause instability. The derivative term is predictive in its action and is sometimes called anticipatory control.

Combining Eqs. (7.19)–(7.21) gives an expression which describes the action of a three-term or PID controller:

$$x_p = K_p \varepsilon + \frac{K_p}{\Gamma_i} \int_0^t \varepsilon \, dt + K_p \Gamma_d \frac{d\varepsilon}{dt} \qquad (7.22)$$

The responses of controllers based on the proportional, integral, and derivative action principles to a step change in the error signal are shown in Figure 7.24. These responses will be strongly dependent on the selection of the gain, the integral time, and the derivative time, and serious oscillations or instability can result from incorrect settings. Dealing with the methods of determining these constants is beyond the scope of this book, but there are numerous texts available[26-28] which provide the techniques required. Manufacturers of controllers also provide guidance to controller settings in their technical literature.

Equation (7.22) describes the behavior of an analogue controller, which operates in a continuous manner. However, digital controllers handle information in discrete bits, so the intervals at which the controller reads the transducer signal become an important factor in the precision of control and

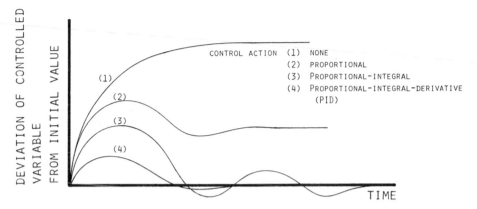

FIGURE 7.24. The influence of the action of controllers on the system response to a step change in the error signal.

Eq. (7.22) must be written in finite difference form in order to describe the behavior of a digital controller:

$$x_p = K_1 \varepsilon_n + K_2 \Delta t \sum_{i=1}^{n} \varepsilon_i + \frac{K_3}{\Delta t} (\varepsilon_n - \varepsilon_{n-1}) \qquad (7.23)$$

where K_1 is the proportional gain, previously identified by K_p, and K_2 and K_3 are equivalent to K_p/Γ_i and $K_p\Gamma_d$, respectively. The integral and derivative times were introduced because the settings on analogue controllers are often labeled in those terms; but digital controllers are essentially computers and it is generally more convenient to refer to K_2 as the integral gain and K_3 as the differential gain. As the sampling interval Δt is made smaller, the action of the digital controller approaches that of the analogue controller but becomes more expensive in terms of the computer power needed. The sampling interval must be matched to the expected rate of change of the variable being controlled. If a slow rate of change is expected the intervals between reading the transducer signal can be lengthy, without undesirable delays between a deviation from the set point occurring and corrective action being initiated.

Digital control[28,42] depends on the describing equations for the desired control actions, usually called algorithms, being programmed into the controller. "Programmable" controllers can be obtained in which the control algorithms are stored on ROM (read-only memory), with only the constants in the algorithms being adjustable, thus giving a digital equivalent of an analogue controller. At the other end of the spectrum, a general-purpose computer can be used for process control, provided it has adequate input and output facilities, giving total freedom in the choice and formulation of control algorithms. Figure 7.25 shows a layout for direct digital control (DDC) of machine conditions (local automatic control). The feedback side of the control loops is similar to that described in Section 7.4 for process monitoring, while the control signals pass through a digital-to-analogue converter (DAC) to an interfacing device, such as a thyristor, for motor speed and temperature control, or a solid-state relay for activating various types of valves. Digital or pulse train outputs are also used for some types of interfacing devices, such as stepper motors. In fact, a number of different types of machine functions can be controlled from the DDC computer.

Direct digital control from a programmable controller or computer is cheaper than analogue control when multiple control loops are needed. Whereas a separate analogue controller is needed for each control loop, it is possible to handle all the control loops of a process from a single digital control device. Logically, there should be a breakeven point in terms of the number of control loops needed to justify a DDC. However, to think of direct digital control as a direct replacement for analogue control would result in most of the advantages of a DDC being disregarded. These are explained in the following sections.

7.5.3. Adaptive Control

The term "adaptive control" is used here in a very general sense to refer to any control system which adjusts the process to maintain a specified optimum performance in an unpredictable environment. This is particularly important for processes such as mixing, extrusion, injection molding, and calendering, where performance is influenced by random changes in the input material properties, the feeding characteristics, and changes in the environment in which the process operates. Adaptive control systems range from simple extensions of closed-loop control to sophisticated learning and self-organizing control systems, which inevitably involve direct digital or computer methods. For this reason, adaptive control has only very recently been introduced into the rubber industry, despite its obvious advantages for overcoming the random variations which continue to plague rubber processing.

A very simple adaptive control system is shown in Figure 7.26, where the internal mixer motor-speed controller set point is being adjusted by the computer, in response to a feedback of the batch temperature, in order to raise the batch temperature to a desired value and then maintain it at that level. In fact, this is hardly different from closed-loop control, except that the computer is supervising the setting of a controller. This involves three

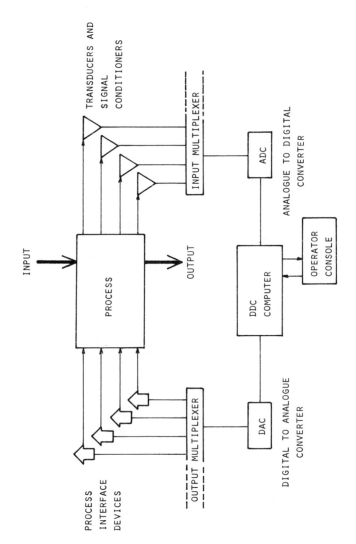

FIGURE 7.25. Direct digital control (DDC) of machine conditions. From: Mikell P. Groover, AUTOMATION, PRO-DUCTION SYSTEMS, AND COMPUTER-AIDED MANUFACTURING, © 1980, p. 418. Reprinted by permission of Prentice-Hall, Inc., Englewood Cliffs, N.J.

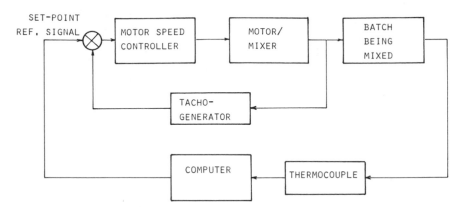

FIGURE 7.26. Adaptive control of batch temperature in internal mixing.

functions which are essential for adaptation to unpredictable time-varying circumstances: identification, decision, and modification.

The identification function involves determining the current performance of the process, by making use of the feedback data from the transducers. Normally, the performance of the process is identified by some relevant index of performance, which can be as simple as a single direct feedback signal, as in the example given in Figure 7.26, or a composite index, which uses the feedback from a number of transducers and includes both quality and quantity elements. The concept of a composite performance index is extended further in *model reference adaptive control*, where a mathematical model of the process is necessary to enable the performance to be quantified.

Once the process performance has been determined, the next function of adaptive control is to decide how the control system should be adjusted, to either maintain a consistent performance or to improve it. The former is concerned with maintaining the performance index at some predetermined value, while the latter involves optimizing the performance index, according to a preprogrammed strategy for adjusting the set points of the local automatic controllers, or their operating characteristics. This is generally known as optimal control and uses some well-documented techniques, such as gradient search strategies.

The third adaptive function is to modify the controller settings, which then completes a cycle of the adaptive control system. When optimal search methods are incorporated into the adaptive control system, each cycle is one step in a series of successive approximations, as the process moves nearer and nearer to its optimum performance, as defined by the performance index.

The main problem with adaptive and optimal control methods is the complexity of the mathematical techniques involved[43] and, consequently, the high level of expertise needed to formulate the control algorithms and commission the control systems. For this reason, monitoring of process performance, with manual adjustment of the set points of local automatic controllers, is often used when it is judged that the cost of developing an adaptive control system is not justified or when the expertise to do so is unavailable.

7.5.4. Sequence Control

The ability to handle complex sequences of operations, such as those which occur during the cycle of a batch process or when any process is being started up, closed down, or changed over from one product to another, is one of the strongest attributes of computer control. This ability is fully utilized in flexible manufacturing systems, dealt with in Section 8.3.2, but it also has much to contribute to the efficiency of conventional processes.

Most processes include a number of elements which can be actuated independently but have to be coordinated in their operation to ensure the safety of the operator, the integrity of the equipment, and the productivity of the process. This is particularly relevant to startup, closedown, and changeover operations, when the responsibility for activating a considerable number of controls in the right sequence and at the right time is often left to the process operator. Ideally, all the necessary operations should be activated by a single "start" command, leaving the operator free to concentrate on those tasks which require manual intervention. This can often reduce the manpower requirements of a complex process. Also, when a standard visual display unit is available at the process, guidance on the manual operations necessary for startup, etc. can be given to the operator in the form of a checklist. Where the monitoring equipment is suitable, the computer can be used to guide operations actively, indicating when operations should be started by audio or visual signals, provided that these signals are helpful.

For sequence control to function, the initiation of a control action must be based upon preceding actions having continued for a preselected time, for example, a warmup time, or having reached preselected values, for example, working temperatures. In the former case the control is open loop, since no allowance is made for the deviation of the process from its predicted behavior; but in the latter case the control is closed loop and relies upon the feedback signals from transducers.

Sequence control acts in conjunction with local automatic, closed-loop, and adaptive control systems. During startup and changeover operations each controller set point may have to be adjusted to suit the product to be run, and the process allowed to approach the set points before production can commence. Provided all the set point information is available in the computer files, the set points can be changed and the sequence of operations initiated by entering the appropriate product code into the computer. During the "warmup" stage the process will be under local automatic control, since closed-loop and adaptive control systems rely on the feedback of process performance data. Once machine conditions have reached preselected values (which may not be the set points), the sequence of operations needed to start production can be initiated and, on completion, closed-loop or adaptive control of the process can commence. Due to the feedback to both closed-loop and adaptive control systems giving a direct measure of the quality of the product, it is possible to achieve useful output before a process reaches its equilibrium operating conditions. Adaptive systems are more powerful than closed-loop systems in this respect; but both systems are capable of achieving useful production much sooner after startup than is possible with local automatic control.

During normal production, sequence control is not needed for continuous processes, except for those elements which are discontinuous in nature, such as batch-off systems, where reels or rolls have to be changed as they are filled. In batch processes, such as internal mixing and injection molding, sequence control is essential throughout the process operation, and if closed-loop or adaptive control systems are used, sequence control must be integrated with them. For example, in internal mixing the sequence of operations for the feeding of a particulate additive may be initiated when the energy input to the process has reached a predetermined value or setpoint. These operations would include raising the ram; metering the required amount of additive into the weigh hopper (itself a closed-loop operation); and when the ram raise is completed, opening the weigh-hopper discharge valve, followed by the lowering of the ram to continue effective mixing. In an adaptive control scheme the set point could be subject to alteration.

7.5.5. Safety and Machine Integrity

Computer control provides an opportunity for setting up systems of checks which ensure that a process will either refuse to operate or shut itself down if the operator is endangered or if expensive damage to equipment will result from further operation. Diagnostic facilities which enable faults to be rapidly identified and located can also be provided.

Interlocks which ensure that a machine cannot be operated unless safety gates are closed and guards in position are already well established in all parts of the rubber industry. These do not depend on computer methods; but it is useful for the location of an open gate or guard to be identified on a visual display unit, particularly with a large installation, such as a mixing facility.

Monitoring of process integrity usually needs to be associated with an automatic system for taking appropriate action in the event of component or system failure, because there is often very little time between the occurrence of a local failure and widespread damage, precluding the possibility of effective manual intervention. Process integrity monitoring and control systems must be justified by comparison of the cost of installing them with the estimated savings on machine repair costs and loss of production over a specified period. In addition, with effective process monitoring it is often possible to reduce the level of routine maintenance.

Damage to equipment is often caused by lubrication failures, which can be detected by a rise in temperature of the bearings affected. Measurement of the temperature of major bearings, such as those on the drive shafts of an internal mixer, is a common practice; but it is less practical on a machine with many moving parts and a multiplicity of bearings. In such cases vibration monitoring is a useful technique, although it is then necessary to locate the cause of the rise in vibration intensity. Measurement of vibration is achieved by attaching an accelerometer to the equipment to be monitored. These are usually low-cost piezoelectric devices. Vibration monitors can often be used to assess the gradual aging of a machine due to wear, which results in a progressive loss of precision of operation, and may provide some early warning of catastrophic failures, such as those resulting from metal fatigue.

Closed-loop sequence controls generally carry their own built-in machine integrity checks, since each step in a sequence is dependent on the ones preceding it. Additional monitoring, such as that described in the previous paragraph, is needed for those aspects of automatic operations which are not protected by sequence checks. However, if a process is stopped because of a malfunction capable of causing serious damage, it is likely that it will need to be emptied of work in progress, particularly if it is dealing with an uncured compound at high temperatures. Manual overrides of the machine integrity systems (but not the operator safety system) must be provided to enable this to be done.

An important feature of computer systems for manufacturing is the provision of diagnostic routines for locating faults within the system, to enable a standard element, such as a printed circuit board, to be identified and changed without a high level of electronics expertise being needed. This enables production to be quickly recommenced, provided the necessary spares are available, without the necessity of calling out either the system manufacturer's service staff or the company's own electronics staff, thus avoiding considerable expense and delay.

7.5.6. Interfacing and Actuation Devices for Computer Control

The use of analogue-to-digital converters, pulse counters, and multiplexers has already been dealt with in Section 7.4.7. These devices are also required for computer process control systems, to interpret the feedback signals to the computer. In addition, devices are needed on the output side to convert the instructions generated by the computer program into adjustments at the process. Digital-to-analogue converters (DAC), pulse generators, and output multiplexers have been dealt with in Section 7.5.2, so this section will concentrate on the devices used to translate the low-voltage and very-low-current signals from the DAC and pulse generator into process adjustments.[28]

Switching operations are frequently required in sequence control, to start motors, open solenoid valves, and so on. This simple off or on control can be actuated through a relay, although it is usually necessary to use transistorized electromechanical or solid-state relays to avoid exceeding the current limits of the computer. When proportional control is needed, as in heater or motor-speed control, the devices used are essentially amplifiers. The power supplied to a heater can be controlled from the computer by the use of thyristors. Variable-speed motors are also thyristor controlled, with manual control via a potentiometer on the control panel. Computer control is simply a matter of replacing the potentiometer, which supplies a control voltage to the thyristor circuits, with a direct line from the computer.

For positional or displacement control, stepper motors and servohydraulic actuators have been introduced. The stepper motor is essentially a synchronous dc motor controlled from a pulse generator, which is, in turn, controlled from the computer. A single pulse will cause the stepper motor to rotate through an angle dictated by the motor design. Step angles are commonly in the range $1°-120°$. If necessary, this rotary motion can be converted to linear motion by an appropriate gearing and transmission system.

Hydraulic actuators with servo, proportional, or digital control valves are widely used in the plastics industry to give precise and programmable control of the rate of injection in injection molding machines. Servo valves adjust the size of the oil flow path to the hydraulic actuator by the movement of the valve spool. This movement is provided by the magnetic core of an electromagnetic positional device. Digital valves have a number of ports of different sizes through which the oil can flow and which can be opened in any order to achieve a combined flow rate. For example, a five-port valve is capable of 31 output steps:

$$2^4 + 2^3 + 2^2 + 2^1 + 2^0 = 31 \qquad (7.24)$$

and by the same argument, a six-port valve gives 63 steps. This system lends itself to computer control due to the binary nature of its operation, but it is unable to give the stepless control of the servovalve. Also, separate digital valves are required for pressure and flow control whereas a servovalve can perform both these functions, although not at the same time. Proportional valves have internal flow monitoring, in addition to regulation, giving the advantage of not needing transducers attached to the actuator to provide the feedback signal for closed-loop control. Mechanical wear and leakage in the hydraulic circuit or actuator, or even a change of oil temperature can alter the relationship between the valve transducer measurement and actuator performance; but in a well-maintained environment they can give a high degree of accuracy and repeatability.

7.5.7. Computer Control Systems

Computer control systems generally start at the process level, being applied to individual machines or to self-contained processing operations, such as rubber mixing or extrusion and continuous vulcanization, in a manner similar to that described for process monitoring in Section 7.4.7. In fact, as the number of functions that process-level microprocessor packages or microcomputers perform is increased, their economic viability is also improved, resulting in progress from monitoring to sequence control and eventually to closed-loop control of process productivity, with increasing levels of automation.

The current trend in microprocessors is toward modular systems designed specifically for instrumentation and control applications. The modules are generally at the plug-in board level, enabling a system to be assembled and commissioned without a deep understanding of microprocessor systems design. Numerous companies are offering systems of this type. Some machines, particularly injection molding machines, are supplied with an integral microprocessor control system. This has the advantage that a working system is immediately available and no process-control expertise is necessary; but it also has a number of disadvantages, which are discussed in Section 7.4.7, including the serious problem of compatibility if a hierarchical system is envisaged.

As the introduction of computer control proceeds, individual processes can be linked with the second-level manufacturing group computer, as shown for process monitoring in Figure 7.20, with the difference that communication will be two way. The basic functions of the secondlevel computer, in addition to continuing to be the "clerk" for process monitoring data, are to "down load" operating and control programs to the first-level computers, together with instructions specific to the product and production run and to coordinate the activities of a number of computer-controlled operations, where necessary.

The hierarchical structure[28] has evolved to be the most effective and efficient arrangement for implementing computers in manufacturing. At one time, before minicomputers, microcomputers, and dedicated microprocessor systems became so predominant, it seemed feasible to use one large plant computer to handle all monitoring and control functions; but the development of computer technology has given the advantage to the computer hierarchy approach. In addition to enabling computer control to be installed gradually throughout a company, spreading the cost over a number of years, the hierarchical system contains redundancy. In the event of a computer breakdown, other computers in the system are programmed to assume the critical tasks of the computer that is down. Software development is more easily managed in the hierarchical configuration; programming for each installation project can be handled separately and once the project is commissioned, changes in software are also more easily accomplished, with less chance of disrupting the system. However, this all depends on planning for a hierarchical system, with communication-compatible computers, right from the introduction of the first-process-level computers.

7.6. QUALITY CONTROL

7.6.1. General Considerations

The development of a quality-control strategy appropriate to a company's needs depends on:

1. The consistency of incoming raw materials and subcomponents in relation to the requirements of the manufacturing operations and product performance.
2. Recognizing the separate requirements of manual and automatic production operations.
3. Conforming with customer's requirements for finished product inspection.

The traditional interpretation of quality control implies "off-line" inspection methods. This definition will be retained, to differentiate it from process monitoring, which is capable of replacing off-line inspection in many cases, with an improvement in effectiveness, due to the increased sampling rate and minimal delay between measurement, analysis, and report. However, it is preferable to consider the task of the quality-control group as being the prevention of circumstances which would lead to customer dissatisfaction and the minimization of manufacturing costs, instead of assigning a limiting inspection role. This emphasizes the "systems" approach to quality control, where the primary objective is to provide the conditions for trouble-free manufacture. The responsibilities of a quality-control group should include the conducting of process-capability studies and the writing of specifications, both of which were dealt with in earlier sections. With the advent of computer process monitoring and control, the validation and calibration of both input and output signals from the first-level computers must also be included. As an increasing reliance is placed on computer information and action, the importance of ensuring that it is accurate and valid also grows. Off-line inspection is then only necessary in well-defined circumstances, such as:

1. Testing of raw materials and other process supplies which cannot be relied upon to be consistently within the requirements of the manufacturing operations or the product.
2. Inspection and testing of the output of operations where a manual element can result in the product being defective.
3. Compliance with legislative, contractual, or licensing requirements.
4. Exceptional circumstances, where it is economically viable to achieve product quality by inspection and selection.

Within each of these areas inspection and testing procedures must be justified by their influence on the economics of manufacture. Goods-inwards inspection is only necessary when a supplier capable of giving the required consistency at a viable price cannot be found. If goods-inwards inspection is deemed to be necessary, its cost must be added to that of the supplies tested and justified by comparison with the estimated savings deriving from:

1. Monies recouped from suppliers for returned defective supplies.
2. Loss of the value added to the supplies by manufacturing operations, prior to their inadequacies being detected.
3. Disruption of manufacturing.
4. Cost of disgruntled customers in receipt of defective products.

Inspection of the output from operations with a critical manual element can be assessed in a similar manner to goods-inwards inspection, with the omission of item (1) in the preceding list, although the emphasis should be placed on deterring defective work, rather than detecting it after it has occurred. Achieving product quality by inspection and selection (or rejection) is generally acceptable only when the cost of the products concerned is structured to include it. Products in this category include small orders, where the cost of the rejects is less than that of developing the manufacturing method to eliminate them and high-precision products, at the limits of current manufacturing technology.

The responsibility for product quality and the authority for implementing corrective action rests largely with the production manager. The quality-control group, despite their title, essentially provide an advisory service, thus avoiding the divisive situation where the production manager is responsible for quantity and the quality-control group is responsible for quality.

7.6.2. Sampling Inspection

Sampling inspection is used for two distinct and separate purposes. First, at goods-inwards inspection, to provide a screening service for rejecting out-of-tolerance supplies and monitoring the suppliers' quality. Second, at any stage during manufacture where a check on the effectiveness of systems and procedures introduced to maintain an acceptable (to the customer) level of product quality is deemed necessary. Feedback of information, both to suppliers and to those responsible for the quality of manufacture, is an integral part of sampling inspection procedures.

Although sampling inspection is economically and conceptually preferable to 100% inspection, it does carry with it a probability or risk that the sample inspected will not be representative of the batch or lot from which it is drawn.[4] This could result in an acceptable batch being rejected (the producer's risk) or of an unacceptable batch being accepted (the customer's risk). To determine and examine these risks, it is first necessary to describe the terminology of sampling plans.

The lot size N is usually taken to be the total number of units delivered by a supplier in a single shipment or, more generally, the total production run of the manufacturer for which the conditions of the system remained essentially unchanged. Thus, it is implicitly assumed that the quality of items within a lot is homogeneous and that the average number of defects did not change during the production run. The average fraction of defective units is called the process average $\bar{p}$, which may alternatively be expressed as a percentage. If a lot consists of separate items the lot size N is simply the number of items, but if materials are to be inspected then a unit can be defined by a mass, which divided into the lot mass gives the lot size N. Similarly, if an extrudate is to be inspected a unit can be defined by length. The sample size n is defined as the number of units to be *drawn at random* from the lot and a sample criterion c is also needed, to define the conditions for accepting or rejecting a batch. This criterion is defined so that when n items are drawn from a lot size of N and k items are found to be defective, then if $k > c$, the entire batch is rejected, and if $k \leqslant c$, the entire batch is accepted.

The risks deriving from a sampling plan can be defined by drawing an operating characteristic (OC) curve, shown in Figure 7.27, where the area marked α defines the producer's risk and the β area defines the customer's risk. In the context of goods-inwards inspection, an OC curve acceptable to both the producer and the customer can be developed by choosing appropriate values of n and c, often by a process of negotiation. The limiting value, defined by the customer, is called the lot tolerance fraction defective (LTFD). It is the upper limit of fraction defectives that the customer is willing to tolerate in each lot. Above this limit the customer would wish to reject all lots but knows that this is impossible if one uses sampling inspection. Therefore the customer compromises by specifying that for no more than β percent of the time should lots with such undesirable quality levels get through the sampling inspection procedures without being detected.

It is important to note that any sampling plan, of specific c and n, completely specifies the α and β risks for given levels of $\bar{p}$ and LTFD. In

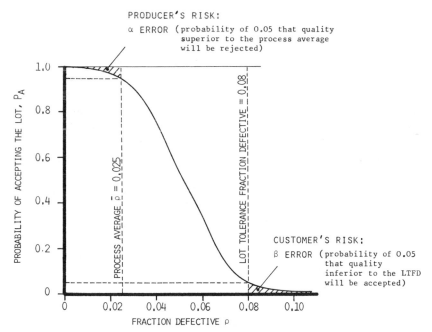

FIGURE 7.27. An operating characteristic curve showing producers' and consumers' risks.[44] Martin K. Starr, PRODUCTION MANAGEMENT: Systems and Synthesis, 2nd ed., © 1972, p. 303, Reprinted by permission of Prentice-Hall, Inc., Englewood Cliffs, N.J.

turn, the sampling plan requires n inspections for every lot. If the process average $\bar{p}$ is truly representative, then α percent of the time the remainder of a rejected lot may need to be 100% inspected, to remove defective units. If this is the case, the average number of units inspected will be $n + (N - n)\alpha$.

As n becomes large and approaches N, the sampling plan becomes increasingly discriminating and α approaches one, reducing the risk to the customer and increasing the cost of inspection for the producer. It is also worth mentioning proportional sampling at this stage, where $n/N = \text{const.}$ Although this seems intuitively sound, it can be shown that as lot size increases, the sampling plan constructed on the basis of proportional sampling becomes more discriminating and should not be used.

One of the most direct approaches to constructing sampling plans is to use tables which have been devised for the purpose.[45,46] These are often defined in terms of an acceptable quality level (AQL), which is the process average $\bar{p}$ redefined as a target to be achieved or surpassed by the producer. Alternatively, OC curves can be derived from appropriate mathematical statements, using the hypergeometric distribution when the lot size is small, so that the effect of sampling n units, $n - 1$ units, etc. can be recognized, and the binomial or Poisson distributions when N is sufficiently large, for example, $N \geqslant 1000$ units.[47]

Sampling-plan tables generally specify three levels of inspection:—normal, reduced, and tightened,—so that the customer's risk is adjusted in response to the producer's performance, to minimize the amount of inspection. Switchover from one level to another is dictated by a specified number of preceding lots being accepted or rejected and, in the case of moving to the reduced level, the total number of defects in the preceding lots.

In addition to the single sampling plans there are the alternatives of using double or multiple sampling plans. Double sampling plans are well documented[45,46] and widely used, usually to reduce costs in situations where the amount of inspection required by single sampling plans appears to be too great. They are also psychologically attractive because they give a second chance to doubtful lots. In single sampling the acceptance decision must be made on the basis of the first sample drawn. With double sampling, a second sample can be drawn, if desired. The double sampling plan requires two acceptance numbers c_1 and c_2, such that $c_2 > c_1$. Then, if the observed number of rejects in the first sample of size n_1 is k_1:

1. The lot is accepted if $k_1 \leqslant c_1$.
2. The lot is rejected if $k_1 > c_2$.
3. If $c_1 < k_1 \leqslant c_2$ an additional sample of size n_2 is drawn, giving a total sample of size $n_1 + n_2$.

If the observed number of rejects in the total sample is $k_1 + k_2$, then:

4. The lot is accepted if $k_1 + k_2 \leqslant c_2$.
5. The lot is rejected if $k_1 + k_2 > c_2$.

All the procedures discussed thus far have been based on classifying items as acceptable or unacceptable. These are termed *attributes sampling plans*. Where an item can be passed as acceptable if a measurement falls within a range of values (the tolerance band) attributes sampling can be used but there is also the alternative of using a *variables sampling plan*.[48] The advantages of variables plans, in comparison with attributes plans, are:

1. Equivalent risks are obtained with smaller sample sizes.
2. Additional diagnostic information is obtained for corrective action.

The disadvantages of variables plans have traditionally been stated as the higher cost of inspection, due to the need to use more complex measuring instruments and the more complicated computations required. The former may still be valid but the latter is certainly not, due to the computer.

The theory governing inspection by variables depends on the properties of the normal distribution and is therefore only applicable when there is reason to believe that the distribution of measurements is normal. If items are considered to be defective when their measurements lie outside the specification limits, the fraction defective will be represented by the areas under the distribution curve beyond the specification limit lines as shown in Figure 7.28(a). When the mean and the standard deviation are known, the fraction defective p relative to the lower specification limit L or the upper specification limit U can be obtained from the tables of the normal distribution, as $p_U = 1 - P_U$ and $p_L = P_L$.

When the standard deviation is known, and thus defines the shape of the distribution curve, and only one specification limit is given, the fraction defective is controlled by the difference between the specification limit and the mean. It is possible to devise a quality parameter Q in terms of this interval and the standard deviation, such that

$$Q_U = \frac{U - \mu}{\sigma} \tag{7.25}$$

if U, the upper specification limit, is given and

$$Q_L = \frac{\mu - L}{\sigma} \tag{7.26}$$

if L, the lower specification limit, is given. The acceptance criteria for a lot or batch would then be to accept if $Q \geqslant k$, where k is the critical value of Q which gives the fraction defective permitted in the specification, and to reject

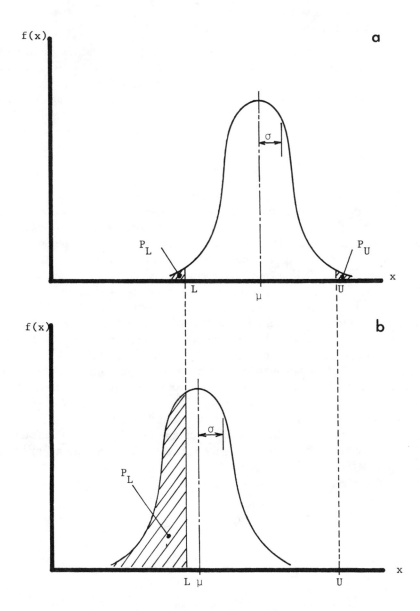

FIGURE 7.28. Influence of a change in value of the mean μ on the fraction of defectives, with constant standard deviation σ.

if $Q < k$. If both upper and lower specification limits are given, they are considered individually and the acceptance criterion is that each should be equal to or greater than its respective acceptability constant k.

The largest value of σ which will just give the fraction defective permitted when μ is midway between U and L is known as the maximum standard deviation (MSD). If σ is found to be greater than this figure, the lot can be rejected immediately. The converse, that a batch is acceptable if σ is less than the MSD, is *not* necessarily true, as the acceptable value of σ depends on the value of μ. This is illustrated by Figure 7.28(b).

In practice μ is not known and has to be estimated from a sample. The best estimate of μ is given by the mean of the sample, $\bar{x}$, and, depending on circumstances, σ may also be unknown, requiring that an estimate of the standard deviation of the lot be made from the sample. If σ is known, then the acceptance criterion for an upper specification limit is

$$Q_U = \frac{U - \bar{x}}{\sigma} \geqslant k \tag{7.27}$$

and for a lower specification limit

$$Q_L = \frac{\bar{x} - L}{\sigma} \geqslant k \tag{7.28}$$

If σ is unknown, it can be estimated from

$$s = \left(\frac{\sum (x_i - \bar{x})^2}{n - 1} \right)^{1/2} \tag{7.29}$$

giving

$$Q_U = \frac{U - \bar{x}}{s} \tag{7.30}$$

and

$$Q_L = \frac{\bar{x} - L}{s} \tag{7.31}$$

Compared with the σ method (where σ is known), this s method increases the probability of error. Taking the σ method, the mean of the sample $\bar{x}$, is not necessarily to the mean of the lot μ; but $(\bar{x} - \mu) n^{1/2}/\sigma$ is normally distributed, with a mean of zero and a standard deviation of unity, so it is possible to show that the difference between them will not exceed a definite figure. For example, for a 95% probability this figure is $\pm 1.96\sigma/n^{1/2}$. When the s method is used, s is substituted for σ, giving $(\bar{x} - \mu)n^{1/2}/s$, which is no longer normally distributed but has a t-distribution with $n - 1$ degrees of freedom. For example, if $n = 10$, there is now a 95% probability that the difference between the sample and lot means will not be more than $\pm 2.26s/10^{1/2}$, compared with the $\pm 1.96\sigma/n^{1/2}$ given by the σ method.

As for the attributes sampling method, tables are available[48] for variables sampling which enable sampling plans to be selected. These plans are also categorized by an acceptable quality level (AQL) and an inspection level, but it is necessary to use the statistical techniques described in this section to implement a plan.

7.6.3. Patrol Inspection and Validation

The need to check the calibration of process monitoring and control systems at regular intervals has been mentioned in previous sections. This is referred to as validation here, to be followed by corrective action if necessary. It represents the updated function of the patrol inspector, alternatively called the line inspector. The traditional role of the patrol inspector is the sampling inspection of work in progress, mainly as a deterrent to substandard manual work.

In comparison with validation, which is undertaken at regular intervals, except when a malfunction is suspected, sampling inspection is best undertaken on a random basis. This is equivalent to taking a sample from random locations in a lot or batch and has the advantage that the arrival of the patrol inspector cannot be anticipated by the operators. This emphasizes the difference between an operation which is under manual control and one under automatic control. The latter is likely to give a progressive change in the quality of manufacture, which is best detected by sequential measurements at regularly spaced time intervals using a process monitoring system, enabling the onset of a problem to be forecast in many instances. Variations in the quality from a manually controlled operation tend to be cyclic in nature, with many short-term irregularities, and are not amenable to forecasting methods.

Sampling plans for patrol inspection can be selected from the tables for either attributes or variables, as is appropriate to the products. It is then necessary to construct a further plan for determining when each item in a sample is to be taken for inspection and, if only a certain proportion of the manual manufacturing operations in a company are to be checked, which operations should be chosen. For short production runs the whole run can be treated as the batch or lot for this purpose, but for long runs the production must be divided up into suitably sized lots, by the number of units produced rather than by time intervals. Random numbers provide the type of unpredictable inspection pattern which is required. They can be produced by computer, reference to random number tables, or even by picking numbers out of a box. At intervals the operations to be checked must be redetermined, taking into account those jobs which will be completed or introduced during the period defined by the selected interval. Also, the inspection pattern for each operation must be redetermined for each separate lot or production run, to maintain the unpredictable nature of the scheme.

7.6.4. Subjective Inspection

A substantial proportion of the inspection currently carried out in the rubber industry is subjective in nature, depending on the judgment of the inspector for an accept or reject decision. In addition to being used to assess products for which visual appearance is important, subjective inspection is often used to determine if a product is functionally acceptable.

Subjective inspection is largely concentrated on finished products, although it can be used at various stages of manufacture by patrol inspectors. A distinction must be drawn between formalized subjective inspection and the detection of an anomaly in a product at any stage of manufacture and by any of the works personnel. The former has to be treated as a further manual operation in the manufacturing route, while the latter is an informal but powerful asset, supporting the activities of the patrol inspector and providing a rapid feedback for corrective action. It can only be obtained by engendering a "quality-conscious" attitude throughout the company.

The setting of standards for subjective inspection is often quite difficult because numbers cannot be used in a precise manner and written descriptions, even when they are accompanied by drawings or photographs, are often open to a range of interpretations. Standards boards are widely used to overcome some of these problems. Defective products which conform to the categories of "unambiguously reject," "marginally reject," "marginally accept," and "unambiguously accept" are pinned to a board for direct comparison with the product being inspected. The main problems of this method are its unwieldiness, a set of defective products is needed for each type of defect in each product, and its maintenance examples tend to "age" in their appearance rather rapidly and require frequent replacement to maintain a viable comparison with current production.

Even in situations where standards boards are well maintained, variations in the standards of inspection will occur unless positive measures are taken to prevent them. Those variations due to fatigue and boredom can be largely overcome by sensible job planning; but there will inevitably be a tendency for a fluctuation which correlates with the number of defects found. When levels of defects are low there is an almost unconscious trend toward tightened standards and a converse trend when defective levels are high. Poor inspection can also be a problem. By treating subjective inspection as an additional manual operation, there is a clear case for including it in the remit of the patrol inspector. Random inspection of both the products being passed as acceptable and the rejects is necessary to ensure that inspection is neither too slack nor too tight.

Subjective inspection is, of necessity, concerned with attributes, rather than variables, and the attributes sampling plans discussed in Section 7.6.2 can be used. If 100% inspection is specified, it should be noted that it is not 100% effective.

REFERENCES

1. Juran, J. M., *Quality Control Handbook*, McGraw-Hill, New York (1962).
2. Kirkpatrick, E. G., *Quality Control for Managers and Engineers*, Wiley, New York (1970).
3. ASTM D2414, "Dibutyl Phthalate Absorption Number," (1979).
4. Turi, E. A., *Thermal Characterisation of Polymeric Materials*, Academic Press, London (1981).
5. Mawer, J. J., *Rub. Chem. Technol.* **42**, 110 (1969).
6. ASTM D445, "Kinematic Viscosity of Transparent and Opaque Liquids," (1979).
7. Brown, R. P., *Handbook of Plastics Test Methods*, 2nd ed., Godwin, London (1981).
8. King, D., "Training within the Organisation: A Study of Company Policy and Procedures for the Systematic Training of Operators and Supervisoes," Tavistock, London (1964).
9. Currie, R. M., *Work Study*, 3rd ed., Pitman, London (1972).
10. Mundel, M. E., *Motion and Time Study: Principles and Practice*, 4th ed., Prentice-Hall, Englewood Cliffs, N.J., (1970).
11. Cox, D. R., *Planning of Experiments*, Wiley, New York (1968).
12. Davies, O. L., *The Design and Analysis of Industrial Experiments*, ICI (Oliver and Boyd), London (1971).
13. Keppel, G., and W. H. Saufley, *Introduction to Design and Analysis*, Freeman, San Francisco (1980).
14. Low, C. W., *Trans. Inst. Chem. Engrs.* **42**, 334 (1964).
15. Spendley, W., G. R. Hext, and F. R. Himsworth, *Technometrics* **4** (4), 441 (1962).
16. Baker, R. J., and J. A. Nelder, *The Glim System—Release 3 Manual*, Numerical Algorithms Group and Royal Statistical Society, London (1977).
17. Box, G. E. P., and K. B. Wilson, *J. Rub. Statist. Soc. B* **13**, 1 (1957).
18. Box, G. E. P., and J. S. Hunter, *Ann. Math. Statist.* **28**, 195 (1957).
19. Derringer, G. C., *Rub. Age* **104**, 27 (1972).
20. Box, G. E. P., and D. W. Behnken, *Technometrics* **2** (4), 455 (1960).
21. Edwards, A. L., *Multiple Regression and the Analysis of Variance and Covariance*, Freeman, San Francisco (1979).

22. Statistical Tables, from White, J., A. Yeats and G. Skipworth, *Tables for Statisticians*, 2nd Ed., Stanley Thornes, Cheltenham, UK (1977).
23. Buxton, J. R., and M. Holt, Paper given at Royal Statistical Society Conference, *Regression Modelling and Data*, Sheffield, England (March 1983).
24. Buxton, J. R., and M. Holt, Internal report, Engineering Maths. Department, Loughborough University, U.K. (1982).
25. Goldsmith, P. L., Paper given at the Royal Statistical Society Conference, Warwick University (March 1974).
26. Haslam, J. A., G. R. Summers, and D. Williams, *Engineering Instrumentation and Control*, Arnold, London (1981).
27. Jones, B. E., *Instrumentation, Measurement and Feedback*, McGraw-Hill, London (1977).
28. Groover, M. P., *Automation, Production Systems and Computer Aided Manufacturing*, Prentice-Hall, Englewood Cliffs, New Jersey (1980).
29. Kane-May, Company Literature, Welwyn Garden City, U.K. (1982).
30. Heimann, Company Literature, Wiesbaden-Doteheim, West Germany (1982).
31. Barnes, Company Literature, Stamford, Connecticut (1982).
32. Vanzetti, Company Literature, Stoughton, Massachusetts (1982).
33. Luxtron, Company Literature, Mountain View, California (1982).
34. Carter Bros., Company Literature, Rochdale, U.K. (1982).
35. Dynisco, Company Literature, Westwood, Massachusetts (1982).
36. Gentron Company Literature, Gentron, Sunnyvale, California (1983).
37. Kistler Company Literature, Kistler, Winterthur, Switzerland (1981).
38. Willshaw, H., *Calenders for Rubber Processing*, Institute of Rubber Industry Monograph, London (1956).
39. Measurex Company Literature, Cupertino, California (1980).
40. Hamblin, D. J., *Manufacturing and Management and Control Systems in the UK Polymer Processing Industry*, SERC, Polymer Engineering Directorate Report (November 1982).
41. B.S. 2564, "Control Chart Technique When Manufacturing to a Specification," (1955).
42. Young, R. E., *Supervisory Remote Control Systems*, Peter Peregrinus, Stevenage, U.K. (1977).
43. Landau, Y. D., *Adaptive Control—The Model Reference Approach*, Marcel Dekker, New York (1979).
44. Starr, M. K., *Production Management: Systems and Synthesis*, Prentice-Hall, Englewood Cliffs, New Jersey (1972).
45. United States Department of Defence, Military Standard, Sampling Procedures and Tables for Inspection 27 Attributes (Mil-STD-105D), U.S. Government Printing Office, Washington, D.C. (1963).
46. B.S. 6001, "Specification of Sampling Procedures and Tables for Inspection by Attributes," (1972).
47. Ott, E. R., *Process Quality Control*, McGraw-Hill, Kogakusha, Tokyo (1975).
48. United States Department of Defense, Military Standard, Sampling Procedures and Tables for Inspection by Variables for Percent Defective (MIL-STD-414), U.S. Government Printing Office, Washington, D.C. (1963).

8

Plant Layout and Operations Methods

8.1. GENERAL CONSIDERATIONS

8.1.1. Defining the Functions of Plant Layout

Plant layout in the rubber industry, where there is usually a number of stages and parallel routes in the manufacturing sequences, is an extremely complex and difficult activity. It is also essential for efficient manufacture. In order to provide a basis for the development of the subject, it is worth attempting a definition:

> The plant layout activity is concerned with the design or modification of a manufacturing system, in which machines and people will be safely and efficiently provided with the services needed for them to perform their tasks, to a standard required by the products being manufactured.

It should be recognized that the *necessary* movements of people, materials, and work in progress in a manufacturing unit dictate the form of the plant layout. In fact, some plant layout engineers choose a definition of the subject based on materials flow[1]:

> Planning and integrating the paths of component parts of a product in order to obtain the most effective and economical relationship between people, equipment and the movement of materials; from receiving, through manufacture to the shipment of the finished product.

It is important that the need for plant layout expertise is recognized when changes within a company are discussed; and that responsibility for plant layout is given to a specific individual in companies too small to have a plant layout engineer. Otherwise, ad hoc conversions, without reference to overall activities and manufacturing requirements, will speedily result in a reduction of manufacturing efficiency. For this reason it is desirable to identify the circumstances which commonly result in a requirement for plant layout expertise[2]:

1. Introduction of new products.
2. Expansion or contraction of existing business.
3. Adjustment within existing manufacture.
4. Technical progress.
5. Safety.

The design and commissioning of a new factory is not included in the preceding list for two reasons: (1) it is not a common problem and (2) it is the classical plant layout problem, which everyone recognizes.

The introduction of a new product invariably brings with it some changes in working methods and may require new handling and ancillary equipment, irrespective of whether it is made by an existing manufacturing route or requires a new one. The changes may be quite small, as in the case of a new injection molded product; but the manufacturing route, from receiving to despatch, should be examined to identify possible consequences of adverse interaction with other products. These occur when machine and labor schedules are disrupted and can often be rectified by reorganization of the work distribution by the production planning and management staff; but sometimes a plant layout solution is required.

New products which require new equipment give rise to larger and more easily recognized requirements for plant layout studies. These range from the need to accommodate a new machine in an existing manufacturing unit to the new factory case.

Both expansion and contraction of existing business pose their own problems. Expansion can take two main routes: more equipment, people, and the buildings to accommodate them can be introduced, using existing methods to meet the requirements for the increased volume of manufacture; alternatively, the economics of production may indicate that a total change to a new method of manufacture is preferable. The former obviously has the advantage of using known technology and methods, but it generally has the disadvantage of requiring more space, which may be very difficult to provide. The latter obviously carries a substantial learning period and risk factor, which must be balanced against the expected benefits. Integration with existing facilities must also be examined, to ensure that the resources necessary to support the new equipment can be made available.

Contraction of a manufacturing unit often takes place in conditions of economic stringency, when the objectives of plant layout changes are to minimize overheads and maintain or improve manufacturing efficiency. In such circumstances a competent assessment of the benefits of changes is essential, for comparison with the additional resources needed. Implementation of the changes also assumes considerable importance, requiring that the remaining operations be kept running while those nominated for disposal are dismantled and removed. This type of procedure is also necessary when moving a manufacturing facility to a new site.

Adjustment within existing manufacture may need to be undertaken for a number of reasons, the most obvious of these being improvements to an existing layout, to reduce costs or save space. Other reasons include the replacement of outmoded or worn-out equipment and movement of equipment, to accommodate changes in working methods or improvements in the working environment.

Returning to the first reason for adjustment within an existing manufacturing facility, the following list[1] provides a number of indicators useful for identifying an unsatisfactory layout:

1. Unexplainable delays and idle time.
2. Stock-control difficulties.
3. Decreased production in an area.
4. Crowded conditions.
5. Large numbers of people moving materials or products.
6. Bottlenecks in production.
7. Backtracking.
8. Excessive temporary storage.
9. Obstacles in material flow.
10. Scheduling difficulties.
11. Underutilized space—horizontally and vertically.
12. Idle people and equipment.
13. Excessive time required in process.
14. Poor housekeeping.

Technical progress brings new equipment, materials, and techniques, all of which have to be evaluated and, if they prove to be advantageous, incorporated into the manufacturing facility. This can involve adjustment in existing manufacture, due to enhancements of well-established techniques or more sweeping changes, when a completely new technique is introduced. The most obvious example of the former is the changes in equipment control and operation resulting from the introduction of the microprocessor; and the continuous mixing of particulate rubber can be used as an example of the latter.

A characteristic of technical progress is that machines are rarely replaced with exact equivalents. Improvements in efficiency, capabilities, and operating methods will have been implemented by the machine manufacturers to compete effectively in *their* market. Full advantage should be taken of these improvements, although they may require substantial changes in the facilities and resources needed to support them. It is worth emphasizing the value of a well-planned equipment replacement and introduction program here, both as an aid to the plant layout engineer and for the commercial well-being of a company. This program provides the plant layout engineer with the necessary information to plan progressive changes in layout with a view to minimizing the disruption occurring during changes, minimizing the costs of changes and avoiding the operation of new equipment interfering with that of established equipment.

As more becomes known about the deleterious effects of certain working environments on health, mainly due to dust and fume hazards, and the standards of physical safety of people in the manufacturing area continue to be raised, production methods have to be modified to comply with these advances. Unless the modifications required are minimal, such as the fitting of new guards to a machine or changing the method of operation at a manual work station, plant layout expertise will be needed.

8.1.2. Guiding Principles for Plant Layout in the Rubber Industry

The rubber industry is firmly based in "batch manufacture," which traditionally involves siting all the machines which have a similar function together and progressing work in batches through the sections or departments thus created. This has been reinforced by the segregation of some operations, such as mixing, because of the environments they create. Organization of a company into profit centers, as described in Chapter 9, has little influence on the basic separation of manufacturing functions, although there is a trend toward the creation of product groups in the postmixing stages of manufacture.

Batch manufacture is sometimes called group technology; in the rubber industry a typical unit or group is a mixing installation, an extrusion line, or a molding press. Each constitutes a complete stage of manufacture. For plant layout the concept of basic manufacturing units is very useful and is the key to the necessary automation of batch manufacture.

As legislation for exposure to noise, dust, and fumes, which constitute health hazards, is tightened and implemented, the dust and fume extraction measures necessary to proceed with many of the traditional manual operations of the rubber industry have become increasingly expensive and are approaching the limits of their capabilities. As noted in Section 6.4.6, these measures add a substantial amount to the manufacturing overhead and can have a detrimental influence on the process with which they are associated. The alternative solution of automation and removal of personnel from the vicinity of such processes is now being pursued vigorously in a number of companies and it has profound implications for plant layout.

Before proceeding to the details of plant layout and handling methods, a review of the techniques and procedures involved is useful. As with all complex activities, plant layout studies require a well-organized and methodical approach to ensure a successful outcome. Plant layout is a project activity, so the following sequence of investigation will have many similarities with those for process and product development, with which plant layout is associated:

1. Identification of the problem.
2. Initial survey.
3. Specification of objectives, scope, resources, time scale, and expected benefits.
4. Detailed survey.
5. Planning and allocating space for individual activities.
6. Synthesis of layouts, followed by evaluation and selection.
7. Installation and commissioning.

The techniques needed to carry out a plant layout investigation are firmly based on methods study, although the scale of the investigation is generally much larger than the typical methods study, which tends to concentrate on a single work station or activity. The preceding sequence indicates a linear progression from one stage to the next; but there will be a substantial amount of "backtracking" between stages (6) and (5), with an occasional need to return to stage (4) for further details of existing usage of space or working practices.

8.2. TRANSPORT AND STORAGE IN MANUFACTURE

8.2.1. Assessment of Requirements

This section is concerned with storage methods and the movement of powders, liquids, and unit items from storage to work station and also between work stations. The details of handling methods at work stations are dealt with separately in Section 8.3.

The majority of items which require storing and moving in a rubber company can be made up into unit loads, which results in trucks and conveyors being predominant, and enables the separation of transport and handling to be readily made. For bulk powder and liquid storage transport and handling methods it is less suitable, since all three functions are generally integrated in a single system. However, the storage and transport elements of the systems will be dealt with in this section, while metering and weighing will be dealt with in Section 8.3, for comparison with manual weighing.

8.2.2. Unit Items

In-factory transport can be divided into three main groups: vehicles for handling over a variable path, equipment for handling over a fixed area, and conveyors for handling over a fixed path. These are used in conjunction with a number of containers or supports, such as boxes and pallets, which may also form part of the storage system. In this section, the transport of rubber, subcomponents, and semifinished and finished products will be followed, from delivery to despatch.

Offloading from a delivery vehicle generally falls into the category of handling over a variable path, with exceptions occurring when the volume usage of a particular item warrants a specially designed vehicle and the installation of a dedicated handling system; or when a "subcomponent" is very large, as in the case for vessels requiring rubber linings. In the general case, a truck of some type is necessary for transfer to unpacking stations, goods-inwards checking areas, and storage. A variety of truck types are available,[3] with motive power ranging from manual to totally motor powered for each type. The type of truck selected will depend on the shape and size of those items large enough to be delivered individually, and by the unit load size and type where small items have been packed together for efficient handling. The offloading method should take into consideration any precautions necessary to avoid danger to the people involved in the operation and to avoid damage to the item being offloaded.

The offloading of hazardous items is usually covered by codes of practice and legislation. The choice of motive power for the various functions of a truck is influenced by the balance of costs of labor and truck utilization, additional tasks for the truck, and the horizontal and vertical distances through which the delivered goods need to be moved.

Rubber is commonly delivered in bales, which are packed together on standard pallets for handling by forklift trucks. Variations on this theme are the delivery of polychloroprene rubber in "chips," which are approximately 25 mm in diameter and 10 mm thick, and silicone rubber, which is delivered in drums to avoid "cold flow" during storage. The bale form inevitably introduces an element of manual handling, even in the largest company, since the pallet has to be opened and the bales transported to the feed conveyor for the mixing operation. Bale cutting may also be necessary.

Subcomponents for rubber product manufacture generally include "inserts," to which rubber is bonded during a molding operation; wire and cable, which are coated with rubber during extrusion; and fabrics, which are coated with rubber during calendering. Most inserts and some reels of wire and cable are small enough to require transport in unit loads, while large reels and rolls of fabric are transported as individual items. Unit loads may need to be broken down for goods-inwards inspection, storage, and issue to production departments. Ideally, the container in which items were delivered should be suitable for factory use; but this is often impractical, which may require the use of in-factory transport methods different from those used for offloading. The containers and supports into which items are transferred for storage and delivery to the production departments should then form an integral part of the work station to which they are delivered, eliminating a further transfer to work-station storage. They should also be designed to be compatible with the handling/manipulation methods used at the work station and be readily identified as belonging to a specific product, if multiproduct manufacture is practiced.

Process supplies include products such as dust mask filters, mold release agents, and degreasing fluids. All these items fall into the category of "consumables," which are just as essential as raw materials and subcomponents. Process supplies for the rubber industry often carry substantial hazards and require the most stringent precautions to be observed in their transport and handling, with all the relevant personnel being trained to react properly to an accident.

Durable items, such as molds, extruder dies, and ancillary equipment specific to products, can also be considered as process supplies. Small items do not present any special problems, but the transport of large molds must be taken into consideration in the early stages of a layout design, together with any other sizable and frequently moved equipment. If an overhead crane is installed, it provides an ideal method of dealing with mold changing in those presses where the platens are oriented vertically. For presses having horizontally oriented platens and in situations where an overhead crane is not available, or for any other job which requires horizontal movement under an overhead obstruction, a lifting platform is required. A forklift truck provides both this facility and a suitable transport method. However, routes from the storage area to the work stations needing the molds or other items must be allocated in the layout plan, to ensure a safe and obstruction-free transfer.

The transfer of subcomponents and consumable process supplies is also usually best achieved by a variable-path system, and trucks of some type are must commonly used. Again the truck type will be dependent on the form of load, level of utilization, and integration with the layout.

Transport of semifinished articles between workstations can be achieved by a number of alternative methods, which depend on the process and product types to a far greater extent than transfer from storage. As alternatives to trucks, there are chain conveyors, roller conveyors, belt conveyors, and overhead cranes. Unlike trucks, all these methods can operate above or below the work level, opening the way to more effective use of space.

The provision of storage space is necessary at a number of locations within a company. Usually, there is a main storage area, adjacent to the delivery bay(s), in which raw materials, subcomponents, and some consumable process supplies are held. This is followed by a number of specialist stores, ranging from solvent stores, sited at a specified safe distance from working areas, to mold storage, adjacent to the product lines. Although storage is often regarded as a passive function, holding supplies until they are required, the efficient operation of a storage system demands a high level of organization and can only be achieved if the layout and design of the stores permit it.

With rubber and many other raw materials being delivered on standard pallets, the main storage facilities can be similarly standardized. Multilevel racks, into which a pallet can be inserted by a forklift truck, are widely used and offer a high level of flexibility in the stores' organization.[3]

The organization of the main store for subcomponents and consumable process supplies is usually combined with that of the raw-materials store, and is concerned with goods-inwards checking, unpacking, and redistribution of supplies into in-house storage and transport containers. A quarantine system is usually operated for a wide range of supplies, including raw materials, to ensure that they are not released for use until passed as satisfactory by the inspection staff. Goods are then supplied for use on a first-in/first-out basis, with records being kept to maintain an accurate inventory. These activities need an ordered, spacious, and well-lit environment, which the plant layout engineer must provide. Suitable and readily accessible storage units must be provided for each type of supply handled. The arrangement of these units should be compatible with the first-in/first-out flow and permit truck access to areas where heavy items are stored. Provision of a pleasant working environment in the main stores usually ensures adequate conditions for supplies, although special conditions may be required to ensure a reasonable storage life for some items.

From the main store, subcomponents often go to preparatory processes, usually to ensure adequate bonding with rubber, and become work in progress (WIP), while the consumable process supplies are distributed to a number of points in the manufacturing unit. At each of these points it is desirable to have a well-organized checking and distribution system, to maintain accountability. These systems would sensibly be located within the specialized storage areas, pointing to further requirements to be considered by the plant layout engineer.

Ideally, there should be no storage of semifinished work; but in practice, storage of WIP is necessary for a number of sound technical and organizational reasons, although the permissible minimum levels should be determined and used.

The first area in which WIP storage is generally found is postmixing. Many companies operate a quarantine system similar to that used for goods-inwards inspection and it is also common for batches to be stored for a minimum period of 24 hours after mixing, to avoid problems which could result from the rapid changes in processing behavior occurring immediately after mixing. Maximum storage time should also be stipulated and a strict first-in/first-out flow observed. Although the rate of change of material properties decreases rapidly after mixing it does not stop; and mixes stored for long periods can become unusable. Some compounds with chemical ingredients which are active at room temperature require cold storage, unless the storage time can be standardized.

The design of a mixed-compound store depends on the physical form of the mixed rubber. Sheets and strip are very common, for which stacking storage is desirable to minimize the use of floor space, with sheets being carried on pallets and strips coiled on trays or reels. For the first-in/first-out flow, the 24-hour moratorium, and the "quarantine until passed by inspection" routines to be rigorously observed, storage must be orderly and locations in the store clearly identified. Release from the store can then be controlled via the instructions to the truck driver for the location from which to take a batch of material demanded by a product line. Unlike the main store, the mixed-compound store does not need a permanent "storeperson," although it is sensible to operate the inspection (materials testing) procedures adjacent to the store, if the volume of production justifies it, rather than in a remote laboratory.

Throughout a company a number of "buffer" stores will be needed to hold an accumulation of WIP generated by differences in production rates between sequential processes, by a high output machine being used to supply a number of product lines and by a batch process being used to supply a continuous one or vice versa. These stores are set up when the quantity of accumulated WIP is too large to be held efficiently in the storage system integral with a work station. Although some of these storage areas cannot be eliminated easily (such as those needed to store molding blanks or preforms, where the extruder is capable of producing preforms for a number of presses molding different products), every effort should be made during a plant layout exercise to minimize both the number of storage areas and the quantities of WIP held in each. All areas where WIP is likely to accumulate should be identified, otherwise an unplanned store will happen. It is essential for efficient manufacture that each storage area is properly arranged and managed.

Ideally, the storage of WIP should be capable of being integrated into a work station, so that the storage methods support the operation and become part of it. It follows that the design for both input and output sides of the work station can contribute strongly to the ease, accuracy, and efficiency of manufacture, particularly if there is an element of manual skill involved. In addition, the storage methods must be compatible with any mechanical handling aids which are used. Careful attention to ergonomics is necessary for a manually operated work station,[4] which is beyond the scope of this book.

Storage conditions for WIP are often very important. For example, cleanliness is essential for subcomponents to be bonded to rubber and extrudates are susceptible to distortion prior to vulcanization. Even after vulcanization, permanent deformation of products can result from unsuitable storage.

Storage of durable process supplies, such as molds, extrusion dies, and ancillary equipment specific to products, is usually a production services group function, being combined with the fitting or installation of such items for production operation and their maintenance during and after use. Where heavy molds or other equipment are stored, there should be sufficient space for the unhampered access of lifting tackle or trucks. The storage system, which is usually heavy-duty racking, should be of a design which enables easy transfer of an item to the lifting tackle or truck. Most accidents happen during this transition. Floor storage of large items does not usually present any particular problems, provided that the space required has been anticipated and allocated. As usual, storage locations should be clearly identified, to permit an item to be located rapidly and to prevent a jumbled and confusing storage arrangement from occurring. A record is normally kept of the periods of use of an item, as a natural consequence of booking it out of and into the store, together with its maintenance record. The latter is used to record requests for repairs and modifications and note the work carried out, together with the routine maintenance. For molds the cleaning frequency may also be noted. In addition to providing an essential record for accounting purposes, these notes highlight problem areas and give an indication of when replacements will be required.

The storage conditions for process durables are usually very important, particularly so in the case of molds, which have highly polished and accurately fitting surfaces. A warm and dry environment is necessary to prevent corrosion. Most stores for molds and similar items have facilities for maintenance work. In some cases the plant layout permits the workshops of the production services group to be located adjacent to a store. In any layout where transfer between buildings is necessary, it is essential that protection for process durables be arranged.

The issue of process consumables is often best carried out from the process durables store. This is particularly useful when the supplies are issued in small quantities or volumes, enabling an efficient quantity to be transported from the main store and issue to occur near the point of use. This also enables a record and inventory to be readily maintained. An exception to this procedure occurs when hazardous or flammable liquids are used. Codes of practice usually stipulate that the storage points should be at or greater than a certain distance from the working areas of a company and that, even in the event of a total failure of the tanks or containers, the spillage should be contained within a restricted area. This is "bunded storage," where a containing wall is built around storage tanks or, for small quantities, the doors to the storehouse open above floor level.

8.2.3. Bulk Fillers and Liquids

Fillers and particulate additives can be stored, conveyed, and weighed or metered by a range of standard methods used throughout the powder-handling industries. Selection of a system depends on the scale and frequency of use of materials, their handling characteristics, and the safety or environmental problems associated with them. The particulate and pelleted materials used in rubber compounds can be split into two groups,—bulk fillers and "small" powders. Whereas only one or two bulk fillers are used in a compound, in proportions of 20–300 parts per 100 parts of rubber by weight, up to 10 different small powders, in proportions of 0.2–6 parts, is quite common. The strategies of handling these two groups of materials are quite different.

Unless volume usage is very small, some form of mechanical handling is desirable for bulk fillers. This requires some characterization of the handling characteristics of the powder or pelleted material, to provide the basic information for system selection. Most companies concerned with the manufacture of powder-handling equipment operate a testing service; but it is useful to understand the basic principles of powder flow behavior, for effective communication with powder-handling specialists.

The basic properties or characteristics of a powder, for specification of a suitable storage and handling system, are bulk density, flowability, and floodability. A number of tests[5] can be performed to arrive at a suitable measure of these characteristics.

Bulk density is defined as the mass of powder required to occupy a unit volume of space, and is always less than the density of the solid material due to the voids between particles. Bulk density is dependent on the mean particle size, particle size distribution, and particle shape, all of which influence the way in which the particles pack together. It is also dependent on the treatment of the powder prior to measurement, which gives more than one measurement of bulk density. The two measurements generally used are aerated density and packed or "tap" density. The former is obtained by pouring a powder sample of known mass into a measuring jar. The latter is obtained when the powder in the measuring jar is subjected to a number of "taps" in a mechanical rig, causing it to assume its most compact configuration, with minimum void space between the particles. The difference between the two bulk densities, expressed by a percentage of the packed bulk density, is termed the compressibility of a powder and is used in the assessment of flowability and floodability. For the definition of compressibility, it is assumed that the particles do not deform. For fillers this is a reasonable assumption, but particulate rubbers will undergo a continued time-dependent creep or cold flow.

The next step, after measurement of bulk density, is the determination of a number of "material angles." The first of these is the angle of repose, which is defined by the slope of a freestanding cone of powder formed under standard conditions. It is also known as the angle of shear. The results obtained are strongly dependent on the conditions of the test and an angle of fall can be defined which is the material angle formed when a powder comes to rest after flowing under standard conditions. In contrast the angle of repose is determined using conditions which impart minimum kinetic energy to the powder mass. The angle of fall is generally lower than the angle of repose, due to the tendency of a powder to self-fluidize or flood. Other methods of measuring material angles are available, but it must be recognized that all these tests are comparative in nature and are approximately equivalent to empirical quality-control tests.

It is desirable that a powder gives mass flow during hopper discharge and that the rate of discharge is constant and predictable. A powder having these characteristics is said to have good flowability and will generally give a low angle of repose and a low value of compressibility. Floodability describes the tendency of a powder to self-fluidize during hopper discharge, giving high and unpredictable flow rates. If a powder has good flowability combined with a low angle of fall and a large difference between the angle of fall and the angle of repose, it will tend to flood. A material with poor flowability will require a "live" hopper with discharge aid devices to prevent bridging and "rat holing," whereas a material which tends to flood will probably require a rotary seal in the hopper, to control the discharge rate. These tests also provide some guidance on the choice of handling system, for example, screw, conveyor, pneumatic, etc., but interpretation requires considerable experience. For more fundamental and informative measurements of powder flow behavior, shear cell tests are used.

In conjunction with the determination of the powder characteristics, consideration should be given to the form in which the powder is delivered from the manufacturing company, the requirements of the mixing process, and the layout and relative positions of the delivery points and the mixing installation. Some nonblack fillers are delivered exclusively in 25-kg bags stacked on pallets, demanding some manual handling and the provision of a bag-slitting and -emptying station, which could be manual or mechanized. In either case, measures to control atmospheric dust levels are essential. Carbon blacks and some white fillers are delivered in a number of forms,[6] from 25-kg bags through tote bins of 1-tonne capacity to big bags of 2 to 3 tonnes and tanker deliveries, direct to the customers' bulk storage silos.

Fillers delivered in small bags stacked on pallets are usually stored in this form in two-, three-, or four-level racks, using forklift trucks. Storage conditions should ensure that the moisture content of the filler does not rise to levels which could interfere with production. Pallets are then withdrawn from the stores as required and the contents transferred to day hoppers, which feed to the mixing process via a weighing or metering unit. The bag-slitting and -emptying station is usually placed in close proximity to the day hopper, to minimize the transfer equipment.

Tote bins and big bags also require suitable storage space but are used in place of the day hopper. Obviously, two discharge points are required for each powder feed, to enable changeovers to take place without loss of production, unless the big bag or tote bin is used to feed a day or buffer hopper.

With bags and semibulk containers the filler can be moved to a point very close to the mixing operation by an unspecialized transport system, such as a forklift truck. The conveying problem is then quite small and may be solved by the use of a short screw or belt conveyor. When a company is large enough to warrant filler deliveries by tanker, the conveying problem is more complex and requires considerably greater capital investment. The traditional method is to use gravity discharge from the tanker to a subground-level conveyor belt, which then transfers the filler to a bucket conveyor, to raise it to the level of the bulk storage silo. This system is now being superseded by pneumatic conveying, which is fully enclosed and is far more adaptable from a plant layout point of view. These systems are generally only required for carbon black; and in the past the creation of fines by pellet attrition in pneumatic conveying has caused some concern, with respect to filler incorporation during mixing. However, this problem does not now appear to be significant.

The economics of bulk liquid handling are quite different than those of bulk powder handling, with the former being viable at quite low-volume usage levels, provided that the delivery of a part tanker load can be procured. This delivery is followed by direct transfer to bulk storage tanks, which may be heated. However, it is more common to have unheated storage tanks and to heat the supply lines between the storage tanks and the mixing process, to reduce the viscosity and so improve the weighing, metering, and mixing characteristics. The alternative to bulk delivery is to use drums, which are labor intensive and generally result in spillage.

8.3. HANDLING METHODS AND OPERATIONS AT WORKSTATIONS

8.3.1. Mixing

The mixing sequence consists of a number of separate operations, some of which take place in environments which are causing companies considerable concern with respect to legislation on working conditions.[7,8] Some of the problems and possible solutions will be examined here. The operations normally associated with internal mixing are:

1. Cutting and manual weighing of rubber.
2. Manual weighing of minor particulate ingredients.
3. Automatic weighing of bulk fillers.
4. Manual mixer feeding and operation.
5. Manual two-roll milling and sheeting-off.
6. Cooling and loading onto pallets.

Most of the environmental problems arise from the dust hazards created during the manual weighing of minor ingredients and the manual feeding and operation of the mixer. The two-roll mill can also present a problem if loose powder additions are practiced. It is also worth noting that these operations also form the major sources of potential variation in mixing. If batch mixing is replaced by continuous mixing, the minor-ingredient weighing is still required; but manual weighing of rubber is replaced by granulation, which generally results in high noise levels. These points will be considered when each of the operations is dealt with.

The handling, storage, and weighing problems for minor additives are quite different from those encountered with the bulk fillers, due to the number needed for a typical compound and the total number needed to support the mixing of a range of compounds. Except in the few companies able to dedicate a mixing facility to a few compounds, versatility is a prime requirement. This points towards manual methods, but the accuracy of weighing is often very important and the problems of maintaining a safe working environment for operators are usually more acute with the minor ingredients than they are with the bulk fillers. The compaction of powders into "dust-free" pellets[9] reduces considerably the dust problems, provided that attrition in handling does not result in the formation of fines and that the problems of accuracy on manual weighing can be overcome.

Most minor ingredients are delivered in 25-kg bags, requiring sack slitting and discharge stations with efficient dust extraction. The type of storage unit into which the bags are discharged depends on the type of weighing system which is used. In the most basic case, manual weighing by scooping materials from open bins is possible, provided that the operator is protected from the airborne dust created by the scooping action. Properly designed scoop bins, which are hoppers having a discharge arrangement for manual transfer of material to a weighing unit, give a more compact arrangement for increased efficiency and are easier to provide with adequate dust control. In both cases, spillage is an unavoidable problem; and there may be some difficulty in meeting the requirements of dust-control legislation with some additives.[10]

Automatic and semiautomatic systems make dust control much easier and remove the responsibility for weighing accuracy from the operator. However, except in large mixing installations handling a narrow range of compounding ingredients, the costs of such systems cannot be justified. For each ingredient in current use it is necessary to have a day hopper supplying a weighing unit; and the difficulty of handling many minor ingredients requires the use of live hoppers and screw or vibrating feeders, in place of a simple gravity feed system. For most companies a system in which one or two weighing units can be used for the whole range of ingredients is clearly desirable.

The minor ingredients can be separated into those for which the best dust-control methods provided for manual weighing are adequate to meet the requirements of current and pending legislation, and those for which they are not, of which there are usually a substantial number. A sack slitting machine in a sealed enclosure, feeding to a number of bins containing different additives in a carousel arrangement, such as the one shown in Figure 8.1, removes the dust hazard from the feed end of the operation. However, the problem of separating the operator from the material being weighed remains. Either enclosing the bins and weighing station or having the operator wear protective gear will impede operations and is likely to be extremely unpopular; but unless fully automatic weighing is adopted, these measures may have to be taken.

For ensuring that weighing accuracy is maintained in a manual system, a technique best described as computer-aided manual weighing can be used. Some electronic scales are computer compatible, enabling a computer-driven display to be used to provide target weights, in response to keying in the compound number. The operator is then required to match the target weight, within a specified tolerance band, before being allowed to proceed to the next ingredient. In the event of a weight being incorrect, a visual signal can be given and the operator "locked out" until the weight has been corrected.

Bulk or semibulk handling systems are normally used in conjunction

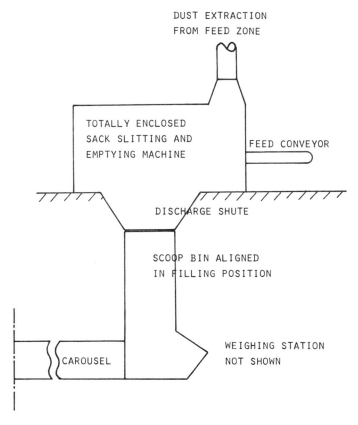

FIGURE 8.1. Arrangement for dust-free slitting of particulate minor additive sacks and filling of scoop bins.

with automatic weighing or metering. Weighing systems are used to feed batch mixers, whereas continuous mixers require a metered feed, which can be provided by rapid cycle batch weighing and smoothing or by a true continuous metering unit. A typical batch-weighing system consists of a weigh hopper, with flexible sides to ensure complete powder discharge into the mixer, mounted on a load-sensing device, which then provides a signal to the system controller, enabling the feed from the day hopper to be stopped when the correct amount of material has been delivered. The load-sensing device can be a mechanical lever arm system, with a dial indicator for manual control of weigh-hopper filling. At the other end of the spectrum, load sensing may be achieved by electronic load cells, with microprocessor control of weigh-hopper filling. A number of hybrid systems are available, mainly with the purpose of providing manual backup if the automatic system should fail. However, increasing reliability in automatic systems, coupled with the mistakes which occur when personnel unaccustomed to manual control attempt to work with short mixing cycles, and the increasing expense of providing manual backup, are causing companies to choose a wholly automatic system.

A number of metering systems are available, to deal with a broad range of material types and metering accuracy requirements. These are mainly based upon feed devices mounted on load cells, where the mass of material in the feed device is compared with a reference value. If this system is used in conjunction with a rapid cycle batch-weighing unit, the interval between discharges is adjusted until the actual value in the metering device coincides with the reference value, which defines the required delivery rate. A different principle is used in the Chronos Richardson loss-in-weight[11] unit. The rate of discharge from a weigh hopper, usually via a screw feeder, is monitored and compared with a reference rate. Adjustments are then made to the speed of the screw feeder as necessary.

Liquid additives can be weighted manually and fed into an internal mixer via the hopper or they can be injected directly into the mixing chamber, under the ram. For large volumes, transfer directly from the storage tanks, by heated pipeline, to an automatic weighing station, followed by direct injection into the mixing chamber can be used.[11] Where versatility is required, a number of liquid additive feeds to a single weighing station can be used, provided that the small amount of cross-contamination which occurs during changeovers can be tolerated.

For small volumes, injection under the ram is not practical and alternative methods must be sought which avoid spillage and the losses which occur due to the "wetting" of the feed hopper during loading. One solution to this problem is to weigh the liquids into low-melting-point polyethylene bags, which can then be heat sealed.

Automatic mixer operation is desirable both for consistent performance and to remove the operator from the environment created by the mixer. The techniques for automatic mixer control are well established,[11,12] although many companies retain manual feeding of rubber and minor ingredients. This can be eliminated, even with sequential additions, if a system capable of recognizing the positions of separate groups of ingredients on an extended feed conveyor is used. Provided that the groups are placed sufficiently far apart for an unambiguous gap to be detected between them, using a photocell or similar detector, the conveyor-belt movement is capable of being controlled by the mixer computer. A number of batches placed on the conveyor would then form a queue.

The preferred method of transfer from the internal mixer to a dump extruder or two-roll mill is direct gravity discharge, requiring that the mixer is sited on an elevated platform. Single-level arrangements using conveyor belts are rarely satisfactory, due to the severe service conditions for the belts and the increased probability of batch-to-batch contamination.

The strip or sheet formed by the extruder or two-roll mill is then fed directly into a cooler, to reduce it to a temperature at which there is no danger of scorch occurring during the slow air cooling which follows.

8.3.2. Extrusion and Calendering

From an operations point of view extrusion and calendering can be treated similarly. They are both continuous processes, for which the feed must be replenished and the output removed without interfering with their running. Also, when provided with a closed-loop control system they are capable of essentially automatic operation, requiring manual intervention only at the feed and output station, except at startup, during a changeover from one product to another, and at shutdown. However, the provision of computer control can aid these functions, as described in Section 7.5.4.

The safety problems with calendering result almost entirely from mechanical sources, with the dangers of running nips predominating. From the long experience of the rubber industry with these dangers, guarding and safety systems are well established and documented.[13] Similar hazards can exist on extrusion lines in the feed and output areas, but they are generally less severe. The major extrusion hazards occur when continuous vulcanization is used, requiring effective enclosure of the vulcanization unit and the extraction of the fumes from it. Because the entry to and exit from the continuous vulcanization unit can be small, this can be readily achieved, but precautions, such as the wearing of respirators and face masks, may be needed if the enclosure is opened under operating conditions.

8.3.3. Molding

The primary factor influencing the layout of molding machines is the need for an operator or handling system to feed the materials and remove the finished products from a number of presses. It is quite unusual for a single molding machine to occupy fully the time of an operator or handling system, even when shuttle molds or swinging cores are used; and the number of presses in an operating group will increase as the cure times of the products increase.

For manual press loading and unloading, the line arrangement has many advantages. Movement from press to press is efficient and containers for molding blanks (compression and transfer molding), inserts, and finished moldings can be sited in front of a press, as shown in Figure 8.2. This enables supplies to be replenished and containers of finished moldings to be removed from the aisle side of the work station, without interfering with the operator's task. Siting presses in long lines also gives the production planner a substantial amount of flexibility for press loading. It is desirable to group together products with cure times which fit into a sensible work cycle. With the line arrangement the number of presses assigned to an operator is not fixed by the layout and can be determined by the cure times of the products being run.

The problems of manual operations in the rubber industry were referred to in Section 6.4.7, where the fume hazard associated with many high-temperature molding operations was identified. The measures necessary to protect personnel from fumes can only be determined by monitoring the atmosphere in the vicinity of the opened presses and the container in which the moldings are placed.[14] Providing local exhaust ventilation intakes as close as possible to the sources of fumes, both at the press and the finished moldings container, can provide a safe working environment for many molding operations. However, as mold temperatures are increased the fume evolution tends to defeat these simple measures and automatic operation may become necessary.

The tire industry, with its standard product, has few problems with automatic loading and unloading of presses; but the technology for automation in jobbing companies, with their short runs and diverse products, has only recently become available, and is only now being explored for application to rubber molding. In Section 6.4.7 the necessity of coupling mold design with the design or selection of an industrial robot manipulator was emphasized; but to enable industrial robots to be used effectively they have to be an integral part of a manufacturing unit, which is referred to as a "flexible manufacturing cell" (FMC).[15] These are now finding an increasing number of applications in the metal-shaping industries. In the context of

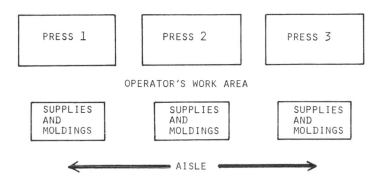

FIGURE 8.2. A typical layout for manual molding press operation.

rubber molding the FMC will consist of a small group of molding machines serviced by one or more robots, plus any ancillary equipment necessary to supply and evacuate the FMC. The layout of the molding machines is dictated by the "space envelope" defined by the reach of the robot(s) and will tend to be of the form shown in Figure 8.3.

Some personnel will still be required to supply the FMC with molding blanks or feed strip, plus any inserts which are required. They will also be required to place these at a designated position on the input pallets so that they can be located by the robot. The evacuation side is less demanding and direct removal of work containers by forklift truck is possible. The key point is that the supply personnel will be working in a fume-free environment due to the total enclosure of the molding presses. This enclosed environment also brings other benefits. Guards can be omitted from the molding machines if their operation is interlocked with the opening of the enclosure, providing improved access to the mold for the robot. It also enables robot movement to be closely synchronized with press opening and closing, minimizing the press open time and the cooling of the mold. Mold cooling will be further reduced, in comparison with manual operation, due to minimal fume extraction being required. In fact there are many technical benefits arising from consistent operation in an environment which is much improved for the process.

With the use of robots and FMCs, computer process control is essential, to cue the robot into the press sequence. Moldings of different types with different cure times can be handled, provided that the number of occasions on which two or more presses reach the end of their cycles simultaneously is not considered to be excessive. While the increased mold open time and cooling associated with this situation can be avoided by programming the computer to delay the start of a cure cycle (with the mold closed but empty), some loss of productivity is inevitable.

Mold changing may also represent a substantial loss of productivity if the whole FMC has to be closed down while one mold is changed. To change a mold "on the run" would require:

1. Removal of panels in the FMC enclosure for entry of mold transport and lifting gear.
2. The erection of temporary guards, to prevent accidental movement into the path of the robot or a closing press.
3. Respirators for the mold fitters, due to the hostile environment.
4. Programming which would enable one or more presses to be dropped out of the cycle of operation.

Clearly the design and economics of FMCs will be taxing production engineers and managers for some years to come; and all their problems and attributes cannot yet be anticipated.

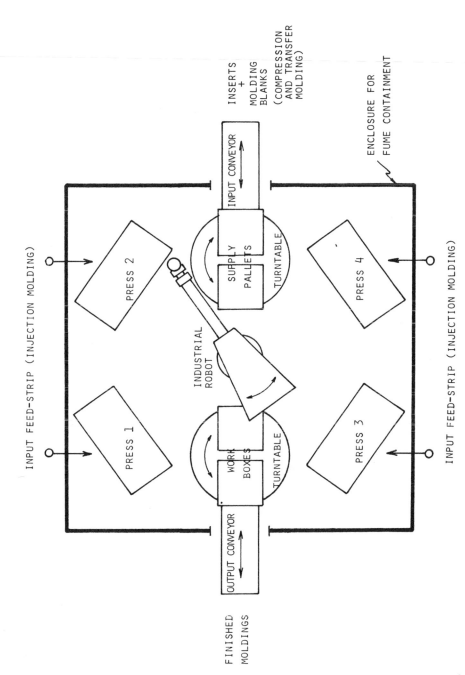

FIGURE 8.3. Plan view of a flexible manufacturing cell for rubber molding.

8.4. PLANNING AND ALLOCATING SPACE

8.4.1. Flow Process Charts and Production Balancing

Flow process charts show the place of a process or operation in the manufacturing sequence and the way in which it interacts with those supplying it or being supplied by it.[2] While a flow process chart can give limited spatial and relational information, these aspects are secondary to its main purpose; and the physical proximity of two processes on a chart can be misleading. A typical flow process chart (FPC) of the type used extensively in plant layout studies is shown in Figure 8.4, together with the five basic symbols, each of which represents a class of activity. The applications of flow process charts can be categorized as:

1. Material charts, where the events occurring to a particular material are followed.
2. Person charts, where the activities of persons are detailed.
3. Machine charts, where the sequence of operations of a machine, such as a forklift truck, are followed.
4. Outline charts, using only □ and ○ symbols, are often used to give an overall schematic of a whole series of processes.

All these are useful for plant layout, although the amount of fine detail will generally be less than that required by a work-study engineer. The nature of the charts will depend on the size and scope of the plant layout problem. However, before a flow process chart can be finalized it is necessary to ensure that a "balanced" situation will be achieved by the proposed processes and operations.

Balanced manufacture is achieved when the processing equipment is selected and operated so that surpluses do not appear at intermediate places in the production sequence, nor do shortages occur.[1] This is a simplified view; it has already been acknowledged in Section 8.2.2 that planned storage of work in progress is necessary. To specify the equipment for a balanced production facility, it is necessary to refer first to the planned pattern and volume of manufacture. Working back from the scheduled delivery of finished products to the despatch area, it is possible to determine the throughput required at a point within the manufacturing unit and, from the estimated work rates of processes and operations, the numbers of each which will be needed. This procedure is necessary for both large and small problems. Replacing a single machine with one of a different type can result in serious disruption if a balanced interaction with machines before and after it in the production sequence is not achieved. At this stage a series of

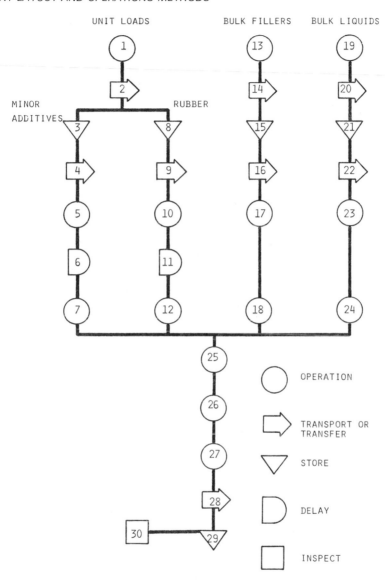

KEY: 1, 13, 19—off-loading from delivery vehicle; 2, 14, 20—transport to storage; 3, 8, 15, 21—storage; 4, 9, 16, 22—transfer from storage to weighing stations; 5, 10, 17, 23—manual or automatic weighing; 6, 11—queuing of manually weighed items prior to mixing; 7, 12, 18, 24—feeding to mixer; 25—mixing; 26—sheeting off; 27—cooling; 28—transfer to mixed compound store; 29—storage; 30—routine testing.

FIGURE 8.4. A flow process chart for rubber mixing, from receipt of raw materials to storage and testing of mixed compound.

alternative solutions to a plant layout problem can be tried and the ones with obvious flaws discarded. The survivors can then go forward to the next stage.

8.4.2. Processes and Operations

After establishing a number of potentially viable solutions to a plant layout problem, it is reasonable to progress to consideration of the physical form and requirements of the individual processes and the operations associated with them, which demands that the conceptual approach is thoroughly three-dimensional. Each process has a "space envelope," which defines the shape and volume of the region required for it to function. Further space envelopes are then needed for the ancillary equipment and operations associated with the process. Generally, space envelopes for equipment are mutually exclusive whereas envelopes for operations, such as manipulation, adjustment, cleaning, and maintenance, are not and can be overlapped. This allows some versatility in the determination of the shape and size of the total space envelope to be carried forward into the layout synthesis stage.

In addition to size and shape, each process will have a number of other constraints and requirements associated with it. These include weight, access, services, and operating environment. Weight or floor loading per unit area will indicate where special foundations are necessary for very large machines, such as an internal mixer, and will dictate the design of elevated platforms or mezzanine floors. Access to a machine is concerned first with the limitations on the position from which it may be loaded and unloaded during production, and second, with the replacement of parts during maintenance, including the total removal of the machine for major work or replacement. Services include electricity, water, steam, compressed air, hydraulic power and, in many cases, dust and fume extraction. Often the position of a process in relation to the source of the services is very important, particularly for the piped supplies, such as steam, water, and hydraulic fluid, where transmission losses are considerable. For each process there will be a distance within which it is preferred to site it and a maximum permissible distance. The operating environment can refer to the need to protect a process handling light-colored materials or medical products from the general environment or, more usually, to the protection of the general environment and operators from the process and the materials associated with it. The normal hazards are fumes, dust, noise, and vibration. Equipment needed to reduce these to safe and acceptable levels will inevitably lead to additions to the total space envelope of a process.

From here, details of the total space envelopes, together with the limitations and requirements just dealt with, go forward to the layout synthesis stage.

8.4.3. Storage Areas

Specifying the capacities of stores for raw materials, subcomponents, process supplies, work in progress, and completed work is often difficult, due to the need to achieve a sensible compromise from a range of conflicting requirements. Close liaison with purchasing and production control groups is necessary to achieve a workable solution.

The provision of storage for raw materials should be geared to the buying strategy and the standard delivery quantities or volumes. The cost of providing storage facilities must be compared with the savings from bulk buying. Also it may be necessary to specify high minimum stock levels for supplies with variable delivery lead times, to avoid disruption of production due to shortages. The selected minimum stock level is an expression of the uncertainty in the rate of use of a material and the lead time from placing an order to delivery.

For particulate materials delivered in tankers and stored in fixed silos or hoppers there are few problems in the management of the storage system, if the correct capacity is specified initially. However, the majority of particulate materials are not used in sufficient quantities to warrant this treatment and are delivered variously in semibulk hoppers, big bags, and conventional small bags on pallets. These represent unit loads and can be dealt with in a similar manner to rubber and other items which constitute unit loads.

For unit loads there are two basic concepts of storage: fixed position and random location. In fixed-position storage a particular item is always stored in a particular place. This means that at any time there must be sufficient space to accommodate the maximum holding of that item, although most of the time it will be well below that level. Random location implies that as a space becomes vacant in the store, through the withdrawal of an item, that space is allocated to the next batch of incoming supplies requiring a similar volume of storage of the type vacated. In this method, as racks and containers become empty they are constantly being refilled by other items, which may be very different. The random-location method can result in 15–30% less storage space being required, compared with fixed position storage.

Despite the saving on space, total random–location storage is not always desirable and most practical storage systems which handle a wide range of materials and supplies operate a hybrid system, partway between the fixed-position and random-location methods. Obvious separations occur when items need different physical and environmental conditions, and when the presence of one material in close proximity to another could cause contamination or deterioration. Random-location storage can also lead to long transit routes from storage position to production unit. In practice it is preferable to have items with a high call-off rate nearer to the store exit than those with a low call-off rate. Also, where forklift trucks are used in conjunction with multilevel racks, it is preferable to have the high call-off supplies near to the ground.

When unpacking is necessary it is often convenient to establish an unpacking area followed by a day store, which is capable of holding materials for 24 hours production or less. Unpacking areas and day stores are usually best sited adjacent to the production lines, particularly if the stores transport system is different from that used in the production unit. An efficient load can be transferred from the stores and then be broken down into its component parts near the process for which it is intended.

8.4.4. Transit Routes for Materials, Products, and Equipment

Transit routes are an important factor in determining a layout and should be approached with the aim of providing a cost-effective solution to the problem of ensuring that materials, products, and equipment can be moved to the locations where they are required, at the right time and safely. The interaction of the selection of transport methods and transit routes with other aspects of plant layout leads to planning for minimization of storage of work in progress and utilization of nonproductive space where possible.

The types of transport system selected will obviously determine the allocation and organization of transit-route space. Referring back to Section 8.2.2, transport methods can be divided into fixed-path systems and variable-path systems, although the distinction becomes blurred where mobile conveyor lines and programmed destination conveyors are used to give a capability for adapting rapidly to a change in the pattern of manufacture. In general, fixed-path systems can utilize nonproductive space, whereas variable-path systems, which usually involve trucks of some type, use transit routes which occupy valuable space which could be used for manufacturing operations.

Both the transport method and transit-route requirements are dictated by the nature of the products, the volume of production, and the rate at which the pattern of manufacture is required to change. The latter two points are extremely important where both the volume and type of products are subject to rapid change. At this stage a detailed breakdown of the pattern of movement between stores and processes and interprocess is necessary, enabling the traffic densities and probable routing directions to be established. This can be achieved in conjunction with the flow process chart.

The prime objective here is to provide sufficient information for a detailed examination of alternative transit routes during the layout synthesis. For fixed-path methods, the space occupied by a system, plus the additional space for the materials or products it carries and for safety guards can be specified, together with fixing and support requirements and maintenance access. The space for loading and discharging a system at work stations should also be included. The performance of a system is an essential part of this specification, since it determines load sizes and often places limitations on routings. Conveyors are a typical example, where the angle of ascent is limited. This information should lead to an adequate provision for the system being made during the layout synthesis and enable the load carrying capacity, working speed, length, and route to be fixed.

Defining an adequate provision of space for variable-path systems is more difficult. Assuming that a variable-path system will involve trucks, the surfaces on which they can run, and the angle of incline they can safely ascend or descend must be defined, as must the predicted floor loadings. Widths of aisles and radii of bends will be dictated by the types of trucks used and by the nature of the traffic: one way or two way. A double-width aisle may be needed for heavy traffic, but passing places at key points are often adequate. Considerable thought must be given to the design and space requirements of junctions or intersections. Visibility is particularly important and the layout in the region of a junction should take into account the need for adequate vision on each approach. The space required by an aisle is further defined by the headroom needed by trucks and their loads, so that the minimum heights at which service lines and conveyors can cross the aisle are fixed. From these considerations, elements of the transit routes are defined, coupling their space requirements with their capabilities. This gives "building blocks" which can be "assembled" in different ways, as demanded by overall priorities at the layout synthesis stage.

Transit routes for materials and products must be designed for continuous use, whereas the moving of equipment is a periodic requirement. Even so, the need for moving any piece of equipment, including very large units, must be anticipated and planned for. At this stage it is reasonable to consider the requirements of equipment too large to be moved using the system being examined for the transport of materials and products. Size is obviously a prime consideration and the feasibility of dismantling an installation into a number of smaller elements must be examined, together with the space requirement for *in situ* dismantling. Weights, viable methods of lifting, and precautions against accident and damage are also important, completing a specification which can again be fed forward for layout synthesis.

8.4.5. Personnel Transit Routes

The routes by which people move about within a company include the mechanistic factors dealt with in the previous section in addition to detailed safety and "work-attitude" considerations. First, it is reasonable to identify processes and work stations where potential hazards exist and to note the nature of those hazards. Transit routes for materials and products have their own hazards, particularly when motorized trucks are used, and these too should be identified. Separate doorways for pedestrians and vehicles should be provided and a similar separation is needed where there are frequent and unscheduled movements of both people and vehicles. In addition, most manufacturing sequences involve jobs where a loss of concentration on the part of an operative can result in an accident, damage to a machine, or defective work. In these instances the disruption which can be caused by people passing close to a work station constitutes a hazard.

Many movements of people can be predicted, such as those at the beginning and end of a work period and at scheduled breaks. Banning vehicular activity at these times may be useful, to allow the rapid transit of a large number of people by routes which are normally "vehicle only." Where mixed routes for both vehicles and people exist it may still be necessary to bar vehicular activity at these times as a safety measure. The use of "vehicle-only" routes by people at preset times can be used to reduce the allocation of space for personnel transit to that needed for people whose movements are unscheduled. Also, by reducing the width of a transit route a greater versatility is gained for layout synthesis. Stairs, bridges over obstructions, and corners that would be impractical in a large aisle can be utilized.

In conjunction with designing personnel transit routes for safe and efficient movement under normal working conditions, the rapid evacuation of people in the event of a fire must also be considered. The provision and positioning of fire doors, in addition to the normal exits, is obviously part of this aspect of plant layout; but the routes leading to the fire doors must also be carefully planned. These routes should be of sufficient size to permit rapid evacuation, should be free from obstacles or changes of direction which could be confusing in conditions of limited visibility, and should avoid passing work stations or storage areas which constitute fire hazards. The latter point will exert a strong influence on the siting of fire doors and the location of operations or stores which involve flammable materials.

8.4.6. Facilities for Personnel

Facilities for personnel can usually be dealt with under the categories of "works" and administration, although a thorough integration between these two should be achieved in the operation of a company. Common canteen facilities can usually be provided in small companies to aid this integration, but distances may prevent it from being achieved economically in large companies.

General toilet and washing facilities are usually the subject of legislation or codes of practice.[8] These approximately define the space which must be allocated for a given work force. The subdivision and location of this space within the company can be dealt with during layout synthesis. In addition to this general provision, rubber companies usually have special requirements in the mixing section, where showers and changing facilities are necessary. Lockers, where outdoor clothing can be exchanged for protective clothing, are also a general requirement. Often, the locker area can be combined with a rest area, where seating is provided and tea and coffee are available, if refreshments at work stations are undesirable.

Office allocation in the manufacturing areas can be limited to providing accommodation for direct line management, but a more reasonable policy is to locate offices near to the sections which their occupants serve. In a small company no office is very far removed from the practical details of manufacture, but a wide range of options are available to larger companies. A number of constraints on office location can be identified. First, the office environment must be conducive to the work performed in it. Second, it is usually more economical of space to group offices together rather than to site them separately. Finally, the offices should not be provided with an adequate environment to the detriment of the manufacturing environment. For example, offices should not deprive a workshop of natural light and ventilation if they make a significant contribution to acceptable working conditions for the operatives. Offices on a mezzanine floor often work very well, enabling line management to monitor the workshop activities easily, as well as conforming to the constraints outlined here.

The allocation of space for administration can be largely independent of the works layout, although the works–administration interface requires careful thought. Also, where a new administration block is being considered, care must be taken to avoid making it seem a remote and separate entity from the works. A prime function of administration and management is communication and the layout should be conducive to the exchange of information. A good starting point is to identify the different types of visitors to the administration group, their business, and the people they will need to see during their visit. Estimates of the numbers of visitors in each of the categories identified are useful for establishing the proportional space allocation.

The basic division of visitors is between employees and others. Employees generally comprise works managers, whose visits are regular and routine, and operatives, whose visits are occasional and are usually associated with specific problems. In the latter case, the usual destinations are the personnel and welfare sections or the wages group. Obviously, convenient entrances from the works "side" of the administration block are needed and, if the company is large enough, a waiting area in a position convenient for personnel, welfare, and wages offices is desirable. This ensures a calm and quiet environment, which gives a good introduction to what may be a difficult interview or meeting. It also provides a sensible separation between the very different aspects of the administration group business. The front entrance of a company reflects the separation. Its function is essentially to impress the visitor and "sell" the company. The design of the approach to the front entrance and the layout of the reception area should reflect the company philosophy and the markets to which the company looks for business. In many instances it is preferable for discussions with visitors to take place in rooms adjacent to the reception area which are designed for the purpose.

Some space must be provided in every company, even the smallest, for medical services. The size and sophistication of the service will depend on the size of a company and the type of processes it operates. In a small company a minimum requirement is a room with a bed, a few chairs, washing facilities, and a comprehensive and well-stocked first-aid kit. Additional medical supplies should be provided if the type of injury which could arise from the processes being operated cannot be dealt with from the contents of the first-aid kit.

Last, adequate car-parking space should be provided for employees, to ensure that access to the company for commercial vehicles and visitors is not obstructed, and to avoid disrupting activities in adjacent industrial or domestic buildings through thoughtless parking. This is also a safety measure.

8.5. LAYOUT SYNTHESIS AND EVALUATION

8.5.1. Layout Synthesis

A layout plan is first arrived at through the judgment of the group responsible for its design. Once this overall layout plan has been created a number of evaluation aids can be used to assess, refine, and compare alternative plans.

The starting point is the flow process chart and the plan of the space in which the layout is to be contained. The type and number (or alternative types) of items of equipment and the space they need to operate should have been previously defined for each element on the flow process chart. The plan of the space for the layout should detail any fixed features additional to those of the building, such as steam lines, ventilation ducts, and drains. However, drawings are two-dimensional and a layout occupies a three-dimensional space. It is unlikely that the available space will be utilized effectively unless the technique used for layout planning is also three-dimensional. The use of models as planning aids in plant layout is well established and versatile commercial systems, such as Lego, enable a layout to be constructed and modified very quickly. In addition to simply modeling the approximate shape of an item of equipment or work station, it is useful to define the "space envelope" required for its operation. This can be achieved using string or wire. By putting this additional information into the model, the initial planning becomes a much more critical process, allowing conflicts of transit and operation space to be identified quickly.

Where the plant layout exercise is concerned with the replacement or reorganization of a small section of a manufacturing unit (which accounts for the largest proportion of layout jobs), the existing transit routes leading into the section and the available transit methods should be utilized, unless they place unreasonable constraints on the layout. Where a whole department is being studied, transit routes into and out of the region still need to be defined, but routes usually become subject to fewer constraints as the size of the exercise increases. The layout model should include these links with the exterior of the region under consideration, defining the type of transit route, the permissible sizes of entrances, and the constraints on their location. The three main entrance types, in ascending order of size, are those for people, industrial trucks, and equipment.

With each process or work-station model there should be a note of its requirements for services, ancillary equipment, and the associated working environment. This could be a note of an undesirable environment produced by a process or, alternatively, could indicate that a certain quality of environment is required by a process or operative for viable manufacture. These factors indicate locations or groupings for processes, which need to be reconciled with the sequence of manufacture, as specified by the flow process chart, and with the need to be able to remove any machine for major repairs or replacement. In the latter case the probability of removal being necessary must be balanced against the disruption which it causes. Machines which can be expected to operate for long periods before replacement can sensibly be situated where movement of work stations is necessary to create a route for removal. Changes in output capacity and the balance between product types should be anticipated, since these are often on a shorter time scale than machine replacement intervals. The layout should be designed to give the possibility of change without major disruption. This often requires that the efficiency of space utilization be traded off against flexibility. However, it must be recognized that many manufacturing units or departments are inflexible by nature and it is often better to consider a new unit, rather than to attempt extending an existing one.

With the concepts outlined here and the provision of the information discussed in previous sections, the plant layout group can make a layout plan. As stated at the beginning of this section, the layout will reflect the quality of human judgment involved. There are no unique solutions to layout problems, since compromises are always necessary. However, the effectiveness of alternative ways of solving a problem can often be assessed quantitatively. The following sections will deal with methods of layout evaluation.

8.5.2. Physical Evaluation: Models

Scale models have been suggested as being the best tool for designing a layout because the visualization is thoroughly three-dimensional. Consequently, use of scale models will avoid the problems associated with working from drawings, where it is difficult to determine if two objects are trying to occupy the same space. Models are particularly useful for assessing if sufficient (or too much) space has been allocated for the movement of materials, products, and equipment. Scale models of trucks carrying their maximum-sized loads, with the space envelopes required for the appropriate maneuvers, can be "run through" junctions and entrances and through loading and unloading sequences. It is also possible to examine all transit routes for potential hazards, in the form of obstacles and those arising from malfunctions and mistakes. Equipment movement sequences can be rehearsed in the same manner as material and product movements. Scale models of the appropriate lifting and moving equipment will be required.

Continuing with the subject of space for movement, maintenance of a machine *in situ* has two requirements: (1) space for the maintenance engineer to perform the necessary tasks and (2) space to remove and replace subassemblies of the machine, preferably without disruption of adjacent operations. This may involve providing the space to maneuver lifting and transport equipment and a route from the machine to the maintenance workshop.

A model can also provide a better understanding of the working environments provided within a layout scheme than is possible using a drawing. An indication of the amount of natural light reaching a work station can be gained and the best position for artificial lights, to avoid glare or shadow, can be determined. The effect of adjacent work stations can be estimated. Enclosure of a work station can create a claustrophobic and isolated situation while, at the other extreme, a work station at the intersection of a busy transit route may need screening to minimize distraction. The separation of noisy and quiet operations and dirty and clean operations can also be decided.

8.5.3. Quantitative Checklists

Quantitative checklists[1] represent an effort to put the qualitative assessment and refinement of alternative layout plans which occurs at the design stage on a formal and organized basis. They ensure that vital factors in layout performance are not ignored and stimulate precise and analytical thinking. Used in conjunction with models, they can be very effective. A typical quantitative checklist is given in Table 8.1. The groupings and the criteria and characteristics (items) defined within each group cover most of the attributes by which a plant layout design can be assessed. However, additions, deletions, and changes to the chart may be necessary to obtain a list which is appropriate to a specific layout or to reflect the changing criteria of the rubber industry.

The technique has considerable versatility due to the weighting which can be applied to all the items in a group or to each item individually. This weighting specifies the importance of each grouping or item within a grouping in relation to the other groupings or items, and is used in conjunction with the numerical score or rating given to each item. This score is determined from:

Perfect	10
Almost perfect	9
Excellent	8
Very good	7
Good	6
Fair	5
Average	4
Poor	3
Unsatisfactory	2
Unacceptable	1

Alternative scoring methods can be used if they are thought to have a more precise meaning within the region of the layout exercise. By multiplying the weighting by the rating, a weighted rating is obtained for the particular item being considered. Adding up the weighted-rating columns for each alternative layout will give a subtotal score for each grouping on the chart and enable

TABLE 8.1
Plant Layout Evaluation Sheet

Project . . . .   Date   . . . .
Region under investigation   . . . .
. . . .
Objectives   . . . .
. . . .
Evaluated by   . . . .

Criteria and characteristics	Weight	Alternative 1		Alternative 2		Alternative 3		Comments and notes
		Rating	Weight rating	Rating	Weight rating	Rating	Weight rating	
(1) General								
(a) Overall appearance								
(b) Crowded conditions								
(c) Excess or duplicate equipment								
(d) Ease of supervision								
(e) Ease of production control								
(f) Provisions for inspection								
(g) Access for maintenance								
(h) Adequate exits								
(2) Flow of materials								
(a) Planned								
(b) Good equipment arrangement								
(c) Good equipment utilization								
(d) Adequate aisle space								
(e) Straight aisles								
(f) Straight flow lines								
(3) Flexibility to meet changing conditions								
(4) Expandability without major disruption								

(Table continued)

TABLE 8.1. *(continued)*

Criteria and characteristics	Weight	Alternative 1		Alternative 2		Alternative 3		Comments and notes
		Rating	Weight rating	Rating	Weight rating	Rating	Weight rating	
(5) Space utilization								
(a) Fully utilized								
(b) Effective use of available space								
(c) Use of "overhead" space								
(6) Materials handling								
(a) Materials handling planned for: stores to preparation preparation to shaping shaping to fabrication finishing to despatch								
(b) Minimum handling								
(c) Short transits								
(d) Mechanized where practical								
(e) Integrated system								
(f) Use of unit loads								
(g) Provision for scrap handling								
(h) Operations during transit								
(i) Materials used from suppliers' containers								
(j) Effective use of gravity								
(7) Storage arrangments								
(a) Raw materials								
(b) Bought-in components								
(c) Process supplies								
(d) Equipment								
(e) Work in progress								
(f) Finished products								
(8) Goods-inwards and despatch arrangements								
(a) Vehicle parking provision								

(Table continued)

TABLE 8.1. *(continued)*

Criteria and characteristics	Weight	Alternative 1		Alternative 2		Alternative 3		Comments and notes
		Rating	Weight rating	Rating	Weight rating	Rating	Weight rating	
(b) Loading bay arrangements								
(c) Receiving close to first operations								
(d) Provisions for goods-inwards inspection								
(e) Mechanical handling where practical								
(f) Despatch area close to last operations								
(g) Provisions for packing								
(9) Service activities								
(a) Convenience of location								
(b) Adequate provision for:								
first aid								
toilet and wash facilities								
rest and smoking areas								
drinking fountains								
lockers or coat racks								
food service								
employee parking								
process supplies issue								
maintenance								
offices adequate and convenient								
rubbish collection								
fire extinguishers and sprinklers								
(10) Building and utilities								
(a) Adequate size								
(b) Practical shape								

(Table continued)

TABLE 8.1. (continued)

Criteria and characteristics	Weight	Alternative 1		Alternative 2		Alternative 3		Comments and notes
		Rating	Weight rating	Rating	Weight rating	Rating	Weight rating	
(c) Sensible subdivision of space								
(d) Sufficient clear height								
(e) Adequate and convenient entrances and exits								
(f) Window position and size								
(g) Ventilation								
(h) Fume and dust extraction								
(i) Type and position of lighting								
(j) Isolation of noisy machines								
(k) Adequate provision for: water sewage drains gas air electricity telephones								
(11) Energy conservation								
(a) Light and heating control								
(b) Efficient and effective space heating								
(c) Recycling cooling water								
(d) Insulation								
(e) Loading bay door seals								
(f) Self-closing doors								

an overall total to be determined, from which the layouts can be ranked in order of excellence.

The key step in this procedure is the determination of the ratings. These are subject to human judgment. If the first item is given a reference rating of

10, the relative importance of the other items in relation to the first one can be specified by assigning a number of less than or greater than 10. It is useful to place limits on the range of ratings, so that no item can be rendered insignificant or disproportionately important. A cross-check on the ratings produced by this procedure can be obtained by comparing a variety of items with each other to establish if the relative ratings are sensible. It is also desirable that the staff who will be concerned with the use of the proposed layout (and the representatives of the operatives) undertake the rating exercise separately and then reconcile their differences by discussion.

8.5.4. Layout Efficiency Indices

Eleven indices have been developed by Gantz and Pettit,[1] which can be used to assess the relative merits of alternative layout plans. Although these indices can be used to assess alternative solutions to a layout problem, it must be noted that the index values which define the best solution to the problem cannot be transferred as "target values" for another job. The reasons for this will be apparent when the indices are defined.

$$\text{Index of indirect materials handling} = \frac{A}{B} \qquad (8.1)$$

where A is the sum of the distances that a part moves automatically from operation to operation without external materials handling. "External materials handling" refers to the manual movement of materials and products from one location to another in boxes, tote bins, and similar containers or supports. B is the total distance that a part travels from goods-inwards reception to the despatch area. This can be rephrased to read "distance from the entrance of the region of the layout exercise to its exit,' for the purpose of dealing with the smaller jobs which constitute the majority of layout problems. This index has been found to give a consistent and sensitive measure of the relative effectiveness of the automation of materials and product handdling.

$$\text{Index of direct materials handling} = B \qquad (8.2)$$

This value, as defined previously, represents the exact distance an item is required to travel during production. It is not properly an index, but it does give a good measure of the efficiency with which a production route is laid out.

$$\text{Index of gravity utilization} = \frac{D}{E} \qquad (8.3)$$

where D is the sum of the vertical distance that gravity feed is used in a multilevel plant and E is the total vertical distance that an item moves up or down, by either machine or human effort, from goods-inwards reception to the despatch area. This index gives a good indication of the efficiency with which gravity is exploited, although it does give unreasonable answers when applied to single-level plants.

$$\text{Primary index of automatic machine loading} = \frac{F}{100G} \qquad (8.4)$$

where F is the sum of the percentages of machine downtime from all cases where the individual percentages of downtime are equal to or less than 50% of the individual work cycles. "Downtime" is defined as that portion of the total machine cycle used for loading and unloading. G is the total number of operators on these machines. This index should only be used when the machine-time portion of the total work cycle is automatic and machines may be left unattended while in operation. A molding machine is a good example of this type. This index is an accurate indicator of the efficient grouping of machines, so that a number of them can be serviced by a single operator. The ratio can also be improved by automating the loading and unloading sequences. For totally automatic operation, the index can go to infinity.

$$\text{Secondary index of automatic machine loading} = \frac{H}{100G} \qquad (8.5)$$

where H is the sum of the percentages of machine downtime from all cases where the individual percentages of downtime are greater than 50% of the individual work cycles. This index is similar to the previous one, but it is useful because it identifies operations where the loading and discharge times are long compared with the machine times. The split at 50% of the total cycle time is completely arbitrary and may be adjusted to suit a specific layout exercise.

$$\text{Index of production line flexibility} = \frac{J_1}{K_1} \qquad (8.6)$$

where J_1 is the number of machines or work stations performing operations on a product under consideration which can be moved to a new position within the same production line during one working shift. K_1 is the total number of machines or work stations in the production line performing operations on the product under consideration. This index has been successfully used as a measure of machine flexibility in relation to the flow of the product in the production line, although the time of one shift allowed for the change to be made is again arbitrary and may be adjusted.

$$\text{Index of work-station flexibility} = \frac{J_2}{K_2} \qquad (8.7)$$

where J_2 is the number of machines or work stations within the region under consideration which can be moved to any other location within one working shift. K_2 is the total number of machines or work stations within the region under consideration. In addition to the time of one shift being adjustable, it can be argued that the capability for changing a die or a mold from one machine to another constitutes work-station flexibility. In both this index and the previous one flexibility can be defined as "the capability for changing the location at which an operation can be performed;" it does not necessarily involve the movement of equipment.

$$\text{Index of floor-area loading density} = \frac{(M+2)(N+2)+P}{Q-(R+U)} \qquad (8.8)$$

where M is the maximum machine length, N is the maximum machine width, P is the total work area normally required by an operator in the performance of his or her job, Q is the total layout floor area, R is the total aisle area, and U is the total floor area occupied by temporary or controlled storage of work in progress and equipment. This index is an accurate indicator of the efficiency with which plant floor is utilized. The word "machine" refers to all production machinery which occupies floor space. Therefore a floor-standing conveyor qualifies, but an overhead conveyor passing over floor areas which are already utilized does not. It should also be emphasized that the region occupied by machines, work stations, and operators may be totally independent of each other. A region under consideration may include automatic operator-independent machines and work stations at which the operators do not utilize machines.

$$\text{Index of aisle space} = \frac{R}{Q} \qquad (8.9)$$

Using terms defined for the previous index, this gives the proportion of floor space given over to aisles. Alternative layouts can be readily compared by using this index, once the viability of the aisle layouts have been determined from model studies.

$$\text{Index of storage space} = \frac{Q-U}{Q} \qquad (8.10)$$

Again using terms defined for the index of floor-area loading density, this index gives the ratio of floor area given over to manufacturing and handling activites to that for the storage of equipment and work in progress.

$$\text{Index of storage volume utilization} = \frac{V}{W} \qquad (8.11)$$

where V is the volume of the space occupied by materials, subcomponents, supplies, and finished products (not work in progress) in the main storage areas. W is the total volume of the space available for raw materials, components, and finished goods.

The indices defined in Eqs. (8.1)–(8.3) and (8.6) refer to single products, so that the performance of a multiproduct layout will take the form of a distribution of indices, rather than a single figure. The relative performance of alternative layouts can then be assessed by inspection of the distribution. Another factor to be considered for rubber companies is the changing form of the product during manufacture. For the indices in Eqs. (8.1) and (8.2) it is reasonable to assess the handling along the whole manufacturing route, ignoring the product form, to gain an overall index. A more detailed measure of performance may then be obtained by calculating an index for each product form, for example, mixed compound, extrudates, calendered sheet, fabricated products, moldings, etc. A similar difficulty arises for the index of floor-area loading density defined by Eq. (8.8). The dissimilar nature of a mixing section, a fabrication section, and a molding shop may result in the index being dominated by an unrepresentative portion of the region under consideration. Again, a more precise comparison of performance can be gained by calculating an index for each grouping of operations.

8.5.5. Cost-Benefit Analyses

In the previous two sections, methods of comparing the relative effectiveness of alternative layouts have been detailed; but the cost of the alternatives has not been included in the various criteria of excellence which have been examined. Except where layout changes are undertaken solely to comply with new safety legislation, a new layout should lead to improved profitability. This can be assessed by a cost-benefit analysis, which compares the cost of taking a new layout through to a fully operational condition with the improvements expected from it, measured by a reduction in the manufacturing cost per item, an increase in the potential range of products, or an increase in the volume of manufacture. Ideally, plant layout studies should be initiated from a well-planned capital expenditure program. This also leads to the assignment of a sensible target period for the recovery of the investment in the new layout, that is, the new layout will be expected to improve the profits of the company so that it "pays for itself" within a certain time and then go on giving an enhanced profitability for a sensible period, before the nature of the market changes or the equipment and techniques become obsolete.

The potential range of products and volume throughput which may be achieved with the efficient operation of a production unit based on a given layout design will be determined by the type of equipment selected, the number of items of equipment, and the effectiveness of the layout design. In most layout studies the volume of production and the range of products are specified at an early stage as objectives, leaving the cost per item produced as the main quantitative attribute by which the cost effectiveness can be judged.

The total manufacturing cost of a product is the sum of the fixed costs plus the variable costs, which must then be compared with the revenue from the sale of the product. The fixed costs are those that are essentially independent of the number of products manufactured, while the variable costs are those which change in some relation with the number made. These concepts are dealt with in more detail in Chapter 10 and will not be reiterated here; but it is worth itemizing the main sources of the fixed and variable costs.

Fixed Costs

Capital expenditure on equipment distributed over investment recovery period or loan repayment schedule.

Civil engineering costs, including site construction or modification and installation of services, distributed over the investment recovery period.

Municipal taxes on layout region.

Wages and salaries.

Heating and light.

Cost of installing and commissioning the layout.

Training costs and learning period for new layout (lost production), distributed over investment recovery period.

Depreciation of equipment (financial control policy).

Insurance.

Operation of personnel facilities.

Technical services.

Variable Costs

Materials.

Bought-in subcomponents.

Energy used in manufacture.

Interest payable on or depreciation on value of work in progress.

Process supplies.

Maintenance.

Employee bonuses/piecework rates.

Changes in sales price in relation to sales volume.

Level of sales/marketing activity in relation to sales volume.

Cost of inspection, quality-control, and production control activities.

For a single product it is possible to draw a simple break-even chart, with the fixed and variable costs of each layout plotted onto it. With multiproduct lines, which represent the majority of cases, the cost effectiveness of the layout will be influenced strongly by the product mix. It is the responsibility of the production control section to optimize the product mix for plant and labor utilization and for conformance with delivery schedules; but it is ultimately controlled by the order book. Where possible, the cost effectiveness of the alternative layouts should be examined using a number of predicted product mixes.

8.6. INSTALLING AND COMMISSIONING A LAYOUT

8.6.1. Planning the Installation

From the models and plans of the finalized layout, detailed drawings and lists of the production equipment, components, and materials can be produced, accompanied by detailed specifications where necessary. Jobs can be divided off for tender to subcontractors and the hire of specialized equipment and services needed during the installation can be investigated. This adds up to a complex package for organization, even when the layout is of modest size. The situation can be further complicated if an existing layout has to be dismantled. Clearly, a powerful tool for installation planning is required to avoid the whole exercise breaking down into confusion and delay.

Fortunately, the need for a planning tool to deal with complex projects was identified a number of years ago and there are now a number of well-established methods available,[16] all based on network analysis. The names of some of them may well be familiar:

PERT—Program Evaluation Research Task

CPM—Critical-Path Method

PRISM—Program Reliability Information System for Management.

PEP—Program Evaluation Procedure

IMPACT—Integated Management Planning and Control Technique.

SCANS—Scheduling and Control by Automated Network System.

The differences between the approaches arise primarily as a consequence of the original job for which the method was developed. All of them share the concept of a critical path and may be referred to collectively as critical-path methods (CPMs). Three steps are required to provide the information for constructing an activity network:

1. All the elements, jobs, steps, tasks, activities, etc. that are required to complete the installation must be detailed.
2. A sequencing order must be determined which is based on technological and administrative dependencies. In other words all the constraints on the sequence of installation must be identified.
3. The time (and cost) to perform each task, activity, etc. must be estimated, including lead times on equipment and material delivery.

Detail is essential for the success of CPMs. Activities cannot be overlooked without adversely affecting the results. Various estimates are required for each activity, with the result that for normally complex projects a vast amount of information is generated. Fortunately, computer programs have been developed for most of the network systems. These determine the critical path (the longest path through the network, which defines the time needed to complete the installation) and the variance to which the completion date is subject, as a result of the variance of the estimates for the individual jobs, tasks, etc. Thus the probability that the actual completion date will occur at a given time before or after the expected or target completion date can be calculated. The latest possible time at which each activity in the network may be started, in order to meet the completion date target, can also be identified.

The details of the application of critical-path methods have been well documented, so there is no need to reiterate them here. However, there are some aspects and extensions which are well worth mentioning.

CPM is a dynamic method. To work effectively it has to be frequently updated to reflect progress, to enable a report of those activities which are ahead or behind schedule, in addition to any shifts of the critical path, to be obtained at any time, via the computer printer or VDU. This tells the manager where to concentrate his or her energies and resources to ensure that the completion date does not slip. Some of the shifts of the critical path, which may create a "firefighting" attitude, can be anticipated by network simulation prior to the start of installation. Because variance of activity times exists, it is reasonable to expect that occasions will arise where the activity times actually experienced along the critical paths will be shorter than those predicted, causing a shift of the critical path. Simulation allows the user to examine the circumstances under which such shifts may occur. If an installation program is to be spread over a long period, then many important decisions will be made as the project evolves. This requires decision points to be inserted in the network, followed by the subsequent alternative paths that the decision implies.

8.6.2. Implementing the Installation

Many plant layout installations are achieved during a company holiday, when the disruption of production does not have to be considered. Alternatively, where changes to an existing manufacturing unit are required, stockpiling and installation in a number of stages may be necessary to minimize the disruption.

The critical-path methods indicate when an activity should be started, but they do not give detailed directions on the method by which an activity is to be carried out, although some consideration of procedure is necessary to arrive at a sensible estimate of the time required for an activity. In order that a layout installation proceed smoothly, the people involved should be given adequate instructions to enable them to carry out their assigned tasks and be provided with the resources needed for their tasks. The infrequent occurrence of layout installation requires that many people are given tasks outside their routine job experience, although these tasks should not be outside their training and competence. Definite guidance is necessary and can be provided through procedure and check sheets and equipment and location lists. Also labeling of machines, equipment, and materials is necessary to identify them and to specify both their original location and their new location. Where office moves are involved, the occupant should be given the opportunity of supervising the move.

Small- and moderate-sized installations are usually carried out by the company production services staff with contracted aid if large and delicate equipment is to be moved. Large jobs are generally beyond the resources of a production service group and extensive contracted aid is required. Large companies are better able to carry out large installations with their own staff, since a considerable number can be mobilized during a holiday.

8.6.3. Commissioning and Startup

During commissioning and startup the plant layout exercise is both brought to fruition and subjected to its most critical assessment by the people who will be responsible for operating it. During this time some unfavorable reactions can be expected, simply because the layout is new. However, these should be minimized if adequate consultation has taken place during the planning stage and if the introduction of the line management, supervisors, and operators to the new layout is carefully organized. The main objective is to prevent anyone feeling confused and inadequate through lack of guidance; and the larger the installation the more important this becomes.

If new equipment is to be installed which requires new skills of the operators concerned with it, then training prior to the startup is desirable. Prior training is essential for the technical and maintenance staff, whose job it is to set up the equipment for manufacture and to keep it in good running order. Manufacturers of large items of equipment are increasingly selling a "package," which includes the training of the personnel who will be concerned with the equipment. Small items of equipment may be purchased ahead of schedule and installed in a training workshop, to which personnel can be seconded at intervals.

A smooth transition from installation to commissioning (putting the equipment in running order, determining control settings, and stocking process supplies stores) will be achieved if the conditions of the handover are clearly defined. Before the transition from commissioning to production operation takes place, all the changes in the responsibilities of the supervisors and line management should be clearly defined and the implications of these understood by the people concerned. Moving into a new layout imposes a very considerable extra load on production management and particularly on the factory-floor supervisors. Positive measures should be taken to relieve some of this burden by direct assistance during startup.

Where there has been a reorganization of the facilities for personnel, as well as changes in the production unit, or where an entirely new production unit is being started up, the first responsibility to the work force is to show them where things are to be found. This initial introduction will be made easier if a model of the new layout has been on display prior to the move, but in any case a guided tour and large notices are very helpful. It is also helpful if all the personnel facilities are working, particularly for refreshments and a schedule of breaks and relief workers (where needed) has been worked out in advance.

It is unreasonable to expect a target production level from a new layout immediately after startup. Attention has been drawn to this by costing for the "lost production" resulting from familiarization and learning time in Section 8.5.5. The larger the new layout and the greater the degree of change, the longer this process will take. Once a pattern of operation has been established, it is generally necessary to assess if the potential of the new layout is being fully exploited. If it is not, some careful questioning will be necessary to establish why not, so that remedial action can be taken. The job is not finished until the efficient operation of the new layout is self-sustaining.

REFERENCES

1. Apple, J. M., *Plant Layout and Materials Handling*, Ronald Press, New York (1963).
2. Pemberton, A. W. *Plant Layout and Materials Handling*, MacMillan, London (1974).
3. Harris, N. D., and J. R. Mundy (Eds.), *Materials Handling: An Introduction*, H.M.S.O, London (1978).
4. Johnson, S., and G. Ogilvy, *Work Analysis*, Butterworths, London (1972).
5. Brown, R. L., and J. C. Richards, *Principles of Powder Mechanics*, Pergamon, London (1970).
6. Watts, R., Paper given at Plastics and Rubber Institute Conference, "Optimisation of Quality and Productivity in Rubber Processing," Burton on Trent, U.K. (1978).
7. *Factories Acts* (22nd ed.), edited by I. Fife and E. A. Machin, Butterworths, London (1972).
8. *Health and Safety at Work Etc. Act 1974: The Act Outlined*, Health and Safety Commission, London (1975).
9. Dolby, P., and P. Jackman, "Dust Suppressed Materials for Rubber Compounding," RAPRA Members Report No. 25 (1979).
10. *Toxicity and Safe Handling of Rubber Chemicals*, British Rubber Manufacturers' Association, London (1978).
11. Chronos Richardson Co. Ltd., Technical Literature, Nottingham, U.K. (1979).
12. Northern Design Ltd., Technical Literature, Bradford, U.K. (1982).
13. *Safe Working of Calenders*, British Rubber Manufacturers' Association, London (1967).
14. *Threshold Limit Values* (*TLV's*), Guidance Note EH15, Health and Safety Executive, London.
15. Groover, M. P., *Automation, Production Systems and Computer-Aided Manufacturing*, Prentice-Hall, Englewood Cliffs, New Jersey (1980).
16. Starr, M. K., *Production Management: Systems and Synthesis*, Prentice-Hall, Englewood Cliffs, New Jersey (1972).

9

Company Philosophy, Organization, and Strategy

9.1. PHILOSOPHY

The formulation and progressive evolution of a coherent company philosophy underpins all of a company's activities. It is responsible for the attitude which employees are encouraged to adopt towards the manufacture of the company's products and towards customers, suppliers, and each other. For a philosophy to be credible, a company's activities must be seen to be derived consistently from it. In its most basic form a company philosophy is a positive statement of the approach adopted towards the challenge of staying in business. Two very simple and different philosophies are:

1. We will operate from a high technical expertise, giving us an outstanding ability to take on high value and personally rewarding work.
2. We will operate conventional, well-established methods in an extremely cost-effective manner, enabling us to offer our products at a lower price than our competitors.

Expanding from the basic philosophy of the company's business platform, there are a number of areas in which a clear statement of company policy is necessary:

1. The company's responsibilities to its beneficiaries.
2. The company's expectations of its employees.
3. The company's attitudes and responsibilities to its employees.
4. The company's policies and attitudes in its dealings with suppliers and customers.
5. The nature and direction of company development.
6. The company's responsibility for the environment in which it is located.

7. The company's responsibilities for the local communities, particularly those from which it draws its employees.

Additional major items include policies on unions and union membership, and relations with governmental and legislative bodies.

Operating a company within the context of a well-formulated philosophy provides a basis for consistency and reliability in business and yields that important attribute—"a good reputation." The lack of a coherent philosophy results in expediency becoming the keynote of company activities, which makes planning very difficult and leads to unpredictable and unreliable dealings, both internally with employees and externally with suppliers and customers. This arises from individuals or groups within a company evolving their own policies for company activities, which may be in conflict with their own interests, those of their colleagues, and those of the company.

The nucleus of the company philosophy is usually dictated by the beneficiaries of a company's activities, tempered by legislative and social constraints. It is then amplified into a practical manifesto, from which a business plan may be formulated, by the executive management, headed by the chief executive.

Publication of a summary of company philosophy can establish a consistent base from which dealings with customers, suppliers, and employees can proceed. It may also have the effect of enhancing the business reputation of the company, provided that the philosophy is praiseworthy and the company's activities are seen to be consistently within the stated philosophy.

9.2. COMPANY ORGANIZATION

9.2.1. The Company's Beneficiaries

A company exists primarily to benefit the owners or shareholders through the making of profit. To do so, it also has to provide adequate benefits for its employees, for them to continue to work for it in a constructive and reliable manner. This in turn benefits the communities from which employees are drawn. In addition, many companies include the support of community projects and activities in their philosophy. However, the shareholders or owners and the employees are the beneficiaries whose actions exert the greatest influence on company activities and viability, and this section will be mainly concerned with them.

If a company does not consistently provide a level of benefit which could reasonably be expected, then the beneficiaries generally withdraw their support. The disadvantage of this type of action is that it merely draws attention to a management failure after it has occurred. To avoid such a catastrophe, the beneficiaries should be organized sufficiently well to state accurately and realistically what benefits they expect from their company and to monitor company performance sufficiently closely to be able to demand remedial action when it becomes necessary.

The role of the beneficiaries in company activities requires beneficiaries' organizations to complement and interact with the management organization. Operatives usually belong to a union and can make their collective views known through their elected representatives. Office and junior management staff are increasingly being represented by unions, as the need for collective action has grown.

The beneficiaries of the profits of a public company are the shareholders, to whom the chief executive is directly responsible. Since the shareholders are often a large group of people, they generally need representatives who will act in their interests and convey their views at meetings of the board of directors. In theory, all the company directors are elected by shareholders, but this is practically very difficult when the shares are distributed among a large number of people who have little knowledge of the business. In practice a board of directors is normally a self-appointed, self-perpetuating body, often drawing new members from the senior management of the company. The resulting weak representation of the shareholders, shown in Figure 9.1, is undesirable because their interests are neglected to such an extent that they only have two courses of action—to continue to support the company or to withdraw support, with the latter often leading to the company's demise. The structure outlined in Figure 9.2 provides a good measure of involvement for shareholders, which is likely to result in their being more willing to support the company through difficult times.[1] The shareholders have elected representatives on the board of directors and also two nonexecutive advisors—the auditor and the monitor—who report back to the shareholders. The auditor's role is expanded from that of annually reviewing the company accounts to performing a similar function at more regular intervals. The monitor's task is to assess the competence and efficiency of the company in achieving the targets set by the board of directors.

Private companies, which do not have shareholders, range from family firms to wholly-owned subsidiaries of multinational corporations. In the family firm the chief executive is often the chief beneficiary, and by filling both roles provides the simplest and most direct form of representation. Alternatively, in companies run by an appointed chief executive, where the number of beneficiaries is small, the structure of the beneficiaries'

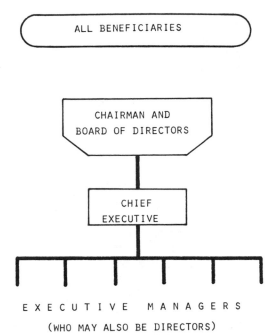

FIGURE 9.1. Indirect representation of beneficiaries.[1]

organization is also very simple and involment in the company is usually direct, through a seat on the board of directors.

The chief executive of the wholly-owned subsidiary usually reports to a small group of professional people who act as representatives of the holding or parent company. These people, through their position and expertise, are able to assign performance targets and monitor achievement against those targets far more effectively than shareholders. In this context the formation of a board of directors from the senior management of the company is quite reasonable; but it should be noted that the function of the board is now quite different, being a management committee and not a forum in which negotiations between beneficiaries and management take place. Their discussions usually center around ways and means of achieving the targets agreed upon between the chief executive and the parent company representatives.

9.2.2. A Basic Management System

A manager may simply be defined as a person who has both a boss and one or more subordinates.[1] Examined in isolation, the basic management

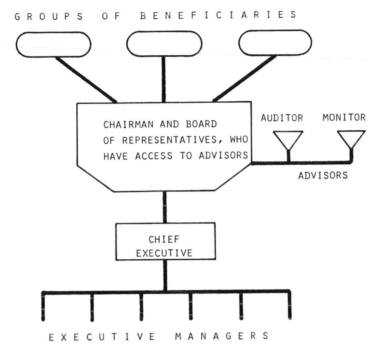

FIGURE 9.2. Group representation of beneficiaries on Board with access to advisors.[1]

process is defined by Figure 9.3. To fit this process into the hierarchical structure of a company, some modifications and communication routes are needed. These are shown in Figure 9.4.

When a task is received from the boss the manager decides how it should be done and then issues instructions to subordinates, who must themselves decide how to carry out the instructions. Taking the simplest case, where only one subordinate is involved, the manager can find from the subordinate's results that:

(A) The task has been completed successfully.
(B) The task has only been partially completed.
(C) The task has not been successfully attempted.

If the subordinate has completed the task successfully the manager must then check (A) that it also enables him or her to complete the task set by his or her boss. The box labeled "check manager's results" in Figure 9.4 is obviously far more important when the task set by the boss has been divided among a number of subordinates and needs synthesis or collation to

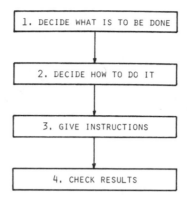

FIGURE 9.3. The four steps of the management process.[1]

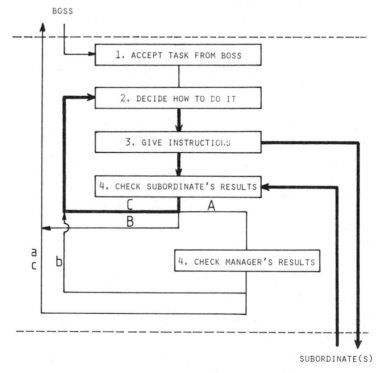

FIGURE 9.4. The management process for one manager.[1]

determine the manager's result. Even so, the manager's result depends on the judgment shown in deciding what should be done. If incorrect instructions have been issued to the subordinate, the manager will have to report failure to the boss (c). When sound judgment is displayed in issuing instructions, which are then successfully completed, the manager can report an unqualified success to the boss (a). Often a manager will find that the results produced by a subordinate are not entirely satisfactory, but that there is sufficient time to do something to improve them (b), involving a return to box 2 and progress through the sequence from that point. If the subordinate has completely failed to carry out the manager's instructions and there is no time to retrieve the situation, the manager has no alternative but to report the failure to the boss (c).

A person who has subordinates but does not have a boss is not a manager but a beneficiary. At the other end of the organization, a person who has a boss but does not have subordinates is an operator. Using this terminology any manager can temporarily become an operator by deciding to carry out a task personally rather than delegating it to subordinates. This should not be taken to imply that a manager is wrong to act as an operator; the contrary is true if the manager is the best person to carry out the particular task. However, assuming that managers only manage, we can now draw a complete management process chart (Figure 9.5).

In practice a manager will generally have more than one subordinate and be involved in many concurrent tasks, which can be of a continuing or project type. Continuing tasks are those for which there is no end point in mind, such as the management of a production department, whereas projects must have a definitive end point and time scale, such as the development of a new product. Figure 9.5 could be expanded to take all these complications into account; but it would be unintelligible. It is useful to bear in mind that the manager's job is a very complex one.

9.2.3. A Management System in Action

The way in which the four stages of the management process identified in the previous section are carried out must satisfy the needs and aspirations of the people concerned, otherwise the resulting frustration will cause a reduction in the quantity and quality of useful activity.

Starting with the giving of instructions, the increasing use of computer methods has accelerated an already rapid trend towards the quantification of objectives. While this is essential to give a subordinate a clear statement of the performance levels at which he or she will be considered to have succeeded or failed in the task, it does appear to run directly contrary to the growing demands by well-educated people to be given an opportunity to

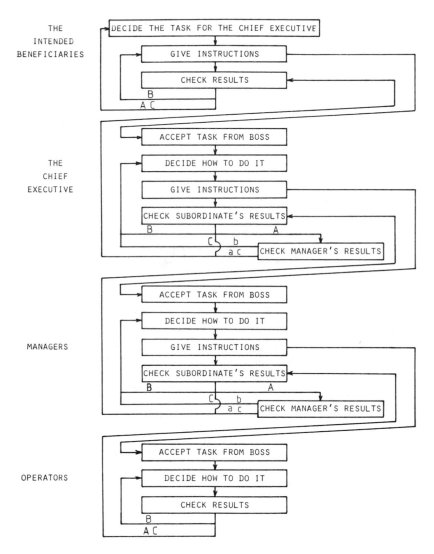

FIGURE 9.5. A complete management system for two levels of management.[1]

exercise their initiative, expertise, and judgment. This conflict can be resolved by quantifying the objectives of a task but not placing unnecessary constraints on the methods by which they are achieved.[1] This requires a positive effort on the part of a manager to ensure that his or her idea of the way to carry out the task he or she is setting is not implicit in instructions to a subordinate. In fact, managers at every level in the company hierarchy will have to become more careful when accepting tasks from their bosses in the future. This is true from the chief executive, who must ensure that the beneficiaries have properly briefed him or her, right down to the junior supervisor or foreman, who needs to be told what results must be achieved but does not want to be told how to achieve them. It is all too easy for a boss to unwittingly frustrate the initiative of subordinates by setting constricting conditions on the way in which a job can be carried out.

Deciding how to carry out a task is really identical to what is often referred to as "management decision making." In fact, most of this book is dedicated to providing a body of knowledge and techniques which will aid the production or technical manager in making decisions. The major trend at this stage is towards an increase in the complexity and number of the alternative courses of action which have to be sorted through before instructions can be issued to subordinates. This trend has been accompanied by a proportional increase in the amount of thinking time necessary for deciding how a task should be carried out. There are now a significant number of people in any organization who spend all or most of their time discussing how to carry out tasks and preparing plans; furthermore, most of these people are either senior managers or highly skilled advisors, and are therefore highly paid. The volume of planning, the number of interacting decisions to be taken, the number of people to be consulted—all have increased.

The importance of the "deciding how to do it" stage has resulted in its subdivision into a number of steps[1]:

(a) Deciding the time and resources which should be committed to each decision.
(b) A positive and imaginative search for alternative ways of carrying out the task.
(c) Evaluation of alternatives and selection of *the* method of carrying out the task.
(d) Preparation of detailed action plans from which instructions for subordinates can be formulated.

Step (a) emphasizes that the resources used in decision making are substantial and must be carefully allocated in proportion to the importance of the task. It also implies that a manager should not try to determine the best method of carrying out a task, but choose the first one which convinces him or her that it is not worth searching for a better one.

Step (b) dominates the decision-making process. The importance of the search for alternatives when dealing with a complex task cannot be overemphasized. When unrelenting commercial pressures have forced companies to increase the efficiency of their existing activities to a maximum in order to survive, the only route available for improvement is innovation.

There is plenty of evidence that many managers have not yet appreciated the importance of committing adequate time and resources to a thorough search for alternatives. This failure to think far enough ahead often results in time forcing a bad decision from an unwilling manager.

A number of aids are available to the manager searching for alternative methods of carrying out the task set by the boss. These include:

> Nonsystematic methods.
> Logical methods.
> Mathematical models.
> Surveys.
> Combined methods.

Nonsystematic methods of generating alternatives include inspiration, intuition, induction, and creativity—in short, having bright ideas. This is certainly the most venerable method of listing alternatives and is probably the most important. It is also the least understood, although there is now a growing body of literature.[2] Creativity has already been identified as a very important attribute in a manager; and it can be reasonably argued that one truly original idea is worth more than a whole list of alternatives generated by the derivative methods which follow.

Logical methods provide a means of working on the terms of reference of a task set by the boss to generate alternatives. For example, a chief executive is concerned about the wastage cost implicit in the difference between the volume of material issued from stores and the volume of material included in saleable products. The chief executive then instructs the works manager to "reduce the cost of material wasted during production by $x\%$ over a period of y months." From this statement "cost" and "material waste" emerge, giving two immediate alternatives:

1. Reduce wastage.
2. Do not reduce wastage.

Expanding on (1) we could obviously explore better control of internal mixing, storage of mixed batches, use of waste-less molding, etc. However, in the context of this exercise it is more interesting to explore the alternatives associated with not reducing waste, such as eliminating high-cost materials, recycling of scrap, etc. These can be built up into a logical "tree" of alternatives of the type shown in Figure 9.6, by performing the various operations of logic on each statement in the three to produce further alternatives. Obviously, missing an important alternative at an early stage can eliminate a

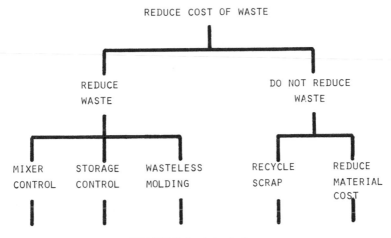

FIGURE 9.6. A logical tree.

whole hierarchy of alternatives; so this is not a substitute for creativity but should be treated as an aid to it.

Operations research methods[3-5] involve setting up a mathematical model of a system, to enable the consequences of a large number of permutations or alternatives to be evaluated in a reasonably short time. Simulation of this type invariably involves the use of a computer, pointing to program development being a major cost and time element, although some commercial program packages are now available for common problem types. Although the main purpose of simulation models is to provide methods of evaluating alternative courses of action for complex and high-cost problems, prediction and optimizing models do actually generate alternatives, once the problem has been formulated in a precise and quantitative form.

Predictive models are essentially forecasting models. A manager faced with a task having a completion date of a year or more ahead will need to produce alternatives which are appropriate to the circumstances at the time of completion. Basing future actions on present conditions is an act of incompetence, which increases in magnitude as the rate of change of conditions and the lead time between deciding upon and completing a course of action increase. Optimizing models, as the name suggests, can produce a "best" solution to a problem. However, the best solution does depend on the validity of the model for the system which is being modeled and upon the criteria of optimality. Despite their exciting potential, optimizing models depend on a deep understanding of their underlying assumptions to avoid inappropriate solutions being used.

Conducting a survey is a well-established method of obtaining information and ideas from external sources. Most major tasks benefit from a literature search, which can include technical journals, trade papers, manufacturers' data sheets, patents, textbooks, etc. Fortunately, the proliferation of information has been accompanied by the setting up of computer data bases, with sophisticated search methods. This makes obtaining relevant literature a relatively quick and simple task, leaving more time for the important job of putting it to use. Collecting ideas and opinions from other people is a very effective way of increasing the list of alternatives. A manager can talk to subordinates, the boss, and to specialists and advisors, both inside and outside the organization. In fact, it is becoming increasingly difficult for one unaided person to draw up a list of alternatives likely to contain one that is better than existing methods.

Combining two or more of the preceding techniques can give the effect of synergism, providing more innovative alternatives than the techniques used either singly or sequentially. One well-known combination is "brainstorming," which consists of a combination of a survey and nonsystematic methods. A small group of people having been previously given a task, meet in an intensive discussion session to appraise fully formed ideas and, more importantly, to join together partially formed ideas into new and viable alternatives. Another combined method is known as Delphi, which is a method of forecasting change, combining survey and predictive methods.

When a manager is satisfied that the list of alternatives is adequate, bearing in mind the time limits imposed by step (a), the manager is faced with the job of eliminating all but one. A systematic procedure known as "the sieve,"[1] has been proposed for this purpose, which has much in common with value analysis. It involves asking six questions about each alternative:

1. Might it achieve the target?
2. Does it require resources beyond those allocated?
3. Is it in line with the organization's ethics?
4. Is the organization well equipped to use it?
5. What trends or events might help or hinder its success?
6. What might go wrong and what would happen then?

This sieve can be very coarse when starting the elimination exercise, to remove rapidly those alternatives which display substantial defects. By adding more detail to the questions on each successive pass, only those alternatives worthy of time-consuming and detailed attention actually receive it. When the final selection is made it will have already become a fairly detailed plan, as a result of the attention it has received during its successive passes through the sieve. Consequently, step (d), is reduced.

Step (d) now involves adding whatever detail the manager thinks is necessary to ensure that subordinates do not misunderstand what he or she wants each of them to do. This defines a plan as "a set of carefully considered instructions for further action."

The role of an instruction is to ensure that subordinates do what is required of them by their boss, requiring that the format of the instructions direct the subordinates only to engage in activities which contribute directly towards the completion of the manager's task. "Slack" occurs when anyone engages upon activities which are to some extent irrelevant to the purpose of their organization. It is not desirable or practical to eliminate slack; indeed, human nature can make any gross attempts to do so entirely counter-productive. However, it can be reduced to proportions which do not endanger the viability of a company if five basic elements of instruction are considered[1]:

1. Define the task—the manager must describe clearly, preferably with figures, exactly what result he or she wants the subordinate to achieve and by when.
2. Allocate resources—the manager must ask a subordinate what resources the subordinate needs to achieve the above results.
3. Describe the method—the manager must discuss how the resources are to be used and how they are not to be used.
4. State checkpoints—the manager must tell the subordinate when he or she intends to check the subordinate's result or progress towards it.
5. Rewards and penalties—the manager must explain the consequences of failure and the rewards of success.

"Defining the task" has already been dealt with in some detail but it is worth examining its importance in the context of the hierarchy of instructions, which should serve to ensure that all the company's activities contribute to the chief executive's task. As one moves up the hierarchy towards the chief executive, the consequences of inadequately defined or incorrect instructions become more serious and affect a greater proportion of a company's activities. Also, the senior managers' tasks tend to be on long time scales, so that the consequences may not become apparent until substantial resources have been committed to inappropriate tasks. It is essential that the chief executive obtains from the company's beneficiaries precise details of the nature and volume of benefit expected, and that each manager ensures that the task set by his or her boss is defined adequately and quantified before proceeding to act upon it. The nature of the targets set will depend on the type of task. Continuing tasks, such as working within a budgetary allowance, need to be defined as performance to be achieved progressively or continuously over a specified period of time, whereas project-type tasks, such as the production of a prototype product, will have a definitive target and a time by which it should be achieved.

A main consideration in the allocation of resources is not to constrain the means of carrying out a task by going into unnecessary detail. At the upper end of the company hierarchy overall resources can be specified in monetary terms, leaving all the detail to the subordinates. As one moves down the hierarchy one finds managers stating resources in terms other than money, simply because the subordinates lack the knowledge and skill to cost resources in money terms.

For a manager to describe the method by which subordinates should carry out a task is to run counter to all the arguments for providing opportunities for the use of initiative and the gaining of job satisfaction. This step should be minimized as far as possible, bearing in mind the capabilities of the individual subordinates. The only area where this step may increase is in vetoing methods which compromise company philosophy and ethics.

Reducing the amount of detail contained in instructions to subordinates places more emphasis on checking their progress towards a defined objective. It is relatively easy to establish checkpoints, which are stated times when a subordinate should report to the manager; it is less easy for the manager to establish what constitutes success or failure at each checkpoint. Again, measures of success or failure should relate to the overall objectives of a task and not limit the methods of carrying it out by implicitly including assumptions about the methods to be used.

In the interests of giving subordinates a wide discretion in carrying out their tasks, a manager should only check their work when significant progress is expected to have been made and at points where corrective action by the manager will be effective. Significant progress carries the statistical meaning of significance, which is important when measuring the progress of continuing tasks. Taking again the example of material waste, day-to-day fluctuations should not be confused with a trend. Simple statistics will indicate when significance will be achieved and there is obviously no point in checking a subordinate's work before that time. However, a manager is quite justified in taking corrective action if no significant results are available at that time. Since a subordinate's task contributes to the manager's task, and onwards up the hierarchy, checking a subordinate's progress in time to take corrective action is necessary, so that it does not jeopardize the manager's task in the event of a failure being reported.

When a manager checks results, he or she is primarily measuring his or her confidence of being able to achieve the target. For short-term tasks this confidence will depend on a subordinate's progress with tasks and the manager's own progress with a task. For long-term tasks, where past results are a poor guide to the long-term future, the manager's confidence depends far more on the continuing validity of the assumptions by which a single plan was selected from the list of alternatives. Depending on the manager's level of confidence of being able to achieve the target, the manager will either do nothing, take action to increase the chances of achieving the target or, if confidence is low, report his or her impending failure to the boss.

In this section a system of obtaining results has been outlined which contributes to job satisfaction, while ensuring that activities which do not contribute to company objectives are kept to a reasonable level. In the following two chapters methods of setting standards against which performance can be judged will be dealt with.

9.2.4. Company Structure

The hierarchy or management structure of a company is a dynamic network of communications and responsibility. Attempts to display this by static charts rarely succeed in showing how an organization really works. The main purpose of a management structure is to provide a framework which directs the activities within a company towards fulfilling the chief executive's task, which should be synonymous with the purpose of the company. The structure must therefore ensure that the chief executive's task can be delegated and subdelegated without gross distortions of instructions or conflicts of interest and responsibility occurring, even though there will be an increasing number of people involved at each stage of delegation.

The chief executive's role as the focal point between the beneficiaries and the management is shown in Figure 9.7, which also establishes the form of most company hierarchies. The pyramid or family tree is a good structure for direct production management; but it becomes unwieldy if service and advisory personnel are included in the "line of command" from the chief executive to the operators. There is also a trade-off between the number of people who report to each manager and the number of levels in the structure. It is desirable to minimize the number of levels in an organization, not only to reduce management costs, but to reduce the slack which results from extended communication lines. This trend is limited by the number of subordinates for whom a manager can assign tasks effectively and check results; and the removal of service and advisory personnel from the production management line of command can aid it. Normally a manager can direct and monitor effectively the activities of four or five subordinate managers.

In the past the high cost of the routine clerical work and recordkeeping associated with in-company transactions has reinforced the adherence to pyramidical organizational structures, by dictating that administration should be centralized and that the majority of "official" transactions should take place along the line of command. The costs of executing in-company

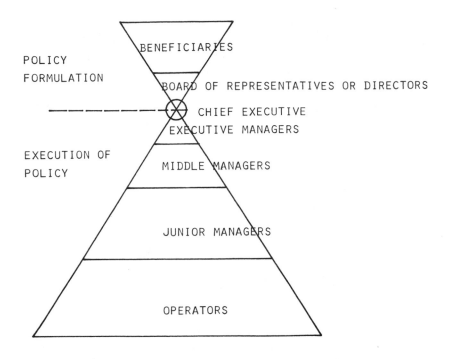

POLICY
FORMULATION

EXECUTION OF
POLICY

FIGURE 9.7. A basic company hierarchy.

transactions and recording them for accounting and other purposes have
been so reduced by the advent of the personal computer and the computer
network that management structures are no longer restricted by the need to
minimize them. This has resulted in the rapid rise of the business center or
profit center as the basic management unit in many companies. A business
center is a semi-autonomous unit which trades with other business centers
within a company and with external organizations, allowing the sound principle of giving managers the greatest possible freedom in carrying out their
tasks to work effectively in practice. A further benefit of business centers is
the reduction of the supervisory burden on senior management, enabling
each senior manager to have more subordinates reporting to him or her, with
a consequent reduction of the number of levels in the hierarchy.

At this stage it becomes necessary to distinguish between administrative
and functional management structures. The administrative structure is the
one which normally appears on management structure charts. It is a
command structure which enables the chief executive to assign tasks and

receive reports of results. It also supports the management seniority and salary structure, office accommodation arrangements, secretarial assistance, and management review procedures.

In Figure 9.8 the administrative structure of a medium-size company, which operates a business center system, is shown. The chief executive has six people reporting directly to him or her, of whom the works manager is generally senior, due to the responsibilities of the position. The works manager could alternatively be called senior executive manager, deputy chief executive, or some other title which indicates this. The command structure has a maximum of five levels, from the chief executive to the manufacturing group operators. In this structure job titles can be readily identified, whereas it is easier to specify functions in the service areas, indicating that staff responsibilities are more difficult to define here and depend strongly on the nature of the company.

The anomalous position of quality control in the administrative structure in Figure 9.8 is dictated by influences originating outside the company. A number of large customers, particularly government departments, demand that the quality manager reports directly to the chief executive, in the belief that this will prevent quality standards being influenced by the pressure on production groups to achieve volume targets. It can be argued that quality can only be controlled at the manufacturing operations, with the function of the quality-control group being to provide the means by which quality can be monitored and controlled *by the production groups*. However, when substantial contracts include the direct reporting of the quality-control manager to the chief executive in their terms, it makes sound commercial sense to comply.

It is the functional structure of the management system in Figure 9.8 which is of the greatest interest, with respect to the trading which occurs between business centers. In addition to each of the manufacturing groups being a business center, the service groups are also business centers, so that the manager of each acts as a business manager. The service groups are separated into those which have an ongoing liaison with the product groups and those which are engaged on specific jobs or projects. The former are included in the domain of the works manager, while the latter report directly to the chief executive. This gives the works manager overall authority for production organization, without any divisive separation of interests occurring.

Whereas the administrative structure often remains stable over a long period of time, the functional relationships between business centers undergo rapid and continual changes. This is true of all companies, and accounting techniques which allow in-company services, such as maintenance, to be "paid for" are well established. The business center concept involves an

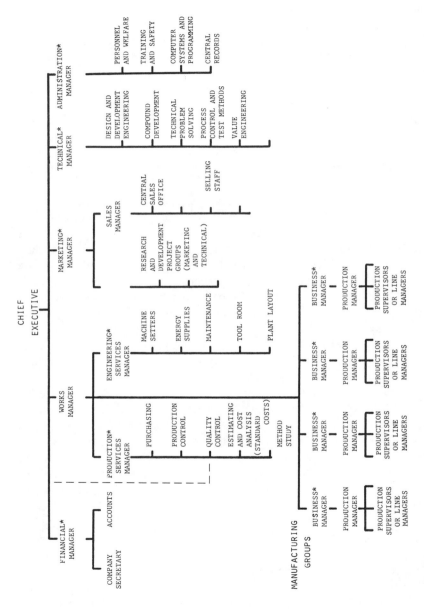

FIGURE 9.8. A management structure based on the business center concept. Business centers are denoted by asterisks.

extension of this accountability and provides a framework within which a complex network of cooperation can function effectively.

The dynamics of the functional network can best be understood by examining the ways in which it functions. These take three main forms:

1. Continuing services.
2. Specific short-term jobs.
3. Medium- and long-term projects.

Continuing services are mainly provided by the purchasing and production control groups in the production services business center and by the maintenance and energy supply (electricity, steam, water, compressed air) groups in the engineering services business center. The cost of these services is arrived at in a similar manner to the cost of engaging services from organizations external to the company. Taking the example of maintenance, "contracts" could be agreed upon for each work station or group of work stations, subject to annual revision, to include both planned maintenance and breakdown services.

Specific short-term jobs, such as changing the mold of an injection molding machine and setting up, so that it can be handed back to the production business center in working order, can be invoiced on the basis of time and resources committed to them. Establishing quality-control procedures for a job and conducting method studies also fall into this category, as do the services of the business center headed by the technical manager.

Projects may be undertaken for the manufacturing groups by any of the service groups; but the research and development project groups operate, as the name suggest, wholly on a project basis. These groups provide the means by which the manufacturing group business managers ensure a continuity of jobs for their operation. However, the introduction of a new product is generally a complex operation which involves a number of groups in a cooperative effort, and will be dealt with in a later section. In all cases the cost of project work has to be met by the manufacturing group for which it is being undertaken. This ensures that projects are only undertaken when they can be seen to have well-defined objectives, programs of work and benefits, against which their costs can be balanced. Projects can be initiated in both manufacturing and service groups. In the former case a business manager will identify an aspect of his or her operation where he or she believes efficiency or profitability can be improved, and will call in the relevant service group(s) to prepare a proposal and a quotation. Alternatively, the head of a service group or the manager of a service business center will identify an aspect of the operation of a manufacturing business center where cost-effective changes are possible. From here à project proposal can be prepared and "sold" to the manager of the business center concerned.

The most important feature of a company organized into business centers is its self-regulating character. Each business center must have an income which is sufficient to pay its operating costs and provide for the ongoing development of its capabilities, militating against the overmanned or overcapitalized group which is a drain on a company's resources rather than an asset. Through the self-regulating character much of the detail of the chief executive's and works manager's tasks, which is essential in a conventionally structured company, can be eliminated, although it must be noted that the command structure still exists and must be used for assigning specific tasks and checking progress. In this way coherence of the company's activities is maintained and the chief executive's task is delegated to the business managers. It is then the responsibility of both the chief executive and the works manager to delegate their tasks in a manner which does not limit the means of carrying them out and does not distort or damage the cost effectiveness of business center activities. This is generally achieved through a business plan, which will be discussed in Chapter 10.

The administrative structure shown in Figure 9.8 could be operated in a conventional manner, without designating any of the departments as business centers, although the functional structure would then be quite different. Trading between departments in monetary terms would be much reduced, resulting in the budget becoming the primary means of ensuring that departmental operating and development costs are appropriate to its contribution to the company's profits. Alternatively, the manufacturing business centers could be made considerably more self-contained than is shown in Figure 9.8, by giving each its own purchasing, production control, and engineering services. Sales facilities could also be attached to those business centers having a saleable product. In fact, if the mixing and compounding business center is allowed to offer a custom compounding service to external organizations and the other business centers are organized into product groups, every business center will need sales facilities. This trend towards autonomy can lead to a logical conclusion in which each business center is a self-contained subsidiary or division of a holding company, comprising only of the chief executive and the administrative group needed to monitor and coordinate the activites of the subsidiaries. This is a common structure for large companies. The level at which subsidiary companies are created and their size depends strongly on financial and legal considerations, which are beyond the scope of this book. However, subsidiary companies can be organized into business centers, which enables a viable and practical functional structure to be created within the majority of administrative structures. It is difficult to define any rules for setting up or modifying a company structure which are more specific than the guidelines laid out in this section. Each case will raise its own problems and require its own individual solutions, although a good concluding remark is, "a sound management structure will include self-regulating features."

9.2.5. Working Patterns

The term "working patterns" is used to describe the way in which working hours are distributed throughout the natural time periods of days, weeks, and years. In recent years people have become far more flexible in their approach to working patterns, provided that they do not feel that they are being disadvantaged for their company's benefit. Since the patterns of working can have a considerable influence on profitability, it is worth examining mutually beneficial working patterns.

From the viewpoint of company profitability, the additional payments to people working unsociable hours have to be balanced against the extra returns on capital investment which come from maximizing the utilization of manufacturing facilities. In simplistic terms 24-hour utilization, compared with utilization over the standard 8-hour working day, can either give three times the productivity, or reduce the capital equipment expenditure, for a production equivalent to that for the 8-hour day, by 67%. It is clear that the gains will be more substantial in capital intensive areas of manufacture than in those which are labor intensive, although in both cases rent, building loan repayments, depreciation, and taxes levied by local government can be substantial and need to be taken into consideration.

The problems associated with working unsociable hours are the disruption of personal life and being unavailable at certain times for essential family duties. If these problems are not circumvented in setting working patterns, a large proportion of the people a company might wish to employ will be unavailable to it, even if an enhanced remuneration is offered.

Traditionally, capital intensive areas of manufacture have operated a 24-hour day, three-shift system, each shift being 8 hours, over a 5 1/2-day week, whereas labor-intensive areas have dispensed with the night shift, which carries the heaviest unsociable working hours payment premium, to give a two-shift system. Staff have generally worked "office hours," usually 8.30–17.00 hours in a 5-day week, although some sections of the industry still maintain a 5 1/2-day week. The main exception to this is the direct management, who have to conform to the operator's working pattern. This traditional pattern is based on largely manual machine operation, requiring strenuous work, even in the capital-intensive areas and depends heavily on operators with children being able to find someone to be responsible for their children's well-being while they are at work. In the past this has been the job of the housewife, who would often never take a paid job; but this is usually no longer economically possible and is certainly not socially desirable. The provision of crèche systems, etc. is beyond the scope of this book; but the social implications of working hours cannot be overlooked by the production manager because of their influence on the availability of people for jobs and their work attitudes.

Many permutations of working hours are possible, but one of the most important developments of recent years has been the introduction of flexible working hours, usually called "flexitime." In this system a person has to work certain core times but is free, within reasonable constraints, to plan when they will work the additional hours which are needed to make up the full working week. Flexitime has been used mainly for office-type jobs where the overhead costs are relatively low and the 7- or 7 1/2 hour working day has been the previous practice. A typical core time would be 10.00–15.00 hours, with the flexing period being from 7.00 to 20.00 hours. Some production jobs could be flexed but the overhead costs associated with many manufacturing operations militate against this approach. Fixed working patterns will always be necessary, but for a diminishing number of people, as the level of automation in manufacture increases.

Looking at the patterns of working as a whole, it can be seen that the use of flexible working hours wherever possible benefits the manufacturer, as well as the employee. Giving couples with children room for maneuver in the working hours of one of them can free the other to do shift work, enabling a company to recruit from a greater number of people. This also diminishes the pressure on a company to create "office hours working" in as many areas as possible and enables it to spread skills and abilities more uniformly over the full manufacturing period.

9.3. MARKET RESEARCH AND COMPANY DEVELOPMENT

9.3.1. Identification of Markets

The identification of markets is closely linked with company philosophy, since a company will need to operate in a manner which will enable it to compete effectively in selected market areas. The progressive rise in the technological requirements imposed on the rubber product manufacturer by customers and the large differences in these requirements which occur in separate market areas, has made the identification of viable markets for company expertise and manufacturing capability an extremely important and critical exercise.

Market research methods are well established and documented.[6,7]

There are also numerous market research companies, whose expertise can be called upon for unfamiliar market types and geographical regions. In this context it is not necessary to deal with the methods of conducting a market research exercise; but it is valuable to look at the considerations involved in establishing the terms of reference of an exercise. A company can establish, change, or expand its market share by developing any of the following:

1. Existing products in existing markets.
2. Existing products in new markets.
3. New products in existing markets.
4. New products in new markets.

Alternatively, expertise or manufacturing capability could be substituted for products in the preceding four areas. In all cases the market research must be directed towards identifying those markets in which the company would wish to become involved. Obviously, market size and potential profit play a large part in the decision to investigate or enter a market; but due consideration must also be given to the compatibility with the philosophy, aspirations, and expertise of the company. Entering a market which requires a radically different philosophy and work attitude can be extremely disruptive. However, if a market is large enough it may warrant setting up a separate production unit in an area where the necessary skills are available.

Markets can also be separated into two groups: those which involve jobbing and those which involve own product manufacture. Jobbing companies manufacture only on receipt of an order and to the specific requirements of a customer. Most of the companies in the rubber industry fall into this category. The main exceptions are the tire companies, who supply the retail replacement trade from stock, and need a much more sophisticated marketing, sales, and advertising organization than the jobbing companies. In essence, jobbing involves manufacturing to a known requirement, while own product manufacture involves producing to a forecast requirement; the former requires a continuity of orders, while the latter requires a continuity of sales.

In times of rapid industrial, social, and economic change a watching brief on current and potential markets and the performance of competitors is essential to enable a company to forecast and react to changes and opportunities in a controlled manner. This broad scan approach can be expanded into a detailed market research exercise when and where it is indicated that the cost of setting up and maintaining a project team is justified by the potential benefits.

Identifying markets for new materials, manufacturing methods, and products and also responding quickly to market opportunities requires expertise in a broad range of areas and an entrepreneural outlook. In

Section 9.2.4, which is concerned with company structures, it was suggested that running a combined market and technical research operation was one way of achieving this objective.

9.3.2. Establishing and Maintaining a Viable Product Range

The overhead costs of a company are closely related to the range of products which it manufactures. A company with only one product can operate at a very high efficiency in terms of utilization of equipment and minimization of manufacturing support services, such as technical, production control, and marketing groups; but it will be extremely vulnerable to changes in demand for its product. A company which supplies a range of products to a single customer, or even to a group of customers in a single market area, also suffers from a similar vulnerability. It is a sound business policy to manufacture a range of products for a number of unrelated markets, although the decrease in manufacturing efficiency which inevitably accompanies an increase in product diversity must be balanced carefully against the gains in security.

Limiting the diversity of machine type, size, and manufacturer is also important, because of the maintenance and spares burdens. It is preferable to produce a range of products from a small number of machine types and sizes. For example, a range of extrudates, intended for a number of markets, can be produced from a single general-purpose extruder and continuous vulcanization line. This policy has resulted in the formation of companies which specialize in certain processes or manufacturing techniques, such as the fabrication of composite sheet products, precision molding, etc.

The overall profit margin of a company is dependent on many factors; but it is strongly influenced by the direct profit margin on the resources employed in manufacture and the level of utilization of the available manufacturing equipment. Ideally, a high direct margin should be coupled with a high utilization level. In less ideal and more normal situations a number of compromises have to be made. For company viability it is necessary to maintain a core of jobs which conform to the ideal; but a low utilization level can reduce or eliminate the overall margin, since resources not being used still contain overhead costs. It is sound policy in such circumstances to raise the level of utilization by accepting jobs which give lower margins, simply to offset the costs which continue to accrue when a manufacturing facility is not used. In these circumstances the work concerned should be additional to normal business. Another source of low-margin jobs is the "package" contract, where a customer offers an attractive job but insists that a minor job of a less attractive nature is accepted as part of the overall deal, knowing that it would be difficult to persuade a company to accept it on its own merits.

The problems of manufacturing versatility and the risks associated with a narrow market outlet usually increase with increasing equipment capitalization, investment recovery periods, and process specialization. They also increase in proportion to the skills needed within a company to ensure the manufacture of a saleable product. Taking an extreme example, if a product can be manufactured by unskilled labor using simple hand tools, the transfer from that product line to another of a similar type is simple, rapid, and inexpensive. The risk involved in concentrating on a single product with a limited sales "life" is minimal and can be justified by the simplification of company organization and the low overhead costs, in comparison with the multiproduct company. Market research can then be geared to establishing a continuity of similar jobs. In times of economic recession such companies can stop and start trading on very short time scales, which is difficult for a high technology or multiproduct company to do because it loses expertise and has continuing comprehensive overhead costs.

There are relatively few areas in the rubber industry where volume demand for low-technology products is sufficient to support an operation based on unskilled labor and hand-tool-type capitalization. Company responsibilities to employees should also lead to efforts to ensure a continuity of employment. The purpose of the example given is to emphasize the causes of company vulnerability to market changes. One of the most important functions of market research is to predict the "life" of a market, so that investment in equipping to supply the market is fully recovered, with an adequate overall profit margin, before the market demand ceases. It is the response of executive management to opportunities and risks of this type which determines the growth or decline of a company.

9.3.3. Introducing a New Product

Although the considerations which precede the decision to embark upon the introduction of a new product or product range into a company have been dealt with in the previous sections, the procedures for doing so and the groups involved have not been discussed. These need to be defined clearly, so that the time and cost of introduction, as well as the disruption to existing production, can be minimized.

The initial idea for a new product is often generated by conventional market studies and technical research; but it can also come from sales staff or directly from a potential customer. In all cases a marketing and research group will undertake an initial study of a proposal and will recommend that a new product be introduced if it appears to be appropriate to their company's activities, and if the return on the investment in development appears to meet company's criteria for profitability. At this stage the resources committed to the new-product proposals are small and the marketing group can be seen to be acting as a sieve, only allowing proposals which appear to have a high probability of eventual success to proceed. After a proposal has passed through this sieve the procedures outlined by the flow chart given in Figure 9.9 can be initiated.

Stage 1 is concerned with elaborating the initial assessment of product viability, calling on the expertise of other service groups for assistance. In a company operating a business center structure, the managers' of the business centers which would eventually manufacture the new product would be involved, if it could be seen that the new product would be manufactured in existing business centers. In the absence of directives from the chief executive or works manager, it would be the responsibility of the business center managers to decide if the product development should proceed or not, since they would have to bear the cost of development. Referring to the terminology of Figure 9.8, quality control and sales and marketing are self-evident in the labeling of stage 1 of Figure 9.9. The abbreviation "Tech" can refer both to the technical research expertise under the marketing manager and to the compound development group, under the technical manager. Similarly, the abbreviation "Eng." identifies two groups, one being the design and development engineering group, under the technical manager and the other being the whole of the engineering services business center, represented by its manager. The latter would only be involved at such an early stage if substantial changes in the manufacturing areas were seen to be necessary for the new product. Stage 1 is complete when the broad manufacturing requirements of a new product have been established and an estimation of the time, cost, and resources needed for its development has been arrived at, forming the basis of the aforementioned decision by the business managers.

In stage 2 a detailed product performance specification is obtained, from which product design and development can proceed. At this stage it may also be necessary to firm up estimates of the number of products required and the incidence of the requirement. In other words, the potential market volume and life of the product should be investigated thoroughly if the initial findings of the marketing group are in any doubt.

Stage 3 is entirely concerned with product design. The two major aspects of this are:

1. Design of a product which will be accepted by the customer.
2. Design of a product which can be manufactured in an efficient and trouble-free manner at a competitive cost.

Design for customer acceptance is simply referred to as "product design" in stage 3, whereas "engineering requirements" refers to design for manufacture. In the former, sales and marketing staff can make a positive contribution, due to their detailed knowledge of the customers' preferences, enabling them to guide the design engineer towards acceptable interpretations of the product performance specification. Design for manufacture requires that an initial selection of the manufacturing methods is made. These often depend on production volume, which was estimated in stage 2. Referring to Figure 9.8, the design and development engineering group undertakes product design both for customer acceptance and manufacturing, avoiding conflicting interests, as far as is possible. During a design exercise it often happens that methods or information essential to its completion are not available to the design engineer and require development or research work. This is referred to as "technical development" in stage 3 and is generally undertaken by the technical research staff.

In stage 4 the final product design or a selection of alternative designs are submitted to the customer. These would take the form of drawings and perhaps a nonfunctional model of the product if it was particularly complex. At this stage the customer may request changes, which result in the feedback loop to stage 3. If substantial changes are requested, as might happen if the customer is uncertain of his or her requirements, the manufacturing viability of the product must be reexamined. If a satisfactory agreement with the customer cannot be reached at this stage, it is preferable to disengage, rather than to commit more resources to an ultimately unprofitable job.

Following customer acceptance of the product design, the manufacturing methods to be used can be established in more detail. In stage 5 this enables the purchasing group to assess the resources needed both for the production of a prototype for customer evaluation and for eventual quantity manufacture. Orders are then placed to allow the former to proceed.

Stages 6 and 7 accompany prototype manufacture. A product specification is derived from the product design drawings and the product performance specification, and enables the prototype products to be evaluated against the customers' requirements. The conditions and procedures which resulted in a product being produced which is acceptable to the customer then form the outline specification. This work is usually undertaken by the service groups responsible to the technical manager, in conjunction with the quality-control group.

When a prototype or sample product is submitted to the customer in stage 8, it is for the evaluation of functional performance and, if the product is to be seen in service, appearance. Provided that the customer's requirements have been embodied accurately in the product specification and that design for manufacture has been adequate, thus avoiding the necessity of design changes in order to achieve prototype manufacture, this stage often

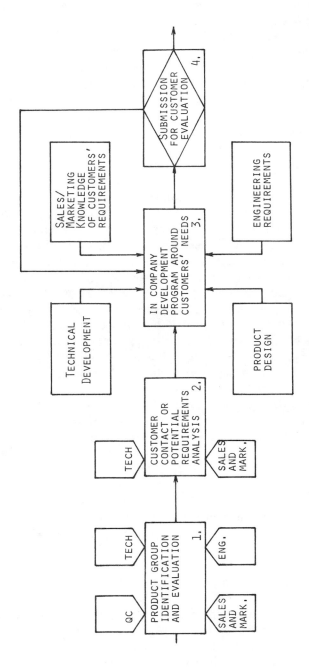

FIGURE 9.9. A functional flowchart for the introduction of a new product. (Illustration courtesy of Woodville Polymer Engineering plc.)

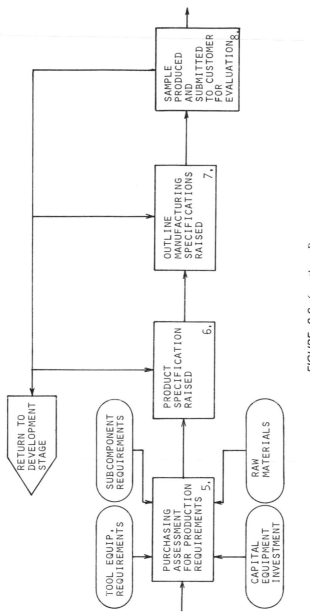

FIGURE 9.9. (continued)

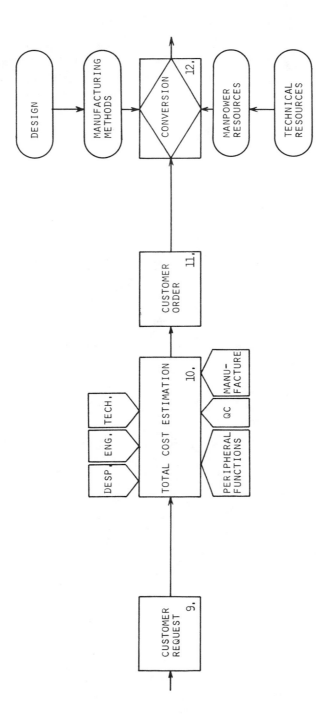

FIGURE 9.9. (continued)

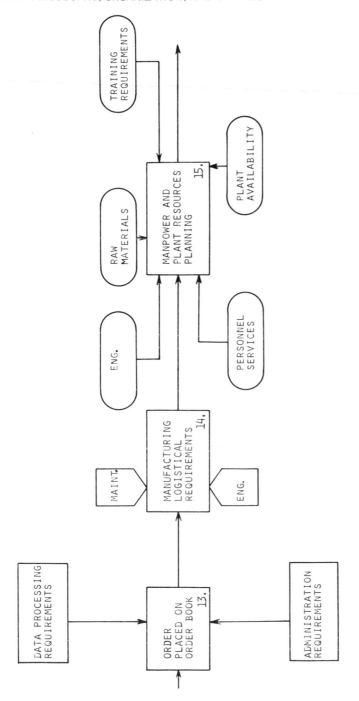

FIGURE 9.9. (continued)

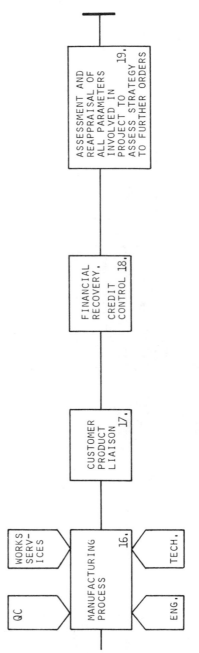

FIGURE 9.9. (continued)

presents few obstacles. However, design of rubber products for arduous applications is often complex, and there are many inadequacies in current design methods. These inadequacies often result in product design being an iterative process, where it is necessary to use the feedback loops indicated in Figure 9.9 in corrective action.

The customer request in stage 9 refers to the final statement of the production volume required and the schedule of deliveries, defining the rate of manufacture and enabling the total cost estimation in stage 10 to be undertaken, so that a price for the product can be determined. "Desp." identifies packaging and despatch and "peripheral functions" refers to any factors which will influence the unit manufacturing cost of the product. These include manufacturing efficiences and general overhead costs which cannot be absorbed onto specific products.

Following the quotation of a product price to the customer and any negotiations which take place over this price, the placing of an order generally follows. On receipt of the order, production equipment specific to the job, such as molds, will be ordered and the development of the production organization necessary to support manufacture initiated. In fact, stage 12 is entirely concerned with "productionizing" the job, involving the groups identified in stage 10. Here all the resources, operations, procedures, and specifications needed for manufacture are finalized.

When "productionizing" is complete the job can be allowed to go onto the order book, enabling it to be planned into production. At this stage the data processing or production performance monitoring requirements are determined and the administrative support for manufacture, including the customer liaison, is organized.

Although the details of all the operations in the manufacturing route have been determined, together with the technical and administrative support, the manufacturing organization and expertise which enables routine manufacture to proceed has not yet been set up. This is tackled in stages 14 and 15, which could be undertaken in parallel. These stages are concerned with:

1. Procedures which will ensure that raw materials and process supplies will be available.
2. In-house routing and transport of raw materials, process supplies, and work in progress.
3. Identification methods and checking procedures.
4. Negotiating new working practices with unions.
5. Training operations.

The infrastructure to support the items raised in the preceding list will generally be available, enabling most aspects of introduction into production to be dealt with using existing practices.

The first production run, in stage 16, will need to be closely monitored for potential organizational and technical problems. This scrutiny is followed through with detailed customer liaison following the first delivery, in stage 17. It then remains to ensure prompt payment against invoice in stage 18.

In practice there will be numerous variations on the product design, development, and introduction route defined in Figure 9.9. When a complex product design is involved, the time scale, from the initial customer contact to delivery of the first production, can be in the order of two or three years. However, in a jobbing situation, where a customer requests the manufacture of a product for which the design has already been carried out, the introduction will generally begin at stage 5, giving a time scale in the region of six months, including the design and manufacture of a production mold. For cases where the required product precision is low, or the customer's confidence in the jobbing company is high, the sampling or prototype manufacture and evaluation may be foregone, enabling the introduction to start at stage 9, with a consequent reduction in time and costs.

The recovery of development costs can also be achieved in a number of ways, although the own product manufacturer generally has no alternative to adding a development cost element to the price of each product. Jobbing companies have more alternatives and generally enter into a development contract with the customer before commencing work on a job. The terms of development contracts can vary widely depending on the nature of the work required, which can be in any of the following three areas:

1. Viability study.
2. Detailed development.
3. Manufacture.

If a customer requests a company to undertake work in areas 1 or 2, without any guarantee of the manufacturing rights, then the customer is generally required to pay in full for the work, so that the company doing it can show a profit on the exercise. In fact, a number of organizations are set up specifically to undertake viability and development jobs. Where a customer does guarantee the manufacturing rights, the situation is more complex and often the subject of intense negotiation. The producing company will usually attempt to share the cash flow and risk burden of development with the customer. If a molded product is involved, the customer will often bear the cost of the mold, which then belongs to the customer.

9.3.4. Manufacturing Capability Development

Savings set aside as a means of ensuring the continuity of a company's activities are known as capital. The sources of capital and the responsibilities associated with them are discussed in Chapter 10. This section is concerned with the ways in which capital can be deployed to meet the objectives of a company, which can be dealt with in six categories:

1. Replacement of existing manufacturing capability.
2. Updating of existing equipment by modifications or additions.
3. Purchase of equipment with capabilites new to the company, to support a move into a new market area.
4. Purchase of dedicated equipment to support the manufacture of specific products.
5. Training of new staff in skills which will enable them to make a positive contribution to the development of the company's activities.
6. In-service training of existing staff in new methods and technologies which will enable them to contribute effectively to company development.

The costs of categories 1 to 4 include installation and commissioning, in addition to the simple purchase price of the equipment. These additional costs are incurred by the design and preparation of the proposed site of the new equipment; the provision of services, such as electricity, water, and compressed air; and the hire of any specialist expertise or equipment. A plant layout study is generally necessary when the purchase of a large capital item is proposed, to determine the impact of introducing the new equipment on existing manufacturing facilities and to enable an accurate total cost to be established. There are also a number of "hidden" costs associated with equipment purchase decisions. A diversity of machine types involves a large amount of capital being "tied up" in spares and results in training and manpower planning problems with operators and maintenance staff. It is unreasonable to expect either an operator or a maintenance technician to maintain a high level of productivity or familiarity with equipment manufactured by a diverse range of companies, even though it may all perform the same function, such as extrusion or injection molding. To avoid reducing the standard of maintenance or the level of productivity while maintaining versatility in assigning personnel to machines, it is desirable to install machines of similar type and having similar controls whenever possible.

Existing equipment can be replaced whenever it can be shown that the return on investment justifies the expenditure. If machine development has not produced significant improvements in productivity since the purchase of the existing equipment, replacement will only be justified when the existing equipment has effectively worn out. This occurs when the mounting costs of maintenance, downtime, and defective work make the purchase of replacements economically viable. However, given the advances in machine capabilities and productivities with the advent of the microprocessor, it is becoming increasingly unlikely that many machines will reach this venerable age, particularly when many existing machine designs are unsuitable for the full exploitation of microprocessor technology. Also, bearing in mind the advantages of machine standardization, there is a strong case for "block replacement" of manufacturing capability, whereby all the machines performing a similar task are replaced by new ones from a single source and having a similar design. There is also the possibility of negotiating advantageous terms on multiple orders, which cannot be achieved on single orders. The implementation of this policy will depend on the existing range of equipment held by a company and the incidence of capital availability; but it should be a guiding principle in the formulation of a capital expenditure program.

Some long-service-life equipment, such as internal mixers, can benefit from modifications or additions to update operating methods, product quality, and productivity, in line with company and market requirements. These can range from the installation of some additional instrumentation for process monitoring to the replacement of major components during reconditioning, as part of a major microprocessor and computer control development scheme.

Purchase of equipment with capabilities new to a company, to support a move into a new market area, represents the most difficult of equipment acquisitions to evaluate and often carries with it a considerable risk. Market research should identify the behavior and requirements of potential customers in the new market area; but new products and new equipment inevitably mean a long lead time between the decision to enter the market and the first returns from sales. Investment in production machinery is usually preceded by the purchase of sufficient equipment for product and process development and evaluation, as described in the previous section, but this is a speculative venture. The probability of recovering the investment in prototype manufacturing facilities will generally be less than in established markets and the recovery time is usually longer. Sound management judgment is of profound importance in a new venture. Moves into new techniques and markets are often forced by the contraction of existing markets, but it is preferable to move willingly, when the rewards appear to justify the risks, rather than waiting to be pushed.

When expansion of manufacturing capacity is necessary to meet increased orders or forecast volume of sales in existing markets, simply installing additional machines of the type currently used has many attractions. The necessary expertise is well developed and there will not be any development problems. However, the effect of the scale of manufacture on the relatiye profitability of alternative methods should be thoroughly investigated. Also, if the service life of the new machines is anticipated to be substantially greater than the expected duration of the increased production volume, then their further utilization or disposal is relevant to the determination of return on investment. Similar arguments can be advanced for methods of dealing with the contraction of manufacturing capacity. Economics may justify a move to methods more appropriate to a reduced volume requirement.

Including staff training in the category of manufacturing capability development is unusual; but it serves to identify training clearly with planning for profitability and company survival, and emphasizes that a high expertise in any area critical to a company's activities cannot be obtained cheaply. Training must be considered as a major source of costs incurred in order to exploit new materials, manufacturing methods, and management techniques fully. For example, an improved control system for an injection molding machine can only yield enhanced profits if products which could benefit from the improved control are identified and programmed onto machines having it, and if the settings of the control system are adjusted to give high productivity.

The expansion of manufacturing capacity often results in a requirement for more space. This can be accommodated by the erection of new buildings, expansion of existing ones, or by renting space. The former two represent substantial investments and are generally undertaken only when it is expected that the increased demand will be sustained. The detailed considerations which enter into the selection of ways and means of making more space available for manufacture are dealt with in Chapter 8.

REFERENCES

1. Argenti, J., *A Management System for the Seventies*, George Allen and Unwin, London (1972).
2. Rawlinson, J. G., *Creative Thinking and Brainstorming*, British Institute of Management, Management House, Parker St., London, U.K. (1970).
3. Tennant-Smith, J., *Mathematics for the Manager*, Nelson, Sunbury-on-Thames, U.K. (1979).
4. Barndt, D. E., and D. W. Carvey, *Essentials of Operations Management*, Prentice-Hall, Englewood Cliffs, New Jersey (1982).

5. Coyle, R. G., *Mathematics for Business Decisions*, Nelson, Sunbury-on-Thames, U.K. (1976).
6. Bellenger, D. N., and B. A. Greenberg, *Marketing Research*: *A Management Information Approach*, Irwin, Homewood, Illinois (1978).
7. Green, P. E., and D. S. Tull, *Research for Marketing Decisions*, 4th ed., Prentice-Hall, Englewood Cliffs, New Jersey (1978).

10

The Economics of Manufacturing Operations

10.1. THE FLOW OF CASH THROUGH A COMPANY

Cash flow management is concerned with controlling the financial resources of a company and thus underpins many of the techniques by which performance is assessed. It is therefore necessary to deal first with cash flow, so that the function and value of the methods of monitoring and controlling the detailed aspects of company performance which follow can be recognized clearly.

A simple diagram of the flow of cash through a company is shown in Figure 10.1, where the difference between the cash withdrawn from the central reservoir to produce the inventory of finished products and the accounts receivable represents the profit, before taxes are deducted. The outflow of cash for interest on loans, taxes, and other unavoidable expenses is a major concern of financial managers. Failure to meet these obligatory external commitments would risk the existence of the company. Historically, many companies have gone bankrupt through mismanagement of cash flows and consequently being unable to generate sufficient funds in the cash reservoir to meet commitments or to provide funds for continuing company development. This situation is known as a liquidity crisis.

The financial resources of a company are generally divided into two categories—liquid and fixed. This section will be concerned mainly with administration of the liquid resources, which include cash, marketable securities, revenue, and inventory. Decisions involving the acquisition and disposition of fixed resources will be dealt with later, in capital budgeting. The amount of liquid cash held by a company, in comparison with the amount invested in fixed assets, depends on the judgment of the executive management. Cash on hand is seemingly idle, not earning a profit for the company and actually losing value in an inflationary economy, even when short-term deposits are used to minimize the loss; but without some cash on hand it is impossible to run a company and meet the bills which accrue

353

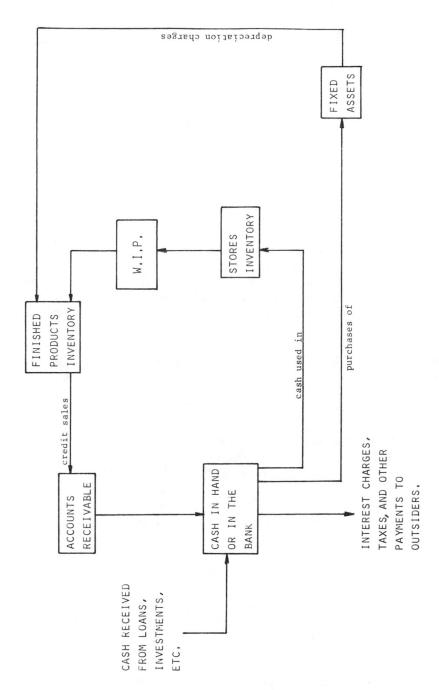

FIGURE 10.1. Company cash flow.

daily. The question is "What is the appropriate amount of cash which should be kept on hand?"

Production managers usually demand a high level of liquidity, to provide flexibility in handling unforeseen outflows of cash, which would otherwise necessitate negotiating a loan at short notice. These cases include taking advantage of unexpected opportunities, such as the availability of a cheap batch of material that requires rapid action to secure and major breakdowns. Financial and marketing managers inevitably demand the minimum practical liquidity, to free more cash to invest in projects which they believe will yield a high return for the company.

As shown in Figure 10.1, the liquid assets of a company will not all take the form of cash in the hand or in the bank. The store's inventory, work in progress, finished products, and accounts receivable all represent reservoirs of resources to which cash values can be assigned. The time for which resources stay in these reservoirs, and out of the cash-in-hand reservoir, is crucial to company profitability and viability. Tying up resources in these reservoirs or having outstanding accounts receivable can all result in a liquidity crisis, even though a sensible cash budget may have been agreed upon. Ensuring that accounts are paid on time is the responsibility of the credit controller, a job which has grown in importance in recent years. The control of the store's inventory, work in progress, and stocks of finished products is mainly the responsibility of the production management, in conjunction with the purchasing and sales groups. The details of preparing cash flow forecasts, to avoid liquidity problems, will be dealt with in Section 10.4.2, on cash budgets.

The simple cash flow diagram of Figure 10.1 becomes considerably more complex when a company structure based on the business center concept is operated. A cash flow diagram similar to the one in Figure 10.1 could be drawn for each business center; but operating in this manner loses the advantages of size for minimizing the fluctuations in cash flow. In fact, it can be shown that the probability of cash outflow due to unforeseen causes exceeding the budgeted amount by a given percentage will diminish as the size of the company and the cash budget increases, provided that the nature of the manufacturing operations do not change. The autonomous nature of business centers can be maintained if a unit with a liquidity problem borrows the necessary cash from those with liquidity in excess of their needs, at an appropriate interest rate. This avoids the need for the company to pay an outside organization for a service it can generate internally.

10.2. COST IDENTIFICATION AND ANALYSIS METHODS

10.2.1. The Classification and Recording of Costs

Selecting methods of recording costs which enable a detailed assessment of company performance to be made and which subsequently provide a basis for decision making, is of prime importance for good management.

The separation of costs into direct and indirect is at the foundation of many costing schemes. Direct costs are those which can be identified with specific products, while indirect costs are those which remain after the direct costs have been identified. The direct material and direct labor costs may be transferred from labor and inventory accounts to a work-in-progress account. The overhead costs, which include all manufacturing costs except those arising from direct materials and labor, are first transferred to a manufacturing overhead account and then to the work-in-progress account. This process is called the absorption of overhead by product and is of crucial importance for capital-intensive companies, but less important for labor-intensive operations. Thus three items are transferred to work in progress:

1. Direct material.
2. Direct labor.
3. Manufacturing overhead.

In labor-intensive operations it is essential for operators to record the work they have done and it is then a common practice to apply a rate of overhead on the basis of the direct labor cost, so that the manufacturing overhead is in some proportion to the direct labor cost. For capital-intensive operations this is inappropriate due to:

1. Overheads which continue to accumulate when a machine is not being used.
2. Machines which operate automatically, without operator intervention or attendance.

For an undistorted assessment of the manufacturing overhead it is necessary to absorb the whole of the overhead accumulated by a process in a given period onto the products manufactured during that period. With computer methods this is not difficult and will emphasize areas of low utilization.

It is worth referring to added value at this stage. As a product progresses through the manufacturing sequence its value increases as a result of the resources absorbed onto it. Thus it is a greater loss to a company to scrap a molding containing a kilogram of rubber than it is to scrap that kilogram of rubber after mixing. This concept is essential to the proper costing of rejected or scrapped work and provides valuable guidance to the quality-control staff.

While direct costs are well defined by their assignment to specific products, indirect costs need to be further categorized in order to differentiate usefully between their origins. This can be achieved through their separation into fixed and variable costs. A variable manufacturing cost is one which will change in some relation with the production output. This definition identifies most direct costs as being variable, although operatives wages are not totally variable, since a good manager relying on a skilled work force will not be able to lay off operatives in response to short-term reductions in orders. A fixed cost is one which does not vary in some relation with output. This rather careful definition is used because fixed costs, such as salaries, often change in proportion to manufacturing capabilities but not in response to short-term fluctuations in output. In labor-intensive operations the variable indirect costs can be quite substantial, due to the difficulty of assigning energy usage and similar costs to specific products, whereas in capital-intensive operations they should be minimal, due to the ease of process monitoring.

To aid decision making, costs may be assigned according to further classifications. These analyses are usually prepared for specific purposes and are not part of the routine recording and analysis used by a company. They include[1]:

1. Out-of-pocket and sunk costs.
2. Avoidable and unavoidable costs.
3. Relevant and irrelevant costs.
4. Opportunity costs.
5. Controllable and noncontrollable costs.
6. Marginal and incremental (or differential) costs.

There is a substantial amount of overlap between some of the preceding classifications; but these and other classifications by which costs are characterized can only be judged by their relevance to the decisions they are intended to aid and expedite.

Out-of-pocket costs require the utilization of liquid resources, whereas sunk costs are those costs for which the expenditure of cash or the incurring of a liability has already taken place. Out-of-pocket costs may be either fixed or variable, such as salaries or direct materials, but sunk costs include the costs of long-lived assets already acquired, allocated to expenses such as depreciation and loan interest.

The terms avoidable and unavoidable in relation to costs are used in their normal sense. Sunk and unavoidable costs are very similar, but the differences between them may be important for guiding some decisions. For example, the depreciation of a building or a machine is a fixed and sunk cost, but it may be avoided if buyers can be found for the building or the machine. The real cost of using the building or machine may be the opportunity cost foregone by not selling or renting them to another user. The main pitfall is in assuming that all unavoidable costs are sunk and that all avoidable costs are variable and out of pocket. Some variable costs, such as taxes on products, are unavoidable if the products are to be made. The chief executive's salary is unavoidable and is a fixed cost, but it is also an out-of-pocket expense.

The purpose of distinguishing between relevant and irrelevant costs is to identify and isolate those costs which are relevant to the decision being made. Failure to do this can result in inappropriate decisions being reached. For example, the book depreciation of a machine is not relevant to the decision to retain or replace it. Only the replacement cost and the extra revenue resulting from the replacement are relevant, unless the machine to be replaced has a resale or salvage value, which is also relevant to the decision. Taking another example, in assessing a company's ability to survive short-term adversity only out-of-pocket expenses are relevant.[1] Thus, for each decision the relevant costs must be determined.

Opportunity costs define the cost of a course of action in terms of the alternative opportunities which have to be given up in order to follow it. The opportunity cost principle is extremely useful in a number of areas. For example, the resources which can be assigned to capital expenditure and to development programs are limited by cash flow considerations, requiring that the available resources are allocated where they will yield the best return for the company. Inevitably there will be more proposals than can be supported and some will have to be abandoned in favor of others. Each proposal will have a potential benefit for the company, usually expressed in terms of forecast profit, which becomes an opportunity cost given up if the proposal is abandoned. By selecting proposals to give the minimum total opportunity costs, the *forecast* benefit from the deployment of the available resources is maximized. The opportunity cost principle can also be used for deciding between alternative uses of production facilities and thereby optimizing the product mix. Linear programming and other operations research techniques[2] implicitly take into account opportunity costs in determining optimal decisions.

The concept of controllable costs can be used as a guiding principle in defining responsibility and determining the distribution of cost information throughout a company. It is generally accepted that a person should only be held responsible for costs which he or she can control. A corollary is that reports to individuals should only contain those costs which they control. This becomes extremely important when the capability of a computer to generate vast amounts of detailed information is considered. To prevent managers from being confused and overwhelmed by the task of identifying the information relevant to them in generalized cost analyses, this task should be assigned to the computer, through sensible programming, leaving managers with time to undertake the tasks for which they are paid.

At this stage it is worth distinguishing between costs and expenses. Cost is a measure of the economic value given up in acquiring an asset. It may represent an actual monetary outlay, as in the case of raw-materials purchase, or a hypothetical comparison with an alternative, as with an opportunity cost. For reporting and analysis purposes, records are kept of actual outlays, and costs are traced through the company as they contribute to the production of assets (added value). Even though the assets may be changed in form, the costs which were incurred in their production are accumulated.

An expense is a cost factor that has been given up by a company in the process of obtaining revenues. A sales transaction could be viewed as simply a change in form among assets in which inventories are exchanged for receivables or cash; but it is more useful to record the receivable or cash at value rather than at its "cost," which would presumably differ from its monetary value. Thus the usual practice is to treat costs as expenses when sales are made.

10.2.2. Marginal and Incremental Costs

The marginal cost is the cost added by manufacturing one more product, whereas the incremental (or differential) cost is the cost added by producing a batch of additional products. Both definitions are extremely important, because many decisions are based on marginal or incremental costs. Two of the most important decisions that senior managers must make are concerned with the price and level of output of a product. Economists have evolved a theory of the viability of a product line that is based on marginal analysis which, despite many simplifying assumptions, provides a rational basis for analyzing pricing and volume decisions. [3]

The relationships between revenue and volume shown in Figure 10.2 are valid both for own product manufacture and for jobbing. In the former case it is assumed that more products will be sold as their price is reduced and in the latter case customers will demand a price reduction for increasing the size of their orders. Thus the average revenue per product (selling price per product) decreases as volume is increased. The marginal revenue, defined as the amount added to total revenues by the sale of one additional product, is less than the average revenue because the price of all the other products must

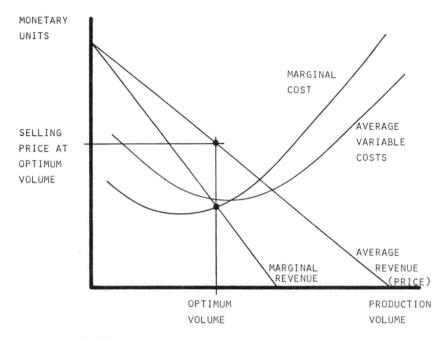

FIGURE 10.2. Average and marginal costs and revenues.

be reduced to achieve this volume of sales. This results in the slope of the marginal revenue curve being greater than that of the average revenue curve.

The average variable cost curve is assumed to have a negative slope initially, reflecting increasing efficiency with increasing volume and then curves upwards as the quantity produced exceeds the volume for maximum efficiency. The marginal cost curve has a similar form and at its minimum point passes through the average variable cost curve.[1] Economists then state that to maximize income on the product line being examined, a company should produce at the volume for which marginal costs are equal to the marginal revenues. If total revenues are less than total variable costs at this level the product should be discontinued. While the revenue from the next item produced is greater than the marginal costs of producing that product, the company should manufacture that product but should cease expanding production when marginal costs equal marginal revenues.

The optimum level of production is identified in Figure 10.2 as being the volume of production at the intersection of the marginal cost and the marginal revenue curves; but the price charged for the product is determined by the average revenue curve. In Figure 10.2 the average revenues are greater than the average variable costs at the optimum level of output, so it is possible that the product line is profitable, but to be certain it is necessary to construct an *average total cost curve*. The location of this curve will then establish whether or not the product line is profitable or not.

The economic analysis of price and volume decisions considers the effect on profits of making and selling *one* additional product. However, most managerial decisions involve the consideration of a number of products. This arises from the indivisibility of production inputs, which generally require that the product is made in batches, and customer orders invariably require specific lot sizes. Moreover, the tolerances in accounting data do not usually permit precise determination of the cost of a single product.

The accounting approach, which must deal with practical manufacture, is to consider the incremental effect of making and selling a number of products. This method focuses on incremental revenues (the net addition to revenues resulting from a specific decision) and incremental costs (the net addition to costs resulting from that same decision). Costs and revenues which are not affected by the decision are omitted from the analysis. A decision will increase profits if the incremental revenues exceed the incremental costs of that decision.

The concept of marginal costing leads naturally to break-even analysis. If the costs associated with a product are separated into fixed and variable costs and the selling price is known, it is possible to calculate the quantity which must be sold in any given period before the manufacturer begins to make a profit. The computation is often displayed graphically, as shown in Figure 10.3 for injection molding and compression molding methods of making a similar product. Compression molding is shown to have a smaller fixed cost (OA) than injection molding (OB) but a greater variable cost per unit, as can be seen for the greater slope of the curve, which represents cost per unit. As a result more products have to be manufactured by the injection molding route before the break-even point is reached (indicated by an asterisk), where total manufacturing costs equal the total sales price. However, if more products than the number indicated by C in Figure 10.3 are to be produced in the given period, then the profitability of injection molding is superior to that of compression molding.

The simple example of Figure 10.3 illustrates a technique which can be used to select among alternative manufacturing methods; but a substantial number of simplifying assumptions have been made which probably not be valid in practice, for the following reasons[4]:

1. Fixed costs are rarely truly fixed and will be influenced in some way by manufacturing volume.

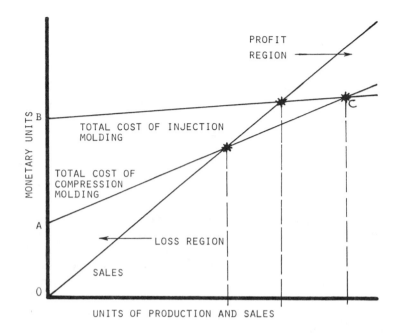

FIGURE 10.3. Break-even chart for compression and injection molding.

2. Variable costs and sales price per unit may both change with volume, producing curves rather than straight lines.
3. The manufacturing operations under consideration will represent one step in a manufacturing route and will be operated in conjunction with a number of other operations, with inevitable interactions.

10.2.3. Manufacturing Overhead

All costs which are connected with the manufacturing operations, other than direct labor and materials, are referred to as the manufacturing overhead. This will include depreciation, loan repayments, energy costs, and maintenance of the equipment used in manufacture, all of which are useful for manufacturing cost control purposes; but it will also include all the expenses associated with the buildings in which the manufacturing facilities are operated and the site on which they are located, which are necessary for financial reporting purposes. In a capital-intensive operation with a computerized production monitoring and data processing system, it is possible to assign the former group of costs to specific products and to treat them as direct costs. The latter group have to be assigned to products on some arbitrary pro rata basis for financial reporting and cannot be used for manufacturing cost control.

For the purpose of allocating and controlling the manufacturing overhead, *cost centers* are established,[1] which can be either locational or functional. Locational cost centers identify manufacturing operations, either individually or as groups. Functional cost centers are generally service groups which operate throughout the manufacturing areas, such as production planning, quality control, maintenance, and internal transport. In comparison with manual reporting and recording methods, computer monitoring and data processing enables much smaller units to be designated as cost centers, giving the capability of very precise control of the direct manufacturing overhead, while simultaneously reducing the clerical burden in some cases. It is important to recognize that a cost center must be a homogeneous unit engaged in a single form of activity if overheads are to be allocated accurately. Also the authority and responsibility for controlling the direct overheads associated with a cost center must be clearly defined, on the basis that people should not be given responsibility for costs they cannot control.

The objectives of analyzing the manufacturing overhead are to facilitate cost control and to ascertain product costs. These objectives are achieved through three successive steps of analysis, known as allocation, apportionment, and absorption. Allocation is straightforward, being the recording of costs incurred by a cost center against that cost center. Apportionment can be divided into primary apportionment and secondary apportionment. Primary apportionment is a method of distributing service costs, such as those incurred by maintenance work or technical problem solving, to the cost center in which the cost was incurred. Provided that the time of the maintenance technician or the process technician and the resources they used in a cost center are recorded, primary apportionment is again straightforward. However, for cost control purposes it is often important to know if the need for the service arose from the machine being used or from the product being manufactured, to differentiate between "problem machines" and "problem products." Secondary apportionment is concerned with distributing service costs which cannot be directly identified with specific cost centers, such as steam, compressed air, space heating, lighting, and depreciation on buildings, in a manner which approximates, as closely as possible, the actual distribution of usage.

Having assigned all the manufacturing overhead costs to production cost centers, the next step is to absorb or apply them to products. The manner in which this is done is extremely important, both for cost control and for accurate financial reporting. Separating them into three categories for each production cost center is beneficial:

1. The direct manufacturing overhead costs.
2. The overhead costs which can be positively identified with the cost center but not with products.
3. The overhead costs which cannot be positively identified with either production cost centers or products and are distributed in some approximate proportion to their estimated usage.

Taking category (1), the direct manufacturing overhead costs include the production equipment overhead, electrical energy usage (if significant), and those services which can be identified with specific products, *for a single recording period*, and can be absorbed onto the products manufactured during that period. When more than one type of product has been manufactured during a recording period, which is usually 24 hours, the *total* production equipment overhead is absorbed onto the products in proportion to their production times. This can give arithmetical problems in analysis if cost centers are established at the single process level and a process is not utilized during a recording period. Nonutilized processes need to be extracted prior to absorption of overheads and then included into an exception report. Identifying underutilized equipment in this way is an essential precursor to taking action to eliminate the cost burden due to it. For the cost determination requirements of financial reporting, the overheads associated with nonutilized equipment can be placed in category (3).

Category (2) overheads arise mainly from maintenance work, including breakdown repairs, and from process adjustments. The level of these provides a good indication of the capability of the equipment in a cost center and its reliability. Wear and general aging will be identified by a general upturn in these costs, as will an inappropriate product/machine match.

Both category (1) and (2) overheads are controllable by the person given the responsibility for and the authority over the cost center. Category (3) overheads are not controllable at the cost center level, although they will be controllable at a higher level. When a business center company structure is operated, each business center can be treated as a supercost center for the purpose of primary apportionment of services, such as quality control, production planning, value engineering, training, and safety, and for allocation of the costs of items, such as steam, compressed air, space heating, and lighting which, for the business center manager, are controllable costs. It can be seen that all the category (3) overheads, a major element of which will be salaries, are controllable at some level in the company, requiring that analyses are available in a number of forms, each of which should be appropriate to the level of responsibility of the people concerned.

By absorbing the costs from each of the three categories onto the products manufactured at a cost center, the product cost can be determined for financial reporting purposes. For effective cost control it is necessary to include in categories (1) and (2) all the significant sources of manufacturing overhead costs which arise from specific products or cost centers. Computer methods are a powerful tool for accomplishing this and are giving an increasingly better return on investment.

10.2.4. Scrap and Rework Costing

For a manufacturing operation to be controllable, in process-control terms, errors must be assigned to their causes in a way which enables corrective action to be taken, otherwise manufacturing is "open loop" and effective control is not possible. Systems set up by the quality-control group should ensure that defective work can be traced to the source; but the assignment of the cost of defective work to the source is often essential to provide an incentive to production managers to take the necessary corrective action.

Defective work can be divided into four categories for costing purposes:

1. Waste material.
2. Rework material.
3. Scrapped semifinished and finished products.
4. Repairs.

In each of these categories the potential savings to be obtained from assigning costs to the source must dictate the resources committed to recording. It will be assumed here that a comprehensive scheme is justified. In each case the source identified will be a cost center and the defective work value will be added to the direct manufacturing overhead costs of a specific product, if possible. Otherwise, it is assigned to the overhead costs "which can be positively identified with a cost center but not with products," referred to in the preceding section. In either case, the responsibility for the defective work is unambiguously defined.

The total amount of material wasted during manufacture can be determined from the store's issue record and the amount accounted for by the products entered into the finished product store, although the variable work-in-progress times will prevent the reporting of an accurate measure of the material wasted on a day-to-day basis. Despite this, a continuous record of the total volume and cost of material wasted is a valuable manufacturing performance index.

There are three sources of wastage associated with the mixing operation:

1. Losses prior to or during mixing.
2. Incorrect weighing of ingredients.
3. Incorrect mixing.

although item (3) more often results in rework than outright wastage. The total weight of material entered into the mixed-compound store subtracted from the amount issued from the stores and the amount sent back for rework determines the losses prior to and during mixing, although work-in-progress times will again prevent an accurate day-to-day report of wastage from being formulated. Knowing the composition of the compounds being weighed, it is also possible to determine which materials are being wasted, which is important for high-cost ingredients.

Mixed compound has added value, in comparison with raw materials, due to the costs of the weighing and mixing operations being absorbed onto it. At this stage the added value could simply be the direct costs incurred in producing the compound, but this raises some problems for cost centers further down the manufacturing sequence. If a product accumulates a very high added value, due to low utilization levels at preceding cost centers, it can be argued that the person responsible for a cost center at which defective work caused it to be scrapped is being unfairly penalized for earlier inefficient operations. Probably the best solution to this dilemma is to base the added value on standard costs for the purpose of allocating the cost of defective work to cost centers. Mixed compound which is scrapped should be weighed and recorded to enable a value to be assigned to it.

Although rework material is not scrapped it does accumulate additional costs, due to the additional operations which have to be performed on it. A cost *per standard rework addition to the mixer* will identify a volume of compound, by which the amount of compound to be reworked can be divided to give a total cost. Rework material from calendering and extrusion operations which results from changeovers can be treated in a similar manner, although it must be apportioned to the calendering or extrusion cost center which produced it. Out-of-tolerance extrudate and calendered sheet can often be reworked, but the value added to it by these operations must be added to the rework cost, in order to determine the total amount to be apportioned to the cost center.

In calendering and extrusion operations material utilization must be considered, in addition to waste and rework. Material utilization measures the efficiency with which the mixed compound used by one of these processes is converted into a product, and is expressed as the length of an extrudate or calendered product produced per unit volume of mixed compound, requiring a weighing operation prior to extrusion or calendering and a record of the amount of product derived from a given weight.

Maximum material utilization is achieved when a process is run as close as is feasible to the minimum product dimensions given in the specification. The savings to be made in direct material costs by installing equipment and process-control systems capable of a much higher level of precision than that required by the overall product tolerance bands may therefore provide a viable return on investment. It must also be remembered that at extrusion and calendering operations the direct material costs will include the value added by the mixing operation.

Material utilization at the vulcanization stage will normally be determined by prior operations, such as forming a feed strip from a two-roll mill, extrusion, calendering, and fabrication, and by the nature of the molding operation. Only in the few cases where an operator controls blank size will there be a variable material utilization efficiency associated with the vulcanization process. The value of defective vulcanized items which are scrapped must be apportioned to the cost center responsible and absorbed onto the appropriate product. However, some nonrubber components, such as metal inserts, may be recovered, reducing the value of the scrapped products. However, the cost of recovery will need to be absorbed onto the product, as will the cost of repairs, where possible.

The main problem of recording the cost of defective work against the responsible cost center, other than the resources needed for recording, lies in identifying positively the responsible cost center. There will be occasions, even in the best controlled manufacturing facility, when problems encountered in a cost center are due to errors incurred at earlier stages in the manufacturing sequence. In such cases it will be the task of the technical problem-solving group to assign the responsibility for the defective work.

In this section the main resources needed to operate the scrap and rework costing procedures are considered to be those associated with recording, based on the assumption that computer data collection and analysis methods will be used.

10.2.5. Monitoring Company Performance

The most useful indicator of overall company performance is the wealth created from the use of limited resources. A realistic measure of performance is therefore obtained by considering the increase in wealth (profits) in relation to the resources available, which are represented by the assets of a company (capital employed). Four indices are needed for adequate monitoring of performance:

1. Direct profit margin.
2. Return on direct capital employed.

3. Overall profit margin.
4. Return on total capital employed.

The direct profit margin is the difference between the selling price of a company's products and the sum of the direct materials, direct manufacturing overhead, and direct labor costs, whereas the overall profit margin is the difference between the selling price and the costs identified for the direct profit margin in addition to all the indirect overheads. These latter items include salaries for production management, service groups, and administrative staff, in addition to site and building maintenance, sales expenses, and many others. The return on direct capital employed is a ratio obtained by dividing the direct profit margin by the direct costs (the direct capital employed) and similarly, the return on total capital employed is obtained by dividing the overall profit margin by the overall company costs (the total capital employed). It is a normal practice to express these ratios as percentages.

The returns on capital employed are required both as ratios and monetary values because separately they do not provide a complete measure of company performance. It is possible for the returns on capital employed, expressed as ratios, to remain static during periods of rapid company growth or contraction. Alternatively, the profit margins can remain stable while substantial improvements or deteriorations in company operating efficiency occur, requiring the employment of less or more capital in order to generate the margins.

While the overall profit margin and the return on total capital employed define the "bottom line" of company performance, it is important for managers to be aware of the amount of capital employed in meeting indirect overhead costs, in comparison with the direct capital employed, and the influence of the indirect overhead costs on operating efficiency, as measured by the difference between the two ratios of return on capital employed. Observed trends in each of the four indices, which are usually determined for each operating period, should reflect company policy and the progressive influence of development through business plans. Deviations from these trends, ignoring short-term fluctuations due to assignable causes, will require corrective action. It should be emphasized that the performance indices cannot be used for cost control, although they are sensitive indicators of the effectiveness of cost control measures. Detailed cost analyses of the types referred to in the preceding sections and in Section 10.3.3 on achievement against standard costs are necessary to identify problems, assign reasons for the occurrence of the problems, and take corrective action.

Historical records of direct and overall profit margins, in either graphical or table form, are only useful if they are all expressed in monetary units of equivalent value. Unless proper corrections are made, inflation can make a deteriorating profit margin look like a progressive improvement. It is preferable to update past records to the current value of monetary units at the end of each company operating period, when the latest values of the indices are entered into the record. This can easily be achieved with a computer data processing system, but it militates against wall charts, which would have to be totally updated at the end of each operating period. However, it can be argued that wall charts and many paper records are outmoded and represent an unnecessary expense when a graph or table can be called up on a computer terminal in a manager's office, as one of many information services available through a distributed terminal computer system.

10.3. STANDARD COSTS

10.3.1. General Considerations

A standard cost is defined as "the cost of performing a given operation under specified conditions,[1] although in some instances it will be an estimated cost instead of one determined from manufacturing practice. The standard costs determined for a product will depend on the specified conditions; normally these will be calculated for the manufacturing route which gives the highest level of productivity for the product being considered, working under conditions which can be expected in a well-managed company running at normal operating levels. A standard cost prepared according to these guidelines is often referred to as an *expected standard*, being approximately equivalent to the expected manufacturing cost.

The use of expected standard costs acknowledges and accepts that there are currently certain ineradicable inefficiencies present in the manufacturing operations, although it must be emphasized that inefficiency is not condoned and standard costs must be changed when techniques for overcoming inefficiencies are introduced. Standard costs relate to the manufacturing costs which, from Section 10.2.3, include the direct material cost, the direct manufacturing overhead, and the direct labor cost. For each of these, allowances must be specified which relate to each manufacturing operation in the company, to ensure that compensation for inefficiencies proceeds from a consistent base during the formulation of a standard cost. It may also be necessary to allocate allowances for manufacturing operations at a number of levels, according to general product group or type, resulting in different levels of total allowances. Clearly, setting up a system of allowances is a complex task, requiring considerable knowledge of the manufacturing operations and reference to analyses of production performance obtained over an extended period of time. However, unless allowances accurately reflect the inefficiencies in manufacture and are updated when these change, all the procedures based on standard costs will lack accuracy and credibility.

Material allowances include scrap, rework, and repairs at levels which are appropriate to normal operating conditions, expressed as a percentage of total material costs for each manufacturing operation. As noted in Section 10.2.4, material will accrue added value as it progresses along the manufacturing route. This must be taken into account in preparing a standard cost. Direct labor allowances include standard breaks in addition to relaxation time and other specific causes of nonproductive time associated with individual manufacturing operations. If the direct labor cost is for an operator whose presence at a manufacturing operation is necessary for production to occur, operator nonproductive time will influence the direct manufacturing overhead, by reducing utilization efficiency, unless relief workers are used. In the latter case, the costs due to the relief worker will be in addition to those of the normal operator. Alternatively, if a worker is servicing a machine which operates automatically, worker nonproductive time refers only to the utilization of the worker, and has no influence on process utilization. However, the nonproductive time of the process *will* add to the nonproductive time of the servicing operator, which again must be taken into consideration in setting up the system of allowances.

Together with the master specifications (which include all the material, manufacturing, product, and performance specifications), standard costs form the essential data bases for initiating and controlling manufacturing operations. Estimated standard costs for a new product determine if an adequate profit margin can be obtained from the expected selling price; and materials requirements planning is generally based on standard cost information. Standard costs also provide manufacturing efficiency targets, against which practical performance can be assessed, and enable cost changes to be implemented quickly and effectively.

10.3.2. Establishing Standard Costs

The procedure of establishing standard costs for a new product starts as soon as a product specification becomes available. It is the responsibility of the estimator to raise an overall initial standard cost, from which the commercial viability of the new-product proposal can be assessed. This standard cost will be based on the direct manufacturing overhead rates and the direct labor rates for each operation on the proposed manufacturing route, in addition to current material costs, with allowances being made on each. The standard cost arrived at for each operation will be based on established standard costs for similar products and consultations with the groups responsible for product development and introduction, thus linking the procedure for establishing standard costs with the sequence for the introduction of a new product described in Section 9.3.3.

The initial standard cost will be subject to some uncertainty, due to much of the information needed to ensure accuracy being unavailable. At this stage accuracy depends largely on the expertise of the estimator, emphasizing the importance of this function. Following the formulation of the initial standard cost, there will be a number of stages in the route to final or production standard costs, which are usually referred to as *the* standard costs. These stages will tend to differ between companies, but the following are desirable[5]:

1. Estimated standard costs.
2. First production standard costs.
3. Production standard costs.

The estimated standard costs are formulated at the end of the product development program and, for jobbing companies, at the time of quoting a price to the customer. With the manufacturing methods being firmly established and many of the problems identified and overcome, considerably greater confidence can be placed in the probability of these standard costs being representative of manufacturing costs. However, there will still be some uncertainty due to the product not having been manufactured under production conditions. The estimated standard costs will be updated at the time of receipt of the customer's order, to account for any changes in the costs on which the standard costs are based. These revised estimated standard costs will be superceded, shortly after production commences, by the first production standard costs, for each operation in the manufacturing sequence. In progressing from the estimated standard costs to the first production standard costs, the responsibility for formulating standard costs passes from the estimator to the value engineer (or value analyst).

The first production standard cost is established from the manufacturing performance on the first production run, after "teething troubles" have been overcome and the learning processes associated with the new product have been essentially completed, thus correcting any inaccuracies in the estimated standard costs. It also forms the first element of a value engineering exercise,[6] being the recording stage of the sequence:

1. Record current methods.
2. Evaluate the efficiency of current methods.

3. Recommend modifications to existing methods or use of alternatives.
4. Implement agreed-upon recommendations.

For effective value engineering, a deep understanding of the manufacturing operations of the rubber industry is necessary, such as is embodied in Chapters 1–8 of this book, coupled with a knowledge of method study techniques.

After the implementation of the recommendations arising from the value engineering exercise, as soon as sufficient time has elapsed for the disruption caused by the changes to have been overcome completely and for normal operating levels to be established, *the* standard costs can be determined for each manufacturing operation. With all the information needed for their formulation now being available, they will represent accurately the *expected* performance of the manufacturing operation.

The formulation of standard costs is based on well-defined procedures and information, indicating that computer methods can provide powerful assistance with the mundane aspects of the job, in order to give the estimator or value engineer more time to concentrate on the areas where judgment is required. In addition to the use of the master specifications data base already mentioned, subsidiary data bases or files will be needed for the material costs, direct labor costs and manufacturing overheads, in addition to the allowances on each of these. With the exception of the latter, all these data bases will be used by other groups, including the production management, production planning, and purchasing groups; and the activities of the estimator and value engineer will result in the creation and maintenance of the standard costs data base. In fact, in a number of instances a service group will be given the responsibility for maintaining and updating a computer data base, for their own use and for use by other groups.

In addition to giving rapid access to information through the data bases, the routine calculations involved in formulating standard costs, which are relatively simple but large in number, can be readily turned over to the computer. However, instead of simply using the computer as a "super-calculator," an approach which could be described as "computer-aided standard costing" provides substantially greater benefits. First, the programs can be written to give the choice of alternative methods and to provide prompts to guide the user through the sequence of operations involved in the formulation of a standard cost, ensuring that important elements are not omitted due to pressure of work. Second, diagnostics facilities can be provided which, when activated, check that information relevant to the product being costed has been used throughout the costing exercise and that the figures entered into the computer are consistent in relation to each other.

The computer methods described here are "interactive" in nature, consisting of a dialogue between the user and the computer, preferably through terminals sited in the offices of the estimator and the value engineer. A distributed terminal computer system is needed to support this type of activity.

10.3.3. Achievement Against Standard Costs

Standard costs identify a level of manufacturing efficiency against which actual performance can be measured, providing targets and incentives for all groups associated directly with manufacture. The difference between an actual cost and a standard cost is termed a variance (which is not the same as the statistical variance used in other parts of this book). The purpose of variance analysis is to identify the causes of differences between standard and actual costs.

Before dealing with variances in detail, it is worth reviewing some of the problems of monitoring and controlling manufacturing productivity through standard costs. First, it is necessary to express variance both as a percentage of the standard cost and as a monetary value, to enable the influence of production volume to be taken into account. For example, the monetary value of some variances can increase, indicating an apparently undesirable decrease in manufacturing efficiency, due to an increase in the number of products manufactured during a recording period. Second, standard costs are formulated for products in isolation. They do not account for the manifold interactions with other products being manufactured, which can influence variances either favorably or unfavorably. Since these interactions start at the orderbook, it is often difficult to assign responsibility for changes in variance. A typical example of an interaction causing a favorable variance is the coincidence of a number of products requiring the same base polymer, enabling the purchaser to obtain a bulk discount.

Both standard costs and operating costs arise at the work-station level, indicating that analyses can be prepared in great detail if necessary. For the person responsible for a cost center, which will include one or more work stations, the availability of variance reports at this level is very important for effective cost control; but the information can be progressively compressed and summarized for people at higher levels in the company hierarchy.

Starting with raw materials, the contribution of the purchasing group to company profits is partially measured by the material price variance, which quantifies their ability to purchase materials at preferential prices:

$$\text{material price variance} = \text{actual quantity} \quad (10.1)$$
$$\times (\text{actual price} - \text{standard price})$$

This relates to *all* of a given material issued from stores during a specified recording period, identifying the purchasing group as the responsible cost center. In contrast, a material utilization variance can be specified which provides a measure of the efficiency with which materials are used at each work station:

material utilization variance = standard price   (10.2)

$$\times \text{ (actual quantity} - \text{standard quantity)}$$

The information needed to calculate this variance is also utilized in the scrap and rework costing procedures described in Section 10.2.4, which are more comprehensive and assign scrap and rework costs to the direct manufacturing overhead and the general profit center overhead. If the simple measure of material utilization defined by Eq. (10.2) is selected in preference to the procedures of Section 10.2.4, then the standard material prices for each work station should include the added value due to previous operations.

    With regard to labor variances, it has been argued[1] that the actual labor cost should be debited to a labor-in-progress account and that the added value transferred from this account to the work in progress should be at the full standard rate for the product concerned, as suggested for the purposes of costing scrap and rework in Section 10.2.4. This means that the credit to the labor-in-progress account will be obtained by multiplying the standard hours of labor for the specified recording period by the standard rate of pay per hour, where

standard hours of labor

$$= \text{number of products} \times \text{standard time per product}   (10.3)$$

$$\times \text{standard labor cost per hour}$$

and the difference between the actual labor cost incurred during the recording period and the standard labor applied is the labor variance.

    If overtime must be worked in order to achieve production targets, it becomes more difficult to achieve a favorable labor variance, due to the premium wage rate which must be paid for overtime. However, if the need for overtime reflects a demand for volume in excess of that obtained by dividing the time in the standard working day by the standard time per product, then it is possible to offset this by an improvement in the direct manufacturing overhead variance, as a result of increased machine utilization. The total variance, to be dealt with later, will verify this. In contrast, if overtime is necessary to compensate for problems or inefficiencies during normal working hours a double penalty is incurred, both from the extra time needed to perform the required operations on the products and the premium wage which must be paid for the overtime.

The determination of labor variances for individual work stations and cost centers does not give any indication of the overall efficiency of labor utilization within a manufacturing group or business center. This can only be assessed by deriving an overall labor variance, which can be defined as the difference between the total actual labor costs for a recording period and the standard direct labor cost content of all the products on which manufacturing operations were performed during that period.

With an accurate direct manufacturing overhead cost, its variance can be treated in a similar manner to the labor variance. The actual direct overhead costs incurred at a work station during a reporting period can be debited to a direct manufacturing overhead account and the added value transferred from this account to the work in progress at the full standard rate for the product concerned. This means that the credit to the direct manufacturing overhead account will be obtained by multiplying the number of products dealt with during the recording period by the standard direct manufacturing overhead cost per product, giving:

direct manufacturing overhead variance   (10.4)

= actual direct overhead costs for the recording period

− (number of products × standard direct overhead cost per product)

In cases where there is a product change during a reporting period, the direct overheads can be apportioned simply on the basis of utilization times. The direct manufacturing overhead variance provides an alternative means of monitoring and controlling the direct overhead costs to that described in Section 10.2.3. However, it must be noted that for financial reporting purposes the actual direct overhead costs must be absorbed onto the work in progress. The other elements of overhead costs will be dealt with later, in Section 10.4.

The direct manufacturing overhead variances will be influenced by the products on the order book, the expertise with which the production planning group control the product mix during assignment of jobs to work stations, and the effectiveness of the production management. It is not possible to gain an assessment of the quality of production planning from overhead variances determined from individual work stations or cost centers. As with the labor variance, this can only be achieved by deriving an overall direct manufacturing overhead variance, which can be defined as the difference between the total actual direct manufacturing overhead incurred during a recording period and the standard direct overhead cost content of all the products on which manufacturing operations were performed during that period.

By adding together the material price variance (assuming scrap and rework costs will be debited from the direct overhead), the labor variance, and the direct manufacturing overhead variance for a specified manufacturing operation on a given product for a single recording period, the total variance can be obtained, where

$$\text{total variance} = \text{actual total direct costs for the product during} \quad (10.5)$$
$$\text{the operating period}$$

$$- (\text{number of products} \times \text{standard cost per product})$$

This may include the performance of a number of work stations and cost centers if the production volume is large. For cost control purposes, the separation of variances at the work-station level needs to be maintained. The total variance is an excellent indicator of the efficiency with which the resources available for manufacture have been deployed.

An overall total variance for a product can be obtained by adding together the variances for each work station on the manufacturing route. This again is an excellent indicator of manufacturing performance but it cannot be used for cost control purposes, due to the factors which influence it arising from a number of separate sources.

Other variances can be formulated readily and their selection depends on the nature of the manufacturing operations. Monitoring, recording, data processing, and reporting at the highly detailed level described in this section is best supported by a distributed terminal computer system. Reports are then available on demand, without the expense of producing routine hard copies. Working on the sound principle of management by exception, highly detailed information is only required when corrective or improving action is indicated. The need for action will be signaled by the trends and fluctuations of the general indicators of production performance, which can be quickly assimilated by the user.

10.3.4. Implementing Cost changes

Increases in material and manufacturing costs and the decrease in the value of money due to inflation erode profits, unless action is taken to pass these on to the customer by increasing the selling price of the products affected. It is necessary to inform customers of changes in selling prices as soon as cost increases are notified or identified, to allow for the lead time involved in negotiation of the new prices and their subsequent progress through the order book to manufacture and eventually to the receipt of monies against invoice.

Material price increases will be notified by the purchasing group and it will normally be the responsibility of the value engineer to identify and evaluate changes in manufacturing costs. These then have to be assigned to the products affected, by adjustment of their standard costs. If the computer data bases referred to in Section 10.3.2 are available, the computer can be programmed to implement the changes and make a report of the changes available to the people responsible for resetting the selling prices.

10.4. BUSINESS PLANS AND BUDGETS

10.4.1. Preparing Business Plans and Budget Forecasts

Business plans and budgets are inseparable. The former is a detailed program of the activities which a company intends to undertake during a future period (usually one year) and the latter defines the allocation of resources to those activities. This is emphasized by the Institute of Cost and Management Accountants' definition of a budget as:[4]

> A financial and/or quantitative statement, prepared and approved prior to a defined period of time, of the policy to be pursued during that period for the purpose of attaining a given objective.

The main objective of preparing a business plan and budget forecasts is to maximize the benefits to be gained from the deployment of a company's resources, within the constraints imposed by cash flow. This differentiates budget forecasts, which are planning tools, from budget targets, which are cost control tools. The latter will be dealt with in Section 10.5.

The preparation of business plans and budget forecasts for a single financial year is often called comprehensive budgeting or profit planning; but all companies must make strategic decisions affecting the long-term future. In large companies the corporate planning function is well defined and in the public sector "planning, programming, and budgeting systems" have been evolved for this purpose.[1] The foundation of planning and setting budgets, both in the short and long term, is forecasting. It is unlikely that a company will be able to finance product development and capital investment programs from resources held at the time of planning, even if this were desirable. Forecasting methods are dealt with in Section 11.2.2, on the subject of inventory planning; market research is discussed in Section 9.3, in conjunction with company development.

The way in which a business plan and budget forecasts are structured will depend on the organizational structure of a company. It is essential that they are compatible with each other, to avoid exerting a distorting effect on the pattern of decision making and to ensure that the resources allocated are used effectively. For example, the categorization and distribution pattern for a traditionally structured company will be substantially different from one for a company organized into business centers. In the former case allocations will be made directly to the manufacturing departments and service groups. In the latter case, the majority of the service groups' income will result from payment by the manufacturing business centers for work carried out; a substantial amount of the business planning and budgeting will occur at the business center level. The main tasks of the central administration executives will be those of coordination and correlation of the individual business center objectives with the overall company philosophy and objectives, and the allocation of the *company's* resources according to those objectives. In fact, the company's business plan and budget allocations are a primary means of management direction in the business center structured company. The forecast resources can be allocated where they will best benefit the company in the long term, rather than simply returning them to currently successful operations. At the business center level, objectives and targets will be specified for the deployment of the budgeted resources; but in keeping with the basic principles of delegation, the means of achieving the targets and objectives should be the responsibility of the business center manager. This will require a secondary planning and budgeting exercise, as previously noted.

The planning, programming, and budgeting systems (PPBS)[1] provide a well-structured approach to the formulation of business plans, irrespective of the time scale. There are many PPBS variants, but they generally have four basic elements:

1. A program budgeting structure oriented towards end objectives.
2. A long-term program and financial plan.
3. Systematic program evaluation.
4. Special analytical studies.

Taking item (1), the budget may only specify lump-sum amounts for each group or business center, but it is normal for the appropriations to be structured further in accordance with some other basis of classification. The principal subdivision is into:

1. Objects of expenditure.
2. Programs.

Objects of expenditure relate to specific categories of input items which may be acquired, such as capital equipment and personnel. Once the budget has been approved, the expenditure in any object class should not exceed the appropriations. Great care should be taken in choosing the objects of expenditure, since it can be difficult to shift resources from one object to another without disrupting the system.

Budgeting for the objects of expenditure is concerned with providing both service and manufacturing groups with the basic facilities needed for them to perform their functions. In contrast, program budgeting is concerned with providing the resources for the utilization and, if necessary, enhancement of the basic facilities for specific goals. There will obviously be a substantial region of overlap, where either of the two methods could be used. In such cases, it is generally preferable to choose program budgeting, which is directed towards output objectives rather than inputs and allows a more flexible approach to the deployment of resources.

Programs often involve a number of groups or business centers, whereas budgets for objects of expenditure normally relate to a single group. For example, process and product development or the installation and commissioning of a computer system depend on the cooperative effort of many groups. Budgeting the distribution of resources can sometimes be problematic, but the mechanisms for doing this are inherent in companies operating a business center structure. All of a program budget appropriation can be assigned to the business center which will derive the greatest benefit from the successful outcome of the program, and which will therefore have the greatest motivation to make it succeed. The business center manager concerned can then subcontract the work involved in the program, including the project planning if necessary, both to other business centers and to outside organizations, in a way which will yield the best return on the budgeted resources.

Moving on to the long-term program and financial plan, the extension of the time scale from a single year to several years in the future is a basic contribution of PPBS. Many long-term projects may involve relatively small current outlays but will require heavy commitments of resources in future years. The implications for the future must be considered in making current decisions.

The financial requirements of projects usually vary with the stage of maturity. In the early stages, the resource requirements may be for the support of market research and for technical research and development. The financial needs at this stage are generally relatively small in comparison with the heavy investment in manufacturing facilities required following the successful completion of the research. In simple terms, unless a company is willing and capable of providing the resources for the implementation of the results of research and development, they should not fund the research for the project. However, this simple viewpoint does not take into account a number of factors:

1. Not all research programs are successful or completed on schedule.
2. A company may decide that it is more beneficial to obtain revenues from licensing the use of research results to other companies than to implement them in-house.
3. With an uncertain future it may be necessary to run several research programs in parallel, to enable the one which is most appropriate to the circumstances at the time of completion to be selected for implementation.

To permit valid comparison of alternatives, data regarding costs and outputs of program must be projected into the future and made available to those responsible for formulating business plans and budgets. A long-term *business plan* presents such data in tabular or graphical form, showing the future implications of current decisions. This business plan is continually revised and extended as decisions are made or new information becomes available.

The selection of the time scale for the long-term business plan is extremely important. The time horizon must be long enough to give adequate weight to recurring operations costs; but in predicting future events and costs, the degree of uncertainty increases substantially as the time horizon is extended. For formal planning purposes, a time scale of five years is often used in projecting future costs. Shorter and longer time scales are used in evaluating individual projects when considered appropriate. In recognition of the increased uncertainty involved with longer time periods, a higher level of aggregation may be used for distant-year projections, and a range of estimates may be used to indicate the degree of uncertainty.

As part of the formal planning and budgeting process, each program, both current and proposed, must be evaluated in terms of its overall impact on the company. This is always done annually and also at shorter intervals in many cases, to provide essential information to guide budget decisions. Computer methods are again very useful in this area, to manipulate and present information in forms appropriate to the decisions being made and also to evaluate the consequences of a number of programs interacting in various ways. The techniques for achieving this are grouped, with many others, under the general name of *simulation*,[2] which is a powerful tool for planning. When the program interaction evaluation is completed, it must be used together with the long-term business plan to determine the desired level of resources to be allocated to each program.

In support of individual program recommendations, special analytical studies may be made of their feasibility and viability. These will vary with the nature of the program but are always oriented towards the decision-making processes of the company and generally take the form of a cost-benefit analysis.

To determine the level at which a program should be supported, it is necessary to obtain information regarding the potential costs and benefits of the program. If the available resources to be spent are fixed, the company might rationally try to maximize the benefits to be gained from spending them. If the level of benefits to be gained from a program is established, it is then rational to attempt to achieve the program objectives at the lowest possible cost. The principal elements of a cost-benefit analysis which can be used for these purposes are:

1. Systematic consideration of alternative means of accomplishing a stated objective.
2. Estimation of the economic costs and benefits to be desired from each of the alternatives.
3. Consideration of the time phasing of the costs and benefits.
4. Assessment of the effects of uncertainty.
5. Consideration of the ancillary costs and benefits of each alternative from the standpoint of the company as a whole.

The essence of cost-benefit analysis is the comparison of alternative programs or courses of action with regard to their relative costs and benefits. It also serves as a model for value analysis or engineering, dealt with in Section 10.3.2.

When cost-benefit analysis is used, great care must be taken in defining the benefits. The simplistic objective of maximizing profit is only a starting point and is subject to many constraints. For example, maximizing short-term profits can result in the demise of a company in the long term, and is often due to a failure to invest adequately in people and equipment. The judgment of a company's senior managers is very important here.

The formulation of business plans and the setting of budgets is usually the responsibility of a planning group which comprises a company's senior managers. In a medium-sized company operating a business center structure, this would include the chief executive and the small number of senior central administration managers in addition to the business center managers. The main objective in determining the composition of the group is to strike a good balance between adequate representation of those who will have the direct responsibility of making the business plan work and unwieldiness due to large numbers. In large companies there will be a number of levels at which planning and budgeting are carried out, to avoid the twin problems of "getting bogged down in fine detail" and large planning groups. The administrative tasks of calling for program proposals from managers, collecting and collating reports on current programs, and then presenting these for the planning group are usually the responsibility of a planning manager, who also has the task of communicating decisions to those managers not directly involved in the planning process. The planning manager will also have a coordinating role when planning and budgeting occur at more than one level.

10.4.2. The Cash Budget and Its Interaction with Other Major Functional Budgets

The formulation of a business plan and the setting of budgets depends on the forecast flow of cash through a company. The incidence of expenditure in *all* areas must be matched with income, to ensure the availability of cash with which to meet the company's financial commitments, otherwise a liquidity crisis may occur, as described in Section 10.1. For this reason, the cash budget is of prime importance. It is usually necessary to prepare a month-by-month forecast of cash flow, as settlements between debtors and creditors are made conventionally at the end of each calendar month.

The accuracy of the forecast cash budget will depend on the forecasts for the other major functional budgets, which interact with the cash budget as shown in Figure 10.4. Starting with the sales budget, this gives the income to be expected from the forecast sales, immediately defining the importance of accurate sales forecasts. The manufacturing budget then gives the forecast expenditure which would be incurred in manufacturing the products demanded by the sales forecast, and is subdivided into the direct materials budget, the overhead budget, and the direct labor budget. It is then the difference between the forecast income from sales and expenditure on manufacturing which generates the cash budget, providing the resources needed for the other functional budgets in Figure 10.4 and hopefully leaving a residue, which is the forecast profit. Figure 10.4 is a gross simplification of a budgeting structure and the interactions are equally simplified; but it does serve to demonstrate the factors which influence the cash budget and the formulation of business plans.

In principle the preparation of a cash budget is simply a matter of adding the budgeted revenues to the opening cash balance and deducting budgeted cash payments. In practice it is not quite so straightforward, since the various functional budgets which must be debited and credited from the cash budget relate to income and expenditure, whereas the cash budget summarizes receipts and payments. This difference arises from the time lag introduced by credit transactions and is further complicated by varying degrees of time lag for different items of revenue and expenditure. The only satisfactory method of dealing with this situation is to prepare a cash flow

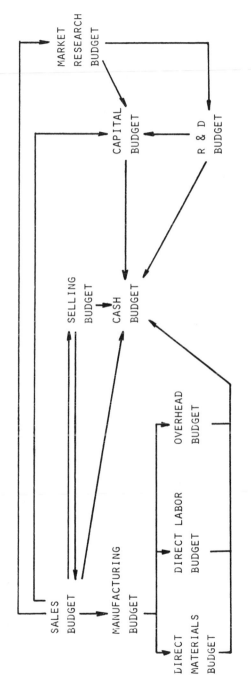

FIGURE 10.4. Interaction between cash budget and major functional budgets.

working sheet, which gives the forecasts for income and expenditure, followed by the known information on time lags, from which forecasts for receipts and payments can be determined, and also capital expenditure finance raised and repaid.

The cash flow working sheet[4] is a valuable planning aid. The natural fluctuations of business will cause a cyclic variation in the cash flow which, if corrective action is not taken, can cause cyclic surpluses and deficits in the cash budget. These can often be avoided by controlling the incidence of expenditure. Alternatively, in cases where interfering with the incidence of expenditure would disrupt some of the company's activities, the need for a short-term withdrawal of cash from deposit or for a short-term loan can be predicted in advance, with obvious advantages.

10.4.3. Plant Purchase and Replacement

Making decisions concerning plant purchase and replacement is a major part of preparing business plans and setting budgets. These are based on three main considerations[6]:

1. The actual financial cost of purchasing, installing, and commissioning the plant.
2. The benefits expected from the investment over a specified period.
3. The useful service life of the plant.

The profitability of plant, which is a fixed asset, will be determined by these three factors, combined with its manner of use and the skill with which it is maintained. An accurate assessment of potential plant profitability will depend on a combination of forecasting and process engineering knowledge. With the pace of technological change ever increasing, the differences between existing equipment and that purchased to replace or extend the capabilities it represents are also increasing. The factors involved have been discussed in detail in earlier chapters, but it must be recognized that they increase the complexity of the plant purchase and replacement decision.

The costs relevant to an equipment purchase decision can be identified using the guidelines provided in Section 10.2.1. There are a number of techniques for dealing with capital expenditure decisions, of which the present-value method of discounted cash flow is probably the most widely used. This takes account of the time value of money. Quite apart from inflation, the current value of a given sum of money is greater than that of the same sum in a year's time, due to its capacity to earn interest or yield a dividend if invested. From the viewpoint of the present-value method, a sum of 100 monetary units receivable in a year's time will have a present value of 93.43 units, if an interest rate of 7% per annum is assumed. Many texts give present-value tables, but it is probably simpler to calculate the present-value Y for the amount receivable X in n-years' time at a return on the capital investment of $100I\%$, where

$$Y = \frac{X}{(1 + I)^n} \qquad (10.6)$$

The benefits of investing money in a plant clearly have to be compared with the return on capital from investing the money elsewhere. For a decision on the purchase of a manufacturing plant, with all the risks and hard work involved, the minimum acceptable rate of return will have to be considerably higher than for money invested in a bank.

Taking a typical example, it has been proposed that four compression molding presses are replaced by a single injection molding machine. The compression presses each have a resale value of £250 after the cost of removal has been deducted and currently occupy two operators full time, at £140 per week each. The presses also take £900 per year each to operate and maintain. The injection molding machine will cost £27000 installed and will only occupy half an operator, whose remaining time can be fully utilized elsewhere, giving a direct labor cost of £70 per week, while maintenance and operating cost will be £3700 in the first year, rising by £600 per year thereafter, in real terms.

The first step is to decide on the minimum acceptable rate of return on capital expenditure. This will take into account the return which is likely to be earned if the money is used for other projects, as well as prevailing interest rates; a reasonable figure might be 15% before tax. Also, a figure must be placed on the time for which the asset will be required or its useful service life, whichever is the shorter, with five years being forecast in this case, after which the salvage value is estimated as £5000, when the cost of removal has been deducted.

The analysis of the problem is shown in Table 10.1, where the net earnings are the difference between the operating and labor costs for the injection molding machine subtracted from those of the compression presses. These figures are adjusted to their present values using Eq. (10.6), inserted in the present value of the cash flow column and then subtracted from the figure for the previous year in the cumulative cash column. From this it can be seen that the payback period for the injection molding machine at the required rate of return on investment is just over three years, indicating that it should be purchased, unless there is a better alternative.

For comparison of alternative proposals it is useful to calculate a

TABLE 10.1

Return on Investment Using the Present Value Method

Year	Capital expense	Net earnings	Present value of cash flow	Cumulative cash at present value
	27,000			
0	−1,000		−26,000	−26,000
1		10,820	9,413	−16,586
2		10,220	7,726	−8,860
3		9,620	6,330	−2,530
4		9,020	5,160	2,630
5		8,420	4,184	6,814
	−5,000			11,814

profitability index,[6] provided that the same level of return on capital has been used in each analysis:

$$\text{profitability index} = \frac{\text{total net income at present value}}{\text{total capital outgoings at present value}} \quad (10.7)$$

which for the analysis in Table 10.1 gives an index of 1.75. When choosing between alternatives, the proposal showing the higher profitability index will normally be chosen in preference to one having the shorter payback period.

In any decision on capital expenditure, whether it is concerned with plant replacement, expansion, or with a product development program, it is essential to take into account taxation, grants, and the likely effects of inflation. Details of company taxation are likely to change at the times when governments review such matters; those who need to be up to date should consult government publications. Grants or other types of financial encouragement are often available for the development and application of certain technologies, of which microprocessors are a prime example.

When taking into account inflation, a clear distinction must be drawn between synchronized inflation, in which wages, materials, energy, and sales prices for the product are expected to advance at the same annual percentage, and the more normal case, where the inflation rate differs between the various classes of expenditure. With a synchronized rate of inflation it will be found that the rate of return on investment increases in approximate proportion to the rate of inflation and, in the case of a 10% inflationary rate applied to the previous example, it becomes approximately 25%. This can be verified by reworking the figures in the analysis presented in Table 10.1, increasing all costs and expenses by 10% annually and using a minimum acceptable rate of return of 25%. It can then be seen that the actual rate of return is in fact slightly above the 25% level, as compared to the 15% level when there was no inflation. This indicates a satisfactory position, the higher return being sufficient to compensate for the falling value of money.

In cases where the inflation rates differ between the various classes of expenditure, the position may be altered considerably. The only way to deal with this is to make an estimate of the inflation in each class of expenditure and then rework the whole analysis. In addition, Gedye[6] has identified two points of great practical importance which influence the expected return on investment:

1. A prospect of wage and salary inflation that is more severe than commodity price inflation and unaccompanied by any improvement in the productivity of manpower will render projects unprofitable and discourage industrial investment. It is therefore a recipe for slump conditions and unemployment particularly in labor-intensive industries.
2. An increase in manpower productivity, even at a rate which is somewhat below the average wage–inflation rate, can serve to make industrial investment much more attractive.

10.4.4. Plant Maintenance and Energy Economy

The reasons for dealing with plant maintenance and energy economy together are:

1. Similar types of cost-effectiveness analysis can be used to determine the level of expenditure on each.
2. Both maintenance and the energy supply are usually the responsibility of the engineering services group.
3. The quality of maintenance can influence strongly the efficiency of energy usage.

It is generally agreed that there are three prime objectives of good maintenance:

1. The safety of those who use the plant.
2. The reliability and efficiency of the plant, as shown by its correct functioning and the minimization of the direct manufacturing overhead costs.
3. The preserving of the plant's asset value, as measured by its useful life and disposal value.

In addition to the above there are a number of intangibles, such as confidence in the plant reliability and the work attitudes engendered both by this and the appearance of the plant.

Traditionally there have been two main strategies of maintenance: preventive maintenance and breakdown maintenance. The former involves replacing components and carrying out work while the probability of failure and subsequent breakdown is still low. Routine maintenance is usually associated with preventive maintenance, in which inspections and certain maintenance tasks are carried out at predetermined intervals. In the case of breakdown maintenance no work is carried out on the plant until breakdown occurs or until imminent signs of breakdown are reported. In practice, most maintenance plans involve a compromise between preventive and breakdown maintenance, although the prediction of the onset of plant failure by computer monitoring and analysis, dealt with in Section 7.5.5, is increasingly influencing maintenance planning in capital-intensive plants.

The costs which must be considered in formulating maintenance plans include the direct costs of breakdowns, covering the work and the items replaced or repaired, and the indirect costs, which include the cost of the time lost at work stations influenced by the breakdown plus the overtime working costs necessary to "make good" the lost production. There may also be a loss of customer goodwill due to late deliveries resulting from a breakdown. In fact, the indirect costs resulting from a breakdown are often substantially greater than the direct costs, although it may be possible to reduce them by a number of measures:

1. The utilization of parallel production routes created by other work stations being able to work on the products affected by a breakdown.
2. The use of a standby plant which is normally idle and has been purchased with the intention of avoiding stoppages of production.
3. The use of buffer stores between work stations.

The preceding items imply that there will be underutilized plant and increased storage of work in progress to provide the safeguards, although each item can also be utilized for routine maintenance jobs which involve stopping operations at a work station, and can also ease changeovers from one product to another.

The optimization (minimization) of costs in both maintenance and energy economy plans can be achieved by comparison of the costs of preventive measures against the losses expected at different levels of preventive measure cost. Figure 10.5 shows a case for which there is a distinct optimum, where the total cost of the maintenance provision plus breakdowns is minimized, whereas for the case shown in Figure 10.6 there is clearly no point in establishing preventive maintenance procedures. Similar plots can be drawn from the savings to be made from energy conservation measures, as compared with the costs of achieving them.

In practice, the information needed to plot curves of the type shown in

Figures 10.5 and 10.6 is rarely available in a coherent form. However, a relatively limited investigation of production records can yield a substantial amount of useful information on breakdown costs, particularly those relating to indirect costs. Similarly, energy usage information may be capable of being extracted. Determining the cost of prevention and direct breakdown costs is generally easier, being contained entirely in the engineering service group records, but carries with it the question of the efficiency and relevance of the methods used. The subdivision of the resources made available for maintenance which give the maximum reduction in breakdown costs will tend to differ as the resource level is changed, with a consequent influence on the shape of the curves in Figures 10.5 and 10.6. The resources budgeted would typically be employed on instrumentation and monitoring equipment, maintenance staff, maintenance equipment, and a stock of spare parts for the plant.

10.4.5. The Cost of Quality

From an economic point of view, all resources allocated to the improvement of product quality are either income-expansion or cost-reduction expenditures.[7] Increasing income through quality improvement is

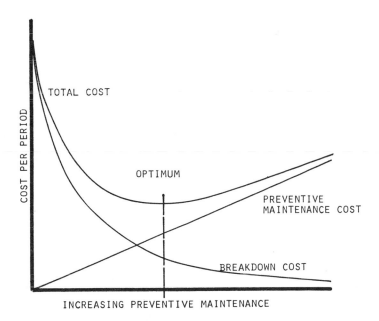

FIGURE 10.5. Curves showing economically justified preventive maintenance.

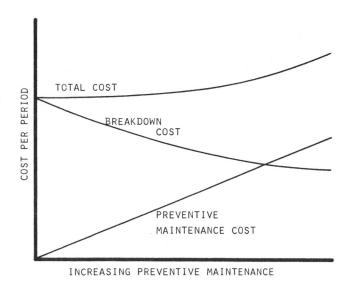

FIGURE 10.6. Curves showing that preventive maintenance cannot be economically justified.

concerned with raising the quality of product design and, consequently, the precision of manufacture. In contrast, cost-reduction expenditures are directed towards improving the quality of conformance with existing specifications for products.

Estimating the increased revenues which will accrue from improvements in the quality of product design and manufacture is difficult, due to supply-and-demand considerations and the response of customers, both present and potential, to the "up-market" movement of the company. It is necessary for customers to be convinced of the monetary advantages of improved product performance, which will be accompanied by an inevitable increase in selling price. Typical benefits of improved quality of product design and manufacture are greater reliability, improved performance, and longer service life.

Evaluation of customer response to improved product quality falls within the responsibilities of the marketing and sales groups; the decision to operate at higher quality levels clearly depends on the judgment of the executive management of a company. These activities are properly an integral part of market research and company development, dealt with in Section 9.3, but it is useful to emphasize here that they are based on a broad view of product quality. The economic benefit of increasing income through improving the quality of product design and manufacturing methods is potentially far greater than the reduction of costs by improving the quality of conformance with specifications for existing designs and methods. However, it is absolutely necessary to maintain a level of quality of conformance with existing product specifications which satisfies existing customers.

The information needed to estimate the cost of achieving and maintaining a certain standard of quality of conformance is readily available in most companies. Similarly, the economic benefit to be derived from the expenditure, in terms of the reduction of manufacturing costs, is also easily obtained. It is more difficult to estimate the cost to the company of handling customer complaints and loss of business from dissatisfied customers, although it is essential to place a monetary value on these for a realistic cost-benefit analysis.

As with the analysis of maintenance costs in the previous section, both expenditure and benefits can be plotted onto a single graph and a minimum total cost determined. In Figure 10.7 the expenditure is divided into prevention and appraisal. The former includes the cost of maintaining the quality-control group activities directed towards devising, implementing, and maintaining methods of avoiding the production of defective work. The latter is concerned with the cost of providing a screening service to reduce the probability of defective work being delivered to the customer. The curve labeled "failure" includes both the internal cost of producing defective work and the external cost associated with allowing defective work to be delivered to the customer.

10.5. BUDGETARY CONTROL

In addition to their planning role, budgets are used widely for cost control, which demands a more detailed approach than that needed for planning. The preparation of budgets is only the setting phase of a formal control system, and is followed by the operating and feedback phases, supported by corrective action where necessary. The Institute of Cost and Management Accountants define budgetary control as[4]:

> The establishment of budgets relating the responsibilities of managers to the requirements of a business plan, and the continuous comparison of actual with budgeted costs to secure, by individual action, the objective of that business plan or to provide a basis for its revision.

There are three main types of budget:

1. Appropriation budgets.
2. Fixed budgets.
3. Flexible budgets.

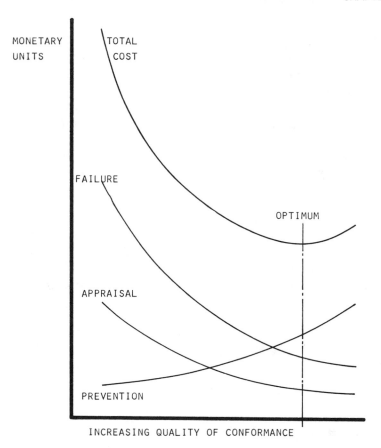

FIGURE 10.7. Curves showing the benefits to be expected from expenditure on improving the quality of conformance.

Appropriation budgets are used in cases where the way in which budgeted resources should be employed in order to obtain the maximum benefit from the expenditure cannot be determined precisely in advance, and also where the determination of the level of necessary expenditure is largely a matter of judgment. Budgets for development programs and for the purchase and replacement of the plant are appropriation budgets. Also, in conventionally structured companies, appropriation budgets are used to allocate the operating expenses of groups such as quality control, production planning, computer systems, and maintenance. In companies having a business center structure the operating expenses for service groups are largely generated by payment for services performed for the production business centers, providing a more direct and accurate method of allocating resources where they are needed than is possible using appropriation budgets.

Although appropriation budgets are not generally highly structured in the sense of specifying how the resources should be utilized, the incidence of expenditure must be subjected to constraints to avoid cash flow problems. For development programs and plant puchase the forecast incidence of expenditure will be contained in the proposal. Service groups also have to indicate how their expenditure is to be distributed throughout the year. Once the total budgeted amounts and the incidence of expenditure are fixed by the planning and budgeting group, they should be regarded as limit values, not to be exceeded, provided that they are sensible and adequate to support the essential activities for which they are intended.

Fixed budgets are designed to remain unchanged, irrespective of the level of activity actually attained. The forecast major functional budgets dealt with in Section 10.4.1 and 10.4.2 are fixed budgets, identifying the purpose of fixed budgets as being planning and coordination. Because they do not change in response to level of activity, they cannot be used for control purposes. For example, if output rises, causing a welcome increase in profits, comparison of the increased manufacturing costs with the forecast fixed manufacturing budget will reveal an unfavorable variance. It must also be noted that appropriation budgets can give the same problem where the level of activity in service groups is related to manufacturing activity, as would be the case for production planning and maintenance. With a business center structure this problem is avoided, since the income of service groups will be related to the work performed for the manufacturing groups. However, the problems of cost control can be largely overcome by using flexible budgets.

The effective control of variable costs can be attained through flexible (or control) budgets, where the actual expenditure can be compared with the budgeted expenditure *for the level of activity actually experienced*. The use of flexible budgets is largely confined to the manufacturing areas, although the budgets of some service groups may be related beneficially to some index of production activity. In manufacturing, flexible budgets are used for establishing targets for overhead costs, indicating their close relation with standard costs. In fact, standard costs take budgetary control to its logical conclusion. Flexible budgets are used where the detailed recording and cost apportionment discussed in Section 10.3.3 is either unnecessary, uneconomical, or simply not possible.

The main difference between flexible budgeting and standard cost methods arises from the way in which the overhead costs are absorbed onto the work in progress. For the former it is accepted that the overhead costs recorded cannot be associated directly with the operations from which they arose, and that a surrogate *measure of activity* must be used. Among the possible choices for level of activity are:

1. Direct labor costs.
2. Direct labor hours.
3. Units of product.
4. Hours of machine operation.
5. Materials used,—costs.

The choice will depend on the nature of the manufacturing operation being examined. Clearly, machine hours will not be a good measure of activity where machines only play a small part in the manufacturing operation and, similarly, direct labor hours will be inappropriate where automation is extensive; although in the latter case detailed monitoring should be feasible, making the use of a measure of activity unnecessary.

All the measures of activity in the preceding list require the calculation of an overhead rate to accomplish the absorption of overhead costs onto the work in progress where

$$\text{overhead rate} = \frac{\text{budgeted overhead for level of activity}}{\text{level of activity}} \quad (10.8)$$

The budgeted overhead can be established from a similar procedure to that described for standard costs in Section 10.3.2. It is then a normal accounting practice to separate the total overhead cost into fixed and variable elements, with depreciation and loan repayments included in the fixed costs, and energy and other operating costs forming the flexible costs.

Once the overhead costs have been absorbed onto the work in progress using the overhead rate, budget variances can be calculated, giving a measure of manufacturing performance. The forms of the variance equations will depend on the measure of activity selected. It will be assumed here that labor hours are chosen, since the use of flexible budgets is most appropriate to labor-intensive operations, where automatic monitoring methods are inappropriate. The variances for the fixed element of the overhead costs then include[1]:

$$\text{budget variance} = \text{actual fixed costs}$$
$$- \text{budgeted fixed costs} \quad (10.9)$$
$$\text{idle capacity variance} = \text{fixed overhead rate}$$
$$\times (\text{budgeted hours} - \text{actual hours}) \quad (10.10)$$

efficiency variance = fixed overhead rate

$$\times \text{ (actual hours } - \text{ standard hours)} \quad (10.11)$$

where standard hours refers to the number of hours in the standard working day. For the variable overhead costs the variances include:

efficiency variance = variable overhead rate

$$\times \text{ (actual hours } - \text{ standard hours)} \quad (10.12)$$

budget variance = actual variable overhead
$$- \text{ (actual hours } \times \text{ variable overhead rate)} \quad (10.13)$$

It must be emphasized that the precision with which these variances represent the true manufacturing performance with respect to overheads depends entirely on the validity of the overhead rate which, in turn, depends on the selection of an appropriate measure of activity. In all cases, the calculated overhead cost per product will be an estimation, and the deviations from actual overhead cost per product can result in variance values which are misleading. For capital-intensive operations, where the overhead costs are a large proportion of the total manufacturing costs, it is very important to obtain an accurate determination of overhead cost per product, pointing to the need for automatic monitoring methods and the use of standard costs.

REFERENCES

1. Beirman, H., and A. R. Drebin, *Managerial Accounting, An Introduction*, Collier MacMillan, London (1972).
2. Buffa, E. S., and J. S. Dyer, *Essentials of Management Science: Operations Research*, Wiley, New York (1977).
3. Copeland, R. M., and P. E. Dascher, *Managerial Accounting*, Wiley, New York (1978).
4. Baggot, J., *Cost and Management Accounting Made Simple*, W. H. Allen, London (1973).
5. Garfield, R. W., Private Communication, Woodville Polymer Engineering, Ross-on-Wye, U.K. (1983).
6. Gedye, G. R., *Works Management and Productivity*, Heinemann, London (1979).
7. Kirkpatrick, E. G., *Quality Control for Managers and Engineers*, Wiley, New York (1970).

Production Management

11.1. PRODUCTION PLANNING

11.1.1. The Production Control Group

The production control group is concerned primarily with scheduling and coordinating all the direct activities which contribute to the manufacture of a saleable product. The Institute of Production Control defines a production controller or planner as a person who

> defines targets, allocates resources, communicates plans, monitors performance, and corrects deviations.

The complexity of a production control group's task depends on the size and nature of the company. Companies manufacturing a limited range of their own products (branded products), such as tire companies, have a less complex task, despite their size, than most general-rubber-goods jobbing companies. The latter manufacture a wide range of product types from an equally wide range of materials; and often present production planners with the additional problem of forward planning being disrupted by customers who place lucrative orders contingent upon short delivery times being met.

It is to be expected that the amount of production planning needed for effective coordination of manufacture will increase as the number of different types of products manufactured in a given period increases. It will also tend to be greater for labor-intensive operations, where the uncertainties of operator performance and attendance demand considerable corrective action. However, it could be argued that breakdowns of capital-intensive operations cause greater disruption of manufacturing schedules.

When a company's production control requirements demand the efforts of more than one person, some thought has to be given to the allocation of responsibilities. This will largely depend on the structure of the company, due to areas of responsibility normally being defined by departmental or business center boundaries. The role of the production control manager is overall coordination, to ensure that the timing of transfer of work from one area to the next will result in the finished products being available for the delivery date.

11.1.2. The Introduction of Work

The introduction of a job into the manufacturing areas is only initiated on receipt of a works order, prepared by the sales office. The activities preceding the issue of a works order depend on the type of company and the status of the product. In a company manufacturing "own products," the works orders are governed by stock levels and selling rates, whereas in a jobbing company the sequence of activities which lead to the issue of a works order are activated by a customer enquiry. These activities depend on the status of the product; the status can be as follows:

1. A new product requiring development work.
2. An irregular or infrequent repeat order.
3. A regularly repeated or ongoing order.

Quotation of delivery times on new products is the responsibility of the company estimators, in consultation with the production control, purchasing, and development groups; but a works order cannot be issued until the job is released from the development groups for production. Irregular or infrequently repeated jobs cannot be included in forward planning in a precise manner, since the quotation of a delivery date depends on the availability of materials and subcomponents, and on the production controller's estimate of the time when it can be introduced into production, based on the forward loading plan (drawn up from the manufacturing budget). For regularly repeated orders, forward planning should ensure that all the resources for manufacture are available as needed; but "exploding the order backwards" into material and subcomponent requirements is still essential for reordering and control of stock levels, and is the basis of materials requirements planning. For all repeat orders this operation can be carried out by computer, using the information contained in the master specification.

When a firm order is received in a jobbing company it first goes into the company order book, which is primarily an invoicing document. From there it is transferred into the works order book, minus the number of products in stock, if any, and is then issued to the production control group. As well as serving as the authority to produce, the works order does the following:

1. Translates the customer's order into terms used in the manufacturing areas.

2. Ensures that all the necessary information is provided.
3. Gives an identifying reference number to the order.

Once the works order is received, the production control group can proceed with their detailed work.

11.1.3. The Principles and Objectives of Production Control

The quality of production control has a direct and immediate influence on profitability. Delays occurring through materials and subcomponents not being available for use at the required time will obviously have an adverse affect on profitability and future orders; but the production planner can make a positive contribution to profitability, through the way in which jobs are loaded onto the various machines and work stations. This is generally referred to as "optimizing the product mix" and is necessary in order to:

1. Maximize machine and labor utilization levels.
2. Achieve economic batch sizes.
3. Minimize the work-in-progress inventory.
4. Achieve standard cost and budget targets.

Item 1 is concerned with minimizing machine (and operative) idle time. Item 2 acknowledges that there will be a setting up or changeover cost associated with each job at each machine or work station, and seeks to maximize the length of production runs, so that the setting-up cost per unit produced is minimized. This must be balanced against the objective of item 3, which is to minimize the capital tied up in work in progress. For example, a customer may order 5000 items, to be supplied at the rate of 1000 per month. Assuming that the items can be produced at the rate of 500 per working day, the production planner then has to decide if two day's production per month, immediately prior to the delivery date, should be programmed, or if several months supply should be produced in a single production run and held in stock until needed. The reduced manufacturing cost per unit resulting from the longer production run must be balanced against the consequences of the loss of liquidity, due to the capital tied up in the stock, the cost of storage, and loss of value due to inflation. In companies where a number of different products are manufactured and use common processes, full advantage has to be taken of the available programming permutations, within the overall limitations of machine availability and delivery dates, to optimize the product mix for minimum manufacturing costs.

The production planner is also the main arbiter on the quotation of delivery times to customers, unless the products are available directly from stock. An order has to be "exploded backwards" along its manufacturing route, identifying the resources needed and the time when they will become available at each stage. The total time, from the confirmation of the order by the customer to delivery of the completed products, is used as the basis of the delivery time. Alternatively, when the gaining of an order is contingent upon meeting a delivery date, the production control group has to reorganize scheduling and loading for this, if possible.

Scheduling and loading are interactive activities which are necessary for the systematic and efficient planning of work. Scheduling is concerned with establishing the start or finish times for each operation in the manufacturing sequence, from consideration of the duration of operations, which will enable an order to be delivered by the due date. Consequently, scheduling is a product-oriented activity. Loading then involves assigning jobs to individual work stations, with the objectives of meeting the requirements of a number of product schedules and optimizing the product mix, showing it to be an operations-oriented activity.

Production control can benefit considerably from the introduction of computer methods. These take three main forms:

1. Batch-operated computer runs to produce hard-copy production analyses to aid manual planning.
2. Interactive computer support to aid and guide manual planning.
3. Computer packages which perform the tasks of scheduling and loading, using batch or distributed system computers.

Batch-operated computers, where a computer technician takes the data to be processed and then returns the resulting analyses in the form of a paper printout at a later time, are being superseded rapidly by distributed systems, which are directly accessible to the user. In these interactive systems the user communicates with the computer through terminals at a number of locations within the company. The terminals can take various forms, depending on the purpose for which they are used, but the most common type is the now familiar visual display unit and keyboard module.

The choice between a manual system and one where the computer does the scheduling and loading depends on the size of the company, the number of product lines, the frequency with which they are changed, the number and type of work stations and, most important, the nature and effectiveness of the current production planning activities. The vast range of options available, deriving from the way in which different permutations of system capabilities can be assembled together, makes general recommendations very difficult. The prime decision is whether to provide computer aid for manual planning methods which are known to be capable of functioning effectively or to replace ineffective or inefficient planning methods with a computer package system which would perform the scheduling and loading tasks. Except in companies with very simple production planning requirements, there is no doubt that computer aid is beneficial, if only to reduce the routine clerical work load. Although the emphasis in the following sections will be placed on medium to large multiproduct companies, simpler requirements can be dealt with using similar techniques on low-cost systems, such as those now being marketed as "personal business computers".[1]

11.1.4. Computer Files for Production Control

There are two types of computer files which are used by the production control group (and others). The first is a data base, which is essentially a library, and the second is an operational file which, as the name suggests, is created specifically for carrying out a task. The primary data bases for production control are:

1. Master specifications.
2. Standard costs.
3. Identification codes.

Master specifications are originated and maintained by the quality-control group and standard costs by the estimators and value analysts. The former give all the information required to manufacture a product, while the latter itemize the time and cost at each stage of manufacture. The computer file system, with its associated program, should enable the information needed for production planning to be extracted, manipulated, and presented in all the ways required. The identification-code data base is essentially a list of the products which can be manufactured by each work station. It therefore determines the possible permutations of jobs and work stations available to the production controller for optimization of the product mix. Although this data base is usually maintained by the production control group, the information for it can be generated by any of the technical-support groups responsible for the introduction of new products, equipment, and techniques into production.

The prime characteristic of operational files for production control is the importance of time scale. This starts with the works order file, which specifies delivery dates and continues into the scheduling and loading files. Such files must be updated frequently, to present an accurate record of the current and planned manufacturing situation, and also to provide a viable basis for scheduling and loading activities. The files will, of course, be amended as the result of scheduling and loading activities, when new works orders are introduced and in response to feedback from production on differences between planned and actual performance.

11.1.5. Scheduling

On receipt of a works order the first step in scheduling is to determine the resources needed, in conjunction with the purchasing group. This can be done by reference to the master specification and standard costs information. The order is then "exploded backwards" into work-station times, transit and storage times, material requirements, and subcomponent requirements. If the incidence of works orders conforms to forecasts, the materials should be available in the stores; similarly, for own product manufacture, subcomponents should be available. For jobbing companies it is a sound policy for the purchasing group to subcontract in advance of a firm order only when the customer is judged to be reliable. Material and subcomponent availability determine the earliest time at which work on an order can commence. The agreed-upon delivery date determines the latest time for work on an order. All this is routine and can be readily performed by a computer program operating from the data bases and works order file. Manual intervention is only necessary when the production planner has to refer to the purchasing group for material and component delivery data. From this point onwards production planning requires judgment, and the computer-aided manual methods diverge from the computer executed methods. This section will follow the computer-aided manual methods route, while Section 11.1.7 will deal with computer-executed scheduling and loading.

The next step is to formulate a schedule which will enable the order to be completed between the imposed time limits. The Gantt sequence chart[1] is a useful aid for scheduling and can form the basis of a computer-aided scheduling method. Figure 11.1 shows a typical rubber industry example, Mixing is followed by a 24-hour minimum storage period, to permit cooling and stabilization of the processing characteristics to occur, and also to allow time for routine testing. Similarly, there is a 12-hour moratorium between the extrusion and use of molding blanks. Transport times between operations are also included. In Figure 11.1 the sequence is arranged for completion at the earliest possible time; by starting each activity in the sequence as soon as the previous one has provided sufficient work for operations to be feasible. Work in progress (and storage space) is minimized by mixing at intervals determined by the demands of the molding operation. Trimming and inspection are done when sufficient products have been accumulated to form an "efficient" work load. This arrangement gives three days to "spare" between completion and the delivery date. This spare time is important because it can be distributed between any of the operations in the sequence, giving the production planner a degree of flexibility when a number of jobs have to be loaded onto the work stations. To provide an ideal schedule as a starting point for loading, all the spare time would be concentrated at the start of the

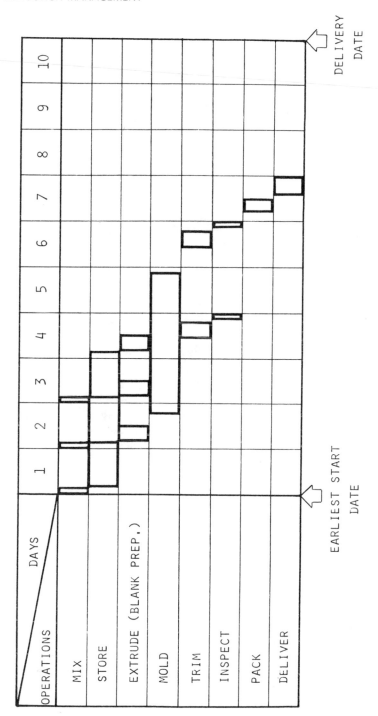

FIGURE 11.1. A sequence chart for job scheduling.

manufacturing sequence, to minimize throughput time and work in progress. In the example given in Figure 11.1, this would involve shifting all the operations forward by three days.

To produce a sequence chart the production planner has to make some initial decisions about the resources he or she is going to allocate to the job. In the example given in Figure 11.1 the molding operation has the greatest influence on the total throughput time. Assuming that three molds are available and the production controller has decided initially to use two of them, the controller then has the option of only using one, which increases the total throughput time but confines the molding operation to one press, or the option of using all three, which increases the spare time but spreads the loading of the job over three presses. This further flexibility, if it is available, is of great assistance to the production planner in achieving the three objectives of meeting delivery dates, maintaining uniform operative and machine utilization levels, and minimizing work in progress. It also emphasizes the importance of interchangeability of jobs between machines. If a large proportion of the products are constrained to a single machine at each stage in the manufacturing sequence, the flexibility of loading will be much reduced, causing a decrease in the efficiency of operator and machine utilization and an increase in storage of work in progress.

When a job can be scheduled to run on more than one machine for a particular operation, the direct costs can be reduced by loading it onto the machine which does it at minimum cost. Standard costs are invariably based on the minimum-cost manufacturing route and products should always be assigned to this route at the scheduling stage. The compromises which have to be made during loading will be dealt with in the next section.

When interactive computer aids for manual loading are being considered it is logical to extend them back to the scheduling stage, to enable a schedule to be recorded directly into an operational file and then retrieved for insertion into the loading plan. Comparisons between the display of ordinary printed information on a computer terminal screen and graphic displays show that the latter methods are more effective because of the perceptual nature of scheduling and loading. The Gantt sequence and loading charts can be adapted readily to a computer graphics terminal method, as can be seen in the next section.

11.1.6. Loading

The task of determining machine loadings from the product schedule is essentially one of trying different permutations until a satisfactory compromise between the main objectives listed in Section 11.1.3 has been reached. A competent production controller can readily achieve a good compromise, provided that he or she has access to effective planning aids and uses them in a systematic manner. The basic aid is the Gantt loading chart, which can be utilized in three mediums: paper charts, proprietary planning boards,[2] and interactive computer graphics.[3] At this stage it is worth introducing the loading chart symbols and Table 11.1 shows those which have become fairly standardized. Additional information is optional and depends on the medium in which the chart is presented.

Paper charts can carry little more than the standard information in Table 11.1, and even with adhesive strips and symbols they are unwieldy and time consuming, bearing in mind that charts need continual updating in order to be useful. Proprietary planning boards, such as the Sched *U* Graph and Product-trol, are far less time consuming to update and can carry more information. They provide a clear overview of the current loading situation but are still rather unwieldy to use during planning, when a large number of permutations may be possible. There is also the further problem of physical size, limiting both the number of work stations and the amount of forward planning which can be displayed. Even so, boards are effective planning aids which are in current use. The Sched *U* Graph system uses a card for each job, cut to the length appropriate to its duration. These are fitted into holders which are then slid into one of the horizontal slots extending to the full length of the board, in a position dictated by the board time scale. Each horizontal slot represents a work station.

Computer graphics provide an opportunity for setting up a more informative and versatile interactive planning aid than the boards. Graphics

TABLE 11.1

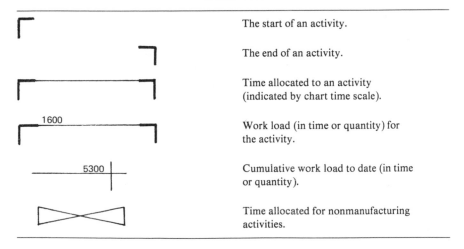

⌐	The start of an activity.
⌐	The end of an activity.
⌐____⌐	Time allocated to an activity (indicated by chart time scale).
⌐__1600____⌐	Work load (in time or quantity) for the activity.
___5300_	Cumulative work load to date (in time or quantity).
▷◁	Time allocated for nonmanufacturing activities.

terminals have advanced remarkably in recent years, as well as becoming available at a very reasonable cost (£3000 for a high-resolution color graphics terminal in 1983). Color graphics are essential to enable the status of jobs to be assimilated rapidly and the keyboard of the terminal must be suitable for making rapid changes to the Gantt chart displayed. Most terminals have "function keys," which can be programmed to implement the specialist operations needed for manipulating the chart, and also have a joystick, which could be used to locate an item on the screen by a movable cursor and then relocate it in a different position. Computer color graphics aid for work-station loading can be implemented in a number of steps, all of which should be within the capabilities of a competent computer programmer. The use of a computer method for exploding an order backwards has already been described. It is logical to continue from this point with the computer, so that a coherent record of the progress of a product through the company is available for immediate recall from the computer files.

As previously mentioned, scheduling and loading are interacting activities. If the minimum-cost manufacturing route allocated to a job during the initial scheduling prevents the efficient loading of work stations, then it is necessary to go back to the scheduling stage to examine the possibilities for changing the production route or the number of work stations to be used. The facility of being able to change rapidly from the loading chart to the sequencing chart on the graphics terminal is extremely useful for this purpose. To enable it to be used effectively two aids are required:

1. A symbol which appears on both the sequencing and loading charts to indicate that alternative routes are available.
2. A simple method of ranking alternative routes by their relative direct costs or predicted standard cost variances.

The symbol denoting the availability of an alternative manufacturing route or the possibility of loading the job onto more than one work station at any stage of manufacture must be an integral part of the data bases and also include a code by which the details of the alternatives can be retrieved. For this operation it is useful to have a normal VDU alongside the graphics terminal, enabling the information to be studied in conjunction with either the sequencing or loading charts. For ranking manufacturing routes by their relative direct costs or by some more sophisticated measure of productivity, a technique originated by SKF, Philadelphia[2] can be used. In the original scheme, an efficiency index is calculated by dividing the direct costs of each work station which can perform a particular operation by the direct costs for the most efficient work station. This can be extended to the whole manufacturing route by adding up the direct costs incurred at each stage of manufacture for each alternative route and then dividing by those for the most efficient route.

The loading of a job onto work stations for the sequence chart is best started at those operations which make the greatest contribution to the manufacturing time and place the greatest constraints on the available loading permutations. In rubber companies these are usually represented by fabrication and molding operations, which are product specific. Most other operations, such as mixing, extrusion, and calendering, are undedicated and can be changed rapidly from one product to another. They are also usually capable of greater output rates than fabrication or molding operations and therefore serve a number of product lines, giving the production controller a good measure of flexibility in formulating loading plans.

The development of the facilities provided by an interactive computer graphics terminal for loading can be built up from those of the basic Gantt loading chart. If simple horizontal bars are used to indicate the duration of a job the main advantages of using interactive graphics are the ease with which jobs can be tried in different permutations and the choice of the region of the Gantt chart to be viewed. This can be limited to the work stations on which a particular job can be loaded or expanded, with a corresponding reduction of the size of each bar on the screen, to achieve a general overview. The symbols defined in Table 11.1 can be used and the jobs identified by their code numbers, as shown in Figure 11.2. At this stage the color graphics are not essential and the techniques could be implemented on an ordinary high-resolution monochrome terminal.

The advantages of color graphics lie in the color coding of bars to indicate the priority and status of jobs. Priority indicates the importance of a job to the company and high-priority jobs are those which:

1. Contribute a substantial proportion of the total production unit profit.
2. Contribute a substantial proportion of the overhead cost absorption.
3. Are for customers who place or who are likely to place large and continuing orders with the company.
4. Enhance the company's reputation.

Item (4) embraces prestigious jobs which may not be very profitable but which receive or can be given wide publicity. Obviously, low-priority jobs are those which do not have any of the preceding attributes.

The term status is used in the context of the differences between the initial schedule, the loading plan, and the practical execution of the job. There are three status categories into which a job can fall:

1. Conforming precisely to the initial schedule.

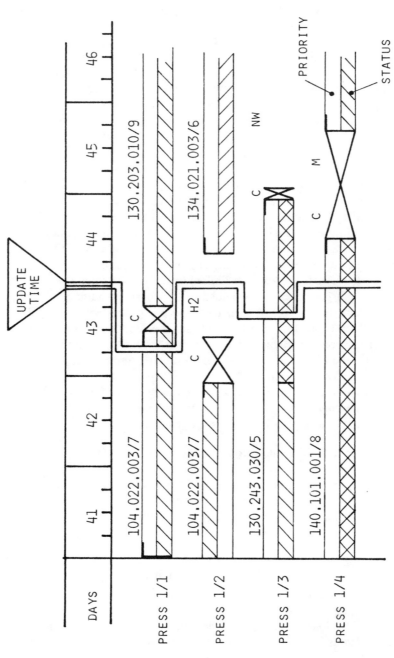

FIGURE 11.2. A portion of a loading chart for molding operations. (*Key:* ▨ Jobs conforming to prediction; ▨ jobs which are currently overrunning their allocated production time or are expected to do so; ▨ ; jobs subject to delays which will cause late delivery; C, changeover and setting up; M, maintenance; H2, held for feed strip; NW, no work; 104.022.003/7, typical product code.)

2. Not conforming to the schedule but, with the adjustment of the spare time and rescheduling of activities, able to meet the delivery date.
3. Unable to meet the delivery date using the resources allocated in the schedule.

Unlike the allocation of priority, which would be a sales office responsibility, status is computer defined. A program can be written to use the information from the sequencing charts to determine, in response to an interrogation command, the status of jobs influenced by a loading decision or by adjustments made to the loading charts on the basis of progress reports from the manufacturing areas. The latter case will be dealt with in Section 11.3, but the principles are similar to those for loading new jobs. When a job falls into category (2) the schedule has to be adjusted, as described in the previous section. The production planner can then proceed to load the job onto the other work stations in the manufacturing sequence. It is possible that at some point in this procedure the spare or flexible time will be totally eroded, whereupon the job will fall into category (3).

There are a number of ways to recover a job from a category (3) situation. The main ones are to allocate more resources to the job or to insert it into the forward loading for a work station at an earlier date than it would have been programmed by simply putting it at the end of the queue. The easiest way of allocating more resources to a job is to authorize overtime working; but this always carries increased costs with it and is therefore undesirable. It can be used as a means of achieving delivery targets on high-priority jobs when alternative methods have failed. Allocating more resources to a job within normal working hours involves either more work stations being used or transferring the job to work stations which are capable of a higher output rate. Often these options will not be available, leaving the reorganization of the order in which jobs are loaded onto the work stations as the only option, other than overtime working, if the products are to be manufactured within the company. Some jobs can be subcontracted or "factored out" to other companies, if the need to do this is identified in time for it to assist in solving the problem of meeting delivery dates.

Returning to the reorganization of the order in which jobs are loaded onto work stations, the color of the bar indicating the job on the graphics screen keeps the controller fully informed of the consequences of his or her reorganization plans. Jobs delayed by the insertion of category (3) jobs at an intermediate point in the queue may themselves acquire category (2) or category (3) status. The planner's task is then one of finding the permutation which gives the best compromise between priority, status, and the other general objectives of production control, such as minimization of work in progress.

Warnings about the violation of other constraints on both scheduling and loading may be necessary. For example, unless a warning is given the production controller may inadvertently plan to commence a job on a work station before an adequate supply of work in progress has been delivered from the preceding work station in the manufacturing sequence. This is a measure to avoid gross planning error and not to deal with the small discrepancies which inevitably occur between planning and execution. The responsibility for dealing with the latter rests with the production management.

So far the emphasis has been placed on situations where the work stations are operating near their maximum capacity. In numerous instances this desirable situation will not exist and there will be a substantial amount of spare capacity. This makes the production controller's task easier, but does not diminish the positive contribution he or she can make to the company's performance. Control of work in progress and achieving economic batch sizes and running periods for work stations become very important. If work stations can be totally planned out of production, the overheads associated with them can generally be reduced. In the extreme case, the equipment may be sold and the staff dismissed, in the interests of maintaining company viability.

In Figure 11.2 a portion of a press loading chart is shown and similar charts will be used for loading all the work stations in the production areas. Each horizontal bar representing a job is split into an upper portion, which indicates the priority, and a lower portion, which indicates the status. Experience with graphics terminals has shown that it is preferable, from the user's viewpoint, to use different shades of a color to denote different levels of an index and then to distinguish between indices by using different colors. For example, high priority could be identified by light red, with mid-red for normal priority, and deep red for low priority. Similarly, a light green could be used for category (1) status, mid-green for category (2), and deep green for category (3). Figure 11.2 shows job status by cross-hatching, in the absence of color. The status of the job on press 1/3 is shown to change from category (1) to (3), indicating that a portion of the products will be delivered on time but that the remainder will be late. Delivery of part of an order on time can often prevent a customer having problems and maintain good relations.

Text information on the loading chart shown in Figure 11.2 includes the job numbers and codes which identify the reasons for nonproductive time on the work stations. Setting out reasons for nonproductive time is not particularly useful to the production planner, but it is essential information for the production manager.

The function of the update time bar in Figure 11.2 is to show when the last report on a job was fed into the computer, so that decisions will not be taken to the assumption that all the information on the chart represents the current situation in the production areas. The ability to recall the number of products completed at the update time and the total number in the order is necessary for both the production controller and the production manager. This information can be displayed on the loading chart or, preferably, on an adjacent VDU, as suggested for scheduling. If the VDU is available, then the information can be extended to computer-calculated estimates of the time to complete an order, based both on standard costs information and the production performance in the period from the start of the job to the update time.

Additional aids for the production controller can be developed as the need for them is identified. As with any new or unfamiliar system, it is better to start by implementing the basic facilities and then adding to them progressively. Also, the use of the loading chart as a key aid to the production manager for monitoring manufacturing performance has been identified; this will be developed further in Section 11.3.4.

11.1.7. Computer Packages for Scheduling and Loading

A number of computer hardware and software companies, including the large ones, such as ICL, IBM, and Hewlett-Packard, offer program packages which include scheduling and loading elements. These take two basic forms:

1. Computer information aids to manual scheduling and loading.
2. Programs with algorithms for performing scheduling and loading operations.

The information aids to manual scheduling and loading are similar in character to those described in the previous sections concerned with graphics aid. Data bases have to be established for the master specification and standard costs information. In conjunction with the works order files, the current loading plan, and the progress reports from the production areas, data bases enable the production planner to do his or her job. The presentation of all this information in text form, either at a VDU terminal or in the form of hard copy, has the disadvantage that it cannot be so quickly assimilated or comprehended by the production planner, in comparison with the color graphics methods, and is far less efficient as an interactive aid to scheduling and loading. This will tend to result in fewer permutations being tried, with a consequent reduction in manufacturing productivity. Despite these disadvantages, commercial computer program packages from reputable companies will be reliable and free from irritating idiosyncrasies. A commercial package for using color graphics methods is not yet available at the time of writing (1983) and their implementation requires a substantial in-house programming effort.

Computer program packages which use algorithms to perform the scheduling and loading operations do not require a display of information for the solution of the problem. These systems are usually based on a set of user-selected priority rules for the allocation of jobs to work stations, using data-base and other information similar to that for the previous methods. The opinion that program packages of this type are less efficient than computer-aided manual scheduling and loading has been expressed, particularly for jobbing companies with fluctuating priorities and criteria. Certainly, the computing power needed for the algorithmic approach is far greater than that for the computer-aided manual methods; but the cost of computing power is dropping very rapidly and the versatility and scope of algorithmic methods will inevitably improve. Such programs are generally beyond the capabilities of in-house programmers and commercial packages must be purchased.

11.2. PURCHASING AND INVENTORY CONTROL

11.2.1. Purchasing Strategy

The purchasing group can exert a strong influence on the profitability of a company, since it is responsible for one of the major budgets. On the negative side, materials and components not available when they are needed will certainly reduce profitability; but the buyer can make a substantial positive contribution by minimizing the total cost of raw materials, subcomponents, and supplies to the company. However, before going into any further detail it must be emphasized that cost minimization does not mean the purchase of inferior or variable-quality supplies. This is usually a false economy, since the saving in purchase price is often more than offset by manufacturing problems. A strict policy of proving trials prior to a change of supplier should be followed.

The basic guide to purchasing strategy is the "economic order quantity,"[1,4] which is the order quantity giving the minimum unit cost for purchasing from a particular supplier. There are a number of factors which influence the economic order quantity and many constraints on its use. The major factors which can be written into a cost equation are:

1. The average stock level.
2. The cost of storage.
3. The cost of ordering, delivery, and handling.
4. Quantity discounts.

The average stock level x_a will be a function of the economic order quantity x_0:

$$x_a = \tfrac{1}{2}x_0 + x_b \qquad (11.1)$$

where x_b is the buffer stock shown in Figure 11.3. The cost of storage is generally determined as a percentage per annum of the purchase price. Table 11.2 details the main sources of variable costs for a situation where adequate space is available in the company stores.

The costs of ordering, transport, and handling can be dealt with collectively for the purpose of determining the economic order quantity, although they will need to be separated for cost assignment purposes. The cost of raising an order will be similar for small and large orders, but in the latter case its ratio to the order value will be smaller, indicating a more efficient use of purchasing group resources. Similarly, the ratios of delivery and goods-inwards handling to order value will become more favorable as the order size increases. For simplicity, the collective cost for a particular supply item can be represented by a standard variable cost per order C_r, irrespective of order size. This average cost is adequate for many cases; the effort of establishing a mathematical relationship between order size and the collective variable cost is generally only justified for high-value orders, where it is apparent that it will exert a strong influence on the economic order quantity.

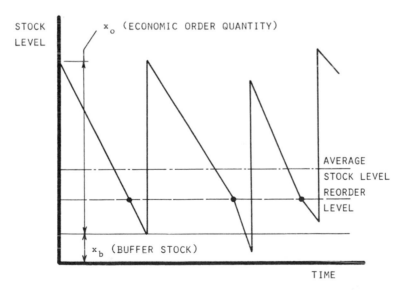

FIGURE 11.3. A typical pattern of stock usage and reordering.

TABLE 11.2
A Sample Determination of Variable Storage Costs

	Percent of purchase price per annum
Loss of bank interest or other revenues as result of capital tieup	14
Obsolescence	3
Deterioration	3
Insurance	0.5
Total storage cost C_s	20.5% of purchase price per annum

Quantity discounts are generally expressed in terms of a series of steps, rather than as a continuous function. For example, rubber is often sold at a standard price for 0–1000 kg, with progressively lower prices for 1000 to 10,000 kg and 10,000–50,000 kg. This gives a number of unit costs $C_u(1)$, $C_u(2)$, $C_u(3)$, etc., one of which will be appropriate to the economic order quantity.

Collecting together the factors that influence ordering cost, a total cost per annum TC can be defined by

$$TC = \left(C_u \frac{x}{2}\right) C_s + \left(\frac{z}{x}\right) C_r \qquad (11.2)$$

where z is the total demand per year and x is the number of units purchased per order. This function takes the form shown in Figure 11.4, with a different curve being obtained for each level of unit cost. Each curve is shown as a full line in the quantity range to which it applies and as a dashed line where it is inapplicable. There is an economic order quantity for each price range, although it can be seen that a lower unit cost can be obtained by moving into the next price range to the right.

It is obviously impractical to move into a higher-volume and lower-cost price range if the demand for the item will cease before the total volume is used; it is often impractical to purchase the economic order quantity for the same reason. Also, it may not be practical to purchase economic order quantities because of restrictions on storage space or, more importantly, because the capital tied up in the stock would seriously reduce company liquidity. Deterioration of items, such as some of the accelerators used in rubber compounds, place further limitations on the size of the order, by imposing a maximum permissible storage time.

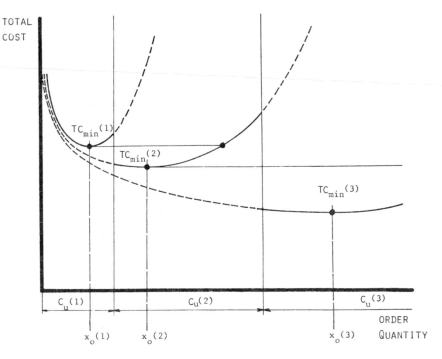

FIGURE 11.4. Curves showing the influence of price discounts for bulk buying on economic order quantities.

For items which are in frequent or continual demand by production, a reorder point and buffer stock level must be determined. The former is fixed by the mean (or quoted) lead time between ordering and delivery, and the actual rate of consumption or usage. An order is placed when sufficient stock remains to cover the mean lead time, as shown in Figure 11.3. The actual lead times form a distribution about the mean, so a buffer stock is essential to avoid frequent shortages. When deliveries are first taken from a new supplier the fixing of the buffer stock level depends on the judgment of the purchasing staff. This estimate can be refined after a number of deliveries have been made by using their distribution, as shown in Figure 11.5, to determine the probability of going out of stock. Assuming a normal distribution, the probability or likely frequency of going out of stock and the time for which the out-of-stock situation is likely to last can be estimated as functions of the buffer stock size. This must be compared with the cost of the disruption which will occur if the item being considered goes out of stock. Some guidance on buffer stock size is given by the condition for minimum cost over a specified period, which occurs when the cost of maintaining the

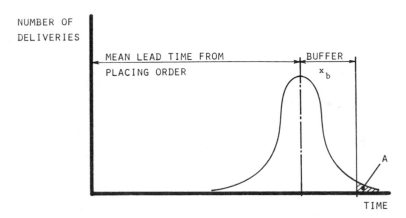

NUMBER OF
DELIVERIES

MEAN LEAD TIME FROM   BUFFER

PLACING ORDER x_b

A

TIME

FIGURE 11.5. A plot showing the probability of going out of stock.

buffer stock equals the estimated cost of the loss of production due to being out of stock, for the specified period.

Limitations on total inventory resulting from a shortage of storage space or a restriction on the total inventory investment must be considered in the context of all the supplies held in stock.[1] Some supplies will have dedicated storage facilities, such as silos or tanks, for which the capacity will have been determined during a plant layout study. Storage space can only be increased by the provision of larger or additional facilities, which are subject to capital expenditure procedures (Section 10.4.3). The majority of supplies are delivered on pallets and go into general-purpose pallet stock storage, where the strategy of the purchasing group can exert a strong influence on inventory costs and the availability of supplies.

For the purchasing group, storage space only becomes a problem when it is inadequate for optimum order quantities and buffer stocks to be maintained or is inadequate to allow the permissible inventory investment to be maintained. The two courses of action immediately available are to maintain stocks at levels which can be accommodated in the existing storage space, or to rent additional storage space. In the former case there is an opportunity cost, which is the extra cost incurred by not being able to buy the quantities which yield the minimum cost per unit. Comparison of opportunity costs for the different inventory items will indicate those which should be given preferential treatment to minimize the increase in inventory costs. There will also be an opportunity cost associated with silos or tanks which are too small, providing the basis for a capital investment decision. When additional storage space A_e is rented its total cost C_e, including the cost of operating it, must be apportioned to each order, such that the additional storage cost C_a is given by

$$C_a = \frac{A_e}{A_e + A_i} C_e A_0 x \qquad (11.3)$$

where A_i is the original storage area available in the company and A_0 is the storage area occupied by one unit x of the supplies being considered. The term which defines C_a must be added to Eq. (11.2), to specify the effect of the rented storage space on the economic order quantity and on the overall unit cost. For practical use, A_e needs to be defined in terms of a "proportion of the effective storage capacity of A_i" and A_0 can be expressed in a sensible lot size, such as a pallet, divided by the number of units in the lot.

An upper limit on the value of inventory which may be carried is usually expressed as a maximum total average monetary inventory (TAMI), recognizing that there will be fluctuations above and below the target figure during normal operation. When the target figure is less than the total average monetary inventory of all stock items, calculated from the economic order quantities plus the buffer stocks, a rational ordering strategy for reduced stock levels has to be implemented. The strategy used should not influence buffer stock levels, since any reduction from their optimum levels would result in an unacceptably high probability of shortages occurring.

Taking a numerical example, where company policy calls for a maximum TAMI of £4000 (including buffer stock) but the TAMI obtained from the sum of the average monetary inventories calculated using the economic order quantity for each item is £5200, as detailed in Table 11.3, the following rule for a rational ordering strategy can be used[5]:

$$\frac{\text{average monetary inventory of an item } j}{\text{total average monetary inventory of all items}} = \frac{(C_j Z_j)^{1/2}}{\sum_j (C_j Z_j)^{1/2}} = \frac{a_j}{A} \qquad (11.4)$$

This important rule states that for each item j, a monetary volume relationship exists with the total monetary volume of all items held in inventory, and that the proportionality, contrary to intuition, is in terms of square roots. It should also be noted that A is not required to be an optimum value, allowing it to be set to £4000 for the example being dealt with. The rule defined by Eq. (11.4) is implemented in Table 11.4 to give order quantities appropriate to the TAMI of £4000, which can be compared with those calculated for the economic order quantities in Table 11.3.

Hitherto it has been assumed that inventory is paid for at the time of delivery, but there is an important exception to this rule known as consignment stocking. Some suppliers operate a system whereby payment is made at the time when supplies are transferred from the stores to production.

TABLE 11.3
Total Average Monetary Inventory (TAMI) Calculated from Economic Order Quantities[a] [Eq. (11.2)]

Item number j	Unit cost c_j	Yearly demand z_j	Economic order quantity x_0^j	Economic order quantity value $c_j x_0^j$	Average monetary inventory $c_j x_0^j/2 = a_j$
1	7.00	1400	400	2800	1400
2	4.00	5000	1000	4000	2000
3	2.00	2500	1000	2000	1000
4	4.00	800	400	1600	800

A_0 = total average monetary inventory = 5200

[a] $C_s = 0.24$ per annum; $C_r = £96.00$ per order.

At intervals the supplier will restock the inventory up to the maximum agreed-upon level and charge for the quantity used. This system reduces the supplier's storage requirements and creates customer goodwill through its obvious advantages.

   Identification of suppliers offering advantageous terms, such as consignment stocking, depends on the knowledge and judgment of the purchasing group. They must also assess supplier reliability and use more than one source of supply for critical items if there is any doubt about their availability when required. An assessment of a supplier obviously starts with the unit price, quality, and delivery lead time, but when a new supplier is being considered for the supply of items for a long period an investigation of financial reliability and viability is also indicated. Low prices may be

TABLE 11.4
Average Monetary Inventories and Order Quantities Calculated from a Maximum TAMI of £4000

Item number j	Annual cost $c_j z_j$	$(c_j z_j)^{1/2}$	Average monetary inventory $A(c_j z_j)^{1/2}/\sum_j (c_j z_j)^{1/2} = \frac{1}{2}c_j x_j$	Order quantity x_j
1	9,800	99	$4000 \times 99/368 = 1076$	307
2	20,000	141	$4000 \times 141/368 = 1533$	767
3	5,000	71	$4000 \times 71/368 = 771$	771
4	3,200	57	$4000 \times 57/368 = 620$	310
		368	$= \sum_j (c_j z_j)^{1/2} \quad A = 4000$	

indicative of an unrealistic pricing policy which will eventually result in the supplier going out of business. When viable suppliers of an item have been identified, the one quoting the lowest unit price can be selected or an attempt to negotiate a better deal with others can be made, when a number of factors such as transport costs, technical service, and goodwill through local buying can be considered.

11.2.2. Inventory Planning

In all companies the pattern of inventory demand by production will be changing constantly, in respect of both the volume and type of supplies required, but the rate of change will generally be faster in jobbing companies than in those manufacturing own products. Inventory planning or materials requirements planning encompasses the techniques used to ensure that the inventory carried is appropriate to current and planned or forecast production.

Forecasting production demand is a basic element of inventory planning. By enabling supplies to be purchased in advance of production demand for many orders, the lead time between a customer placing an order and taking delivery of the goods can be reduced, improving company performance in the marketplace. Anticipating production demand carries with it the risk that the forecast will prove to be inaccurate, leaving the company with unwanted inventory. The rubber industry is fortunate in having numerous supplies which are common to a wide range of products, so that the major inventory items will serve the demands of a number of product lines. Inaccurate forecasts may well result in an increase in inventory costs, but they will not lead to a major loss. The exceptions to this versatility are rubbers and compounding ingredients only used in a small number of products, or even a single one, and product-specific nonrubber items, such as those used in rubber-to-metal bonded components. In these cases the only alternatives available to the purchasing group are to wait until a demand manifests itself before ordering the item or to order in advance of demand, accepting the risk implicit in the forecast. This emphasizes the importance of coordinating activities with the marketing and technical groups in order to reduce the range of supplies used and to avoid creating nonstandard products whenever possible.

Since the purpose of forecasting for inventory planning is to enable inventory holdings to be changed to satisfy production demands and to minimize inventory costs, they only need to be projected forward by a period which is sufficient to enable those objectives to be achieved. This period is usually defined by the time between obtaining the forecast and having the supplies ready for use by production, the majority of which will be the lead time between ordering and delivery of the item being considered. This short period reduces the uncertainty in the forecasts to a level which is sufficiently small for them to provide a sound basis for inventory planning provided that:

1. The forecasting technique is appropriate to the situation in which it is being used.
2. The information base for the forecast is reliable.
3. The forecast is updated frequently.

Forecasts for inventory planning are largely based upon time-series analysis, unlike market forecasting which projects much further into the future and employs a variety of techniques.[4] Time-series analysis uses the historical demand for an inventory item to predict future demand, and is usually accomplished by decomposing the historical data into five components—level of demand, trends, seasonal variations, cyclical variations, and random variations. The level of demand is derived from the slope of the cumulative demand curve which is the running total of supplies received, with respect to time, from the beginning of the recordkeeping period. Both these are shown in Figure 11.6, where the actual demand is shown as a full line and the trend as a dot–dash line. The trend can be obtained by regression analysis, using a commercial computer program package to search through a number of equations which give smoothly changing relationships between either cumulative demand or level of demand and time. Generally simpler equations can be used to fit the level of demand data and, in the absence of strong changes in the circumstances influencing the demand, the equation which gives the best fit to the historical data will yield the best forecast. When the recording period is sufficiently long, seasonal and cyclic variations of the demand above and below the trend curve can be quantified, either by further regression analysis or by examination of a demand plot, using paper or, preferably, a computer terminal. Those variations can be added to or subtracted from the trend analysis forecast, as appropriate, to improve the forecast accuracy. It should be possible to assign reasons to both seasonal and cyclic variations; apparent systematic variations which cannot be explained must be suspected of being artifacts of the analysis method. Random variations occur as a result of nonrecurring external influences, such as strikes, accidents, and mistakes, and can exert a strong influence on the trend curve. For this reason random variations are often disregarded.

The goodness of fit of a regression equation to the data can be simply assessed by reference to the correlation coefficient or by a more rigorous statistical analysis, to determine the significance of the equation via an *F*

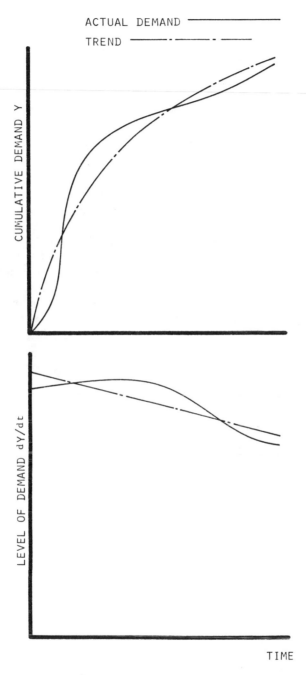

FIGURE 11.6. Historical data for time-series analysis.

ratio. Control limits on the expected accuracy of a forecast can also be defined from the standard deviation of the distribution of the data about the regression curve, at one, two, or three standard deviations from the mean, depending on the practical circumstances. If the actual demand then falls outside the control limits set on the forecast demand there is good reason to suspect that the conditions controlling demand are subject to new influences. The adequacy of the model (type of regression equation) being used should then be examined and, if necessary, replaced with a new one.

The statistical methods described in the preceding paragraph can only be applied practically in a simple manner if the seasonal and cyclic variations are small compared with the trend. If they are not, the goodness of fit, as indicated by the correlation coefficient or the F ratio, will be poor and the control limits will not provide a measure of forecast accuracy. This problem can only be overcome by quantifying and removing cyclic and seasonal variations from the data prior to the statistical analysis.

Regression analysis is a very powerful technique for situations where discontinuities in the conditions controlling demand are relatively rare. Discontinuities can be caused by events such as the sudden upsurge in demand when a new market for a product is opened up and historical data are no longer a useful guide to future demand. If there is a marked lack of continuity in the demand data then time-series analysis will be totally inadequate for forecasting; but methods are available which can cope with a mildly discontinuous situation.

Techniques which discard or give a diminishing importance to data as they age include the last-period demand, the moving average, and the exponentially weighted average. The last-period-demand technique simply forecasts that the demand in the next period will be the same as that recorded in the one preceding it. This technique responds fairly well to changing conditions but it overreacts to random influences. This latter fault can be overcome by using a moving average, defined by

$$\bar{y}_t = \frac{1}{n} \sum_{i=1} y_{t-i} \qquad (1.5)$$

where $\bar{y}_t$ is the forecast demand for period t and y_{t-i} is the actual demand in period $t - i$. The number of time periods n over which the actual is averaged must be determined by experience. If too few are used the moving average suffers from the same fault as the last-period demand. If too many are used, the forecast will not respond well to changing conditions of demand.

With the simple moving average all the data used are given equal weighting, but it is preferable to reduce the influence of data on the forecast as it ages. This is accomplished by using the exponentially weighted moving average:[4]

$$\bar{y}_t = a[y_{t-1} + y_{t-2}(1-a) + y_{t-3}(1-a)^2 + \cdots + y_1(1-a)^{t-2}] + (1-a)^{t-1}\bar{y}_0 \tag{11.6}$$

where $\bar{y}_0$ is the forecast demand for period 0 and y_{t-1} is the actual demand for period $t-1$. The exponential smoothing constant a can take values between zero and one, with equal weighting to all data, as in the moving average, when $a = 0$ and only the y_{t-1} demand being used in the forecast when $a = 1$, as in the last-period-demand method. Guidelines for a range from 0.1 to 0.3, with larger values being advocated for dealing with rapid changes of demand.

Alternatives to time-series methods must be sought when demand is periodic or, as previously mentioned, is subject to frequent discontinuities. In the former case probability analysis can be used,[6] provided there is a sufficient history of previous demand available. Distributions of the frequency and volume of demand can be converted to probability distributions by dividing each category's observed frequency by the total number of observations, for each distribution. The probability distributions then provide estimates of the likelihood of a specified frequency or volume of demand occurring. In other words, the probability distribution quantifies the risk inherent in a forecast of periodic demand. This approach can also be used for forecasting the seasonal and cyclic demands superimposed upon a trend which is amenable to time-series analysis.

When probability analysis indicates that past demand is unlikely to provide a reasonable forecast of future demand, two courses of action are available. Waiting for a demand to occur before ordering is very simple and has the advantage that it is a low-risk strategy, but it can only be used when its advantages are not offset by the occurrence of production shortages and orders being lost through the quoting of long delivery times. Generally ordering on notification of demand is viable only for the minority of jobs which arise unpredictably and need supplies which are not held in stock. This implies that the majority of demand must be forecast, using experience, knowledge of the company's markets, and personal judgment where statistical methods are inadequate.

The techniques described in this section should be considered as aids to judgment and must be used in conjunction with market information. Prior notice of many discontinuities or major perturbations in demand can be obtained by feedback of information from the sales group and inventory levels adjusted accordingly, to supply the demand when it occurs. This feedback of information is generally essential to jobbing companies, where the majority of demand may not be amenable to quantitative forecasting methods.

11.2.3. Inventory Management

Proceeding from inventory planning and ordering, inventory management is concerned with:

1. Receipt and checking of supplies delivered.
2. Allocation of storage space.
3. Operating the supplies identification and withdrawal system.
4. Reporting changes in the quantities stored as they occur (perpetual inventory).

The checking of deliveries is usually separated into an immediate validation by the store's personnel prior to unloading, followed by a quality inspection after unloading. Validation starts with identification of the delivery against an outstanding order, usually by matching an order number quoted on the delivery document (normally a bill of lading) against the "open" purchase orders file. Once this initial identification has been made, the itemization of the supplies on the delivery note can be checked against that on the purchase order. If these match, the validation procedure is completed by a physical check on the delivery, for condition, type, and quantity if possible, either prior to or during unloading. This physical check is intended to detect gross mistakes in time for the goods to be returned to the supplier by the transport used to deliver them. Often, it is limited to checking labels and noting the number of pallets or containers.

If a discrepancy is detected at any stage in the validation procedure, a decision to accept the delivery or not has to be made rapidly. In the case of gross mistakes in the type of supplies delivered, the decision to return them is unequivocable. However, when part of an order is incorrect, or the physical quantity does not tally with the purchase order or the delivery document, the goods-inwards reception personnel need to seek advice from someone competent to make a decision. This can be the purchasing, production, or technical staff, depending on the nature of the problem. To enable a decision to be obtained with the necessary speed, it is desirable for the store's super visor to have a list of people to contact for particular types of problems.

Bulk supplies, such as oils and powders which are fed directly into tanks and silos, and those which are delivered in standard pallet loads, such as rubbers, cans, and bagged powders, generally need no further checking for quantity or type after the initial validation. Items which are not delivered in these forms generally require transferring to in-house containers for storage and transporting, during which time a check on quantity can be made.

Some supplies have to be quarantined until they are passed as being acceptable for production by the quality-control group. For this purpose supplies can either be placed in a quarantine store prior to inspection or be placed directly into the main storage area but not released for use until passed by the quality-control group. The system will depend on the nature of the supplies and on company policy; but in either case the inventory management system should have a quarantine category for supplies delivered but not available for use.

In conjunction with the movement of supplies from the delivery vehicle to a storage location where they are available for use in production, there must be a parallel flow of information. For each delivery it is necessary to:

1. Make a record of the delivery time and date, with a report of any discrepancies between the purchase order, the delivery document, and the physical supplies.
2. Note if supplies are available for use or quarantined.
3. Inform the quality-control group of the type, quantity, and location of supplies which require inspection.
4. Enter supplies into the inventory holdings record when they become available for use in production, together with their store's location and position in the order-of-use listing.

For bulk powders and liquids the reference to the store's location and order of use does not apply, unless more than one tank or silo is used for each supply type. For supplies which form unit loads it is extremely important, since a delivery may be distributed to a number of locations in the store, depending on the storage system in use (see Section 8.6.3).

Withdrawal of supplies which form unit loads from the stores is usually carried out on the authority of a requisition, made out in advance of production demand by the production control group. A number of methods of organizing the physical transfer of supplies from the store to the point of use are possible. The prime requirements are flexibility in the timing and order of transfer, to accommodate deviations from the production plan, and a clear definition of responsibilities. It is preferable to have both the entry and withdrawal of supplies and the recording of those activities entirely under the control of the store's personnel, requiring that they are also responsible for a substantial portion of the internal transport. The store's personnel are then responsible for moving supplies from the store to work stations in accordance with the requisition plan issued by the production control group, which can be overriden by production supervisors to accommodate deviations from the work-station loading and job scheduling plans. Supplies in the form of unit loads which are withdrawn from the stores to meet a demand which does not materialize due to, for example, a breakdown, can be booked back into the stores, provided that the unit load is intact or can be reassembled.

When supplies are moved from stores to work stations they are deducted from the store's inventory and added to the work-in-progress inventory. This system of noting all entries and withdrawals of supplies, so that the store's inventory holds are known at all times, is termed perpetual inventory, to distinguish it from other inventory recording and control methods, in which the inventory holdings are only known accurately when they are at certain levels or at certain times (stocktaking). Perpetual inventory[1,4,5] provides the opportunity for much more precise control of inventory holdings than other systems, but in the past it has not always been justified on a costs basis because of the clerical burden it has imposed. With the wide availability of commercial computer inventory control and management program packages this argument has become invalid, except in the case of low-cost items.

It is frequently advantageous to divide inventory into three classes[4] according to the monetary usage, which is defined as the product of annual demand and unit cost. This is called ABC analysis. As a general guide the A class includes high-value supplies, with a monetary value of 75–80% of the total inventory holdings value, while representing 15–20% of the inventory volume. The B class accounts for 10–15% of inventory value and 20–25% of the volume, while the C class includes low-cost items totaling 5–10% of the inventory value but accounting for 60–65% of the volume. For ABC analysis the entire inventory is listed in descending order of monetary usage and then designated by the ABC system. This concentrates attention on those items which represent the largest annual expenditures, and can therefore influence decisions concerning order quantity, reorder point, safety, or buffer stock and stocktaking frequency.

When supplies become work in progress they accumulate added value as they pass through the various stages of manufacture. For example, the mixes held in the mixed-compound stores have a higher monetary value than the sum of the compounding ingredients because the cost of mixing has been added to the raw-material costs. The added value accumulates until the full manufacturing cost has been absorbed into the product. This can either be its value when it enters the finished product store or, if transport costs are included as an integral part of the manufacturing cost, its value when it is delivered to the customer. The value of the work-in-progress inventory is therefore dependent on its distribution through the manufacturing sequence.

The perpetual inventory system is equally applicable to the finished products store as it is to the production supplies store, and in each case the flow of costs associated with the physical movement of items is accounted for using two main methods[4]: FIFO (first in, first out) and LIFO (last in, first out). Under FIFO, which is the most widely used system, the value of inventory withdrawn from stock is computed on the basis that it is withdrawn from the oldest stock on hand and is valued at a cost appropriate to the oldest stock. In contrast, LIFO involves computing the cost of inventory withdrawn from stock using the prices of the most recently purchased items. The underlying purpose of LIFO is to match current revenues against current costs, but it can result in an unrealistic inventory valuation for balance sheet purposes. FIFO is simple and follows the physical movement of inventory in many companies. The ending inventory for FIFO approximates closely the actual current value as the costs assigned to the stock are the most recent. It must be emphasized that the order of withdrawal of stock which is assumed for accounting purposes need not be related to the actual order of withdrawal.

Mistakes in perpetual inventory systems tend to accumulate with time unless they are corrected by a physical stocktaking at regular intervals. During this exercise the total inventory holding in both the production supplies and finished product stores, in addition to the work in progress, must be determined and subsequently compared with the recorded holdings. Checklists identifying inventory items and the locations at which they are to be found need to be issued and areas of responsibility allocated. The objective is to ensure a coordinated sweep which results in all the inventory being located and recorded.

11.2.4. Computer Methods for Inventory Management and Control

Many of the techniques and procedures described in the preceding three sections have been integrated into commercial computer program packages. These are supported by hardware ranging from microcomputers to distributed system computers, often with specialized terminals, for larger companies. In this section the general features and capabilities of a distributed system package will be dealt with.

The layout of a distributed system is shown in Figure 11.7, with double lines indicating that there is a flow of information to and from each terminal.

For the store's personnel the computer terminals aid the recording associated with the perpetual inventory system and can provide a number of other services which form the basis of an efficient store's operating system. These include the location, type, and capacity of unused storage space, and the location of supplies needed to fulfill a store's requisition, according to the order-of-use listing.

The use of the computer to identify the location of supplies or products

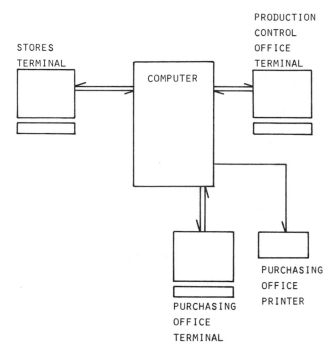

FIGURE 11.7. A simple distributed terminal computer system for purchasing and inventory management.

and available storage space imposes the discipline of setting up a rational store's location identification system and then adhering to its use. Normally, a three-digit code can be used, indicating bay, row, and column. Codes will also be needed for supplies, although there is a good case for retaining a name wherever possible, because of its association with the physical supplies.

When production supplies are delivered and have been accepted, the storeperson will record all the details of the delivery using the computer terminal keyboard and then call up a list of suitable and available storage space. This list can then be compared with a plan of the store's and storage locations selected which result, as far as is possible, in supplies of a given type being in close proximity to each other. The locations and quantity at each location can then be recorded and any quarantined supplies flagged as being unavailable for use. This prevents quarantined supplies from being used inadvertently and enables the quality-control group to call up a list of supplies requiring inspection, together with their locations. The security of the quarantine system is improved if the flagging of incoming supplies is a computer function, and release onto the order of use files is controlled by the quality-control group. The entry of products into the finished goods store follows a similar procedure to that for production supplies but without the quarantine system.

The recording work associated with booking items out of the production supplies store can be minimal, since most of the information needed for the transaction will be in the computer files. The store's requisitions can be entered into a computer file by the production control group and recalled by the store's personnel on their terminal, without the need for any physical documents. Deleting the items requisitioned from the order of use listing can then:

1. Correlate the supplies booked out with the details on the store's requisition (product details).
2. Delete the supplies withdrawn from the total inventory holdings.
3. Return the storage space to the "available-for-use" listing.

The only action required from the storeperson is to key in the time and date of the transaction.

For the purchasing group, the feedback of information from the store's personnel, via the computer, provides the basis for:

1. Forecasting.
2. Determination of economic order quantities (or rational order quantities if a maximum average inventory value limit is imposed), reorder points, and buffer stocks.
3. Listing of ordering requirements.
4. Listing the status of purchase orders and supply requisitions.
5. Listing of the time for which each inventory item will continue to satisfy the demands of production before it goes out of stock.
6. ABC analysis.

These five categories of *current* information and analyses include the main services which a purchasing group needs to function effectively.

In an environment of changing demand, the frequent recalculation of order quantities, order points, and buffer stocks is essential, particularly for A class supplies, to minimize inventory costs and avoid shortages.

The listing of ordering requirements is the only item which requires the creation of a document. Orders can be printed out directly in a form suitable for sending to suppliers, provided that an appropriate computer segment is available.

Listing the status of purchase orders is obviously necessary to enable problems to be identified and chased. For this purpose it is useful to be able to list only those orders which have not conformed to the quoted delivery date or order quantity, including those which have been rejected wholly or in part by the goods-inwards reception personnel or the quality-control group. Most program packages enable orders to be listed by the order date, by quoted delivery date, or in item sequence. Listing by quoted delivery date is useful for planning goods-inwards reception activity.

Listing items in order by the time for which they will continue to supply production demand, starting with the one having the shortest time, is a simple application of forecasting methods, and is helpful to the purchasing staff for identifying those orders which should receive most attention. It is also useful to the production controllers, as an early warning of items likely to go out of stock, enabling them to formulate contingency plans in time to minimize the disruption and loss of production caused. The listing can also include the jobs affected.

Supplies with limited storage life are ordered in quantities which, according to forecasts of usage, will enable them to be used within the storage life limit, with some margin of safety. Even so, differences between forecast and actual demand can occur, and result in waste if the production control group is not fully informed of the life of the supplies in stock. For this purpose a continuously updated listing of perishable inventory holdings, in order of their remaining storage life, is an essential aid to production planning, particularly when the perishable items are A class.

In addition to the listings and analyses needed for the day-to-day control of purchasing and inventory, there are a number of reports which have to be produced periodically, including:

1. Period-end stock status report—with achievement against standard costs.
2. LIFO/FIFO reports.
3. Inventory transaction register.
4. Physical inventory.

The inventory transaction register is a summary of purchase orders issued, deliveries received, and requisitions dealt with. It provides the purchasing group manager with a report of the purchasing and store's activity, from which allocation of tasks and responsibilities can be formulated. It also provides an indication of the effectiveness of the organization and methods of operation of the purchasing group and the store's personnel.

The physical inventory (stocktaking) first involves the printing of a list of current inventory items and their store locations. This list acts as a recording document to be used by the stocktakers. When stocktaking is completed the physical counts have to be reconciled with the perpetual inventory records. After keying in the physical counts at a computer terminal, a report of variances between the counted quantities and those in the perpetual inventory files is produced, enabling supplies with significant variances to be physically rechecked. If the second count confirms the variance, the quantity recorded can be adjusted. A register is then formulated to provide a report and a record of all adjustments to the on-hand balance.

The computer facilities described in this section are fully available in program packages and computer systems produced by IBM, ICL, Hewlett-Packard, and others. The use of interactive distributed terminal systems has been emphasized because they can virtually eliminate the need for written records, instructions, and reports, together with the time and effort spent in distributing them. Programs for single terminal microcomputer systems, which are suitable for small companies, are also widely available. In summary, support to inventory management and control represents an ideal application of computer methods, since they involve large numbers of routine operations. The computer system provides clerical support to the purchasing group and the store's personnel which would be impossible to provide by any other means, and frees them to concentrate on aspects of their jobs which require judgment.

11.3. IMPLEMENTING THE PRODUCTION PLAN

11.3.1. Identification and Routing

Codes are used to identify raw materials, subcomponents, work in progress, and finished products. The main objective is to provide all the information needed to implement and monitor manufacturing in a very compact form, which would not be achieved by quoting a lengthy generic or trade name plus a number of other details. Codes have also been used in an attempt to preserve confidentiality regarding sources of supply, names of suppliers, and the identification of customers. In addition, codes may be used to identify work stations and operators, both for communicating work instructions and monitoring production performance.

During their progress through the manufacturing sequence, compounding ingredients appear in a number of forms. In the raw-material stores they are given individual codes, which are replaced, after mixing, by a compound code to which mixer crew, time, and date codes may be added. A compound can then be used for a number of products and go through a number of preparatory stages before the product is formed. The intermediate form, such as feed strips, can again be common to a number of products. Individual codes can be used at each stage of manufacture, but there is then no facility for tracking work through the various stages of manufacture as an aid to problem solving. A compromise has to be sought between the operational simplicity of individual codings and the utility of codes which grow in a cumulative manner through the manufacturing sequence.

Communication of production plans, resulting from the scheduling and loading activities of the production control group, also relies upon the use of codes. The manufacturing route for a product can be defined by associating the compound, subcomponent, or product codes with the sequence of work stations to be used. In-process stores also have location codes, which are necessary for rapid location of supplies and for operating effective quarantine and first-in first-out procedures. However, these are usually managed by the production groups and do not enter into the routing instructions from the production planning groups.

Routing requires identification of all items in the manufacturing sequence. This is normally achieved by the use of cards or tickets attached to the work in progress or, more usually, to the containers used to transport the work in progress between work stations. Where cumulative codings are used, these tickets will carry the full manufacturing history of the products. If the products are individually coded, as is the case for tires, the manufacturing details can be recalled in the event of a service failure.

All the necessary documentation for manufacture is generally issued by the production control group in the form of shop packets. These include the manufacturing instructions and the work identification tickets.

11.3.2. Manpower Requirements Planning

The task of ensuring that there are a sufficient number of trained personnel available to meet a company's manufacturing targets is mainly the responsibility of the production management. However, the distribution of the product mix between manual and automatic operations can produce a significant short-term variation in direct personnel requirements, indicating the importance of production planning in this area. The work of indirect personnel, such as supervisors, maintenance engineers, and cleaners, is not specific to products and is not so strongly influenced by the production volume or product mix.

The direct personnel work content of production over a specified time period can be found by "exploding" the work station loading plan for that period into working hours, using standard cost information. This procedure is similar, in concept, to the materials requirement planning discussed earlier. However to translate the working time requirements obtained by this method into the number of direct personnel a company needs to employ in order to meet its production targets some further factors have to be taken into account.

A first approximation to the number of direct personnel can be obtained by considering the number of working hours in an employee week and the way in which the employee weeks are distributed through the operating periods into which a year is divided. This takes holidays into account but not absences due to illness and other reasons, which can be defined as a percentage of the available working hours, to be deducted from them. In addition, the working hours will include nonproductive time for which an allowance must be made. The magnitude of this allowance will be strongly dependent on the effectiveness of management and can be treated as a performance index.

An assumption implicit in the preceding analysis is that every member of the direct personnel *can* be fully occupied with productive work and that nonproductive time is simply due to a failure to supply the resources needed for manufacture or that operator productivity is lower than the "normal" level. In fact, a large proportion of nonproductive time arises from the nature of the jobs to be performed. The work of many operators is dictated by machine cycles, which can often include "idle" time in each cycle but not allow sufficient time for another task to be performed. For products with long manufacturing runs, the nonproductive time can be minimized by attention to plant layout and production planning, so that an operator's job can be designed to contain minimal idle time. This is far more difficult to achieve in short-run manufacture, where the product mix is frequently changed. Relief operators also add to the problem of designing jobs for full productive occupation, although they are often essential for processes which cannot easily be stopped and started.

The turnover of direct personnel also decreases productivity, due to new personnel going through a learning curve, even after a well-organized training program. In the absence of such a program the learning period can be greatly prolonged. The loss of productivity from this cause can be expected to be proportional to the turnover rate, which is often in the region of 2–5% of total direct personnel per month. A similar loss of productivity also arises when new methods are introduced into production and retraining is necessary.

Establishing the number of indirect personnel needed to support manufacturing is extremely difficult and whenever "overmanning" of a company occurs, it is usually in this category. One of the major safeguards against overmanning is to set an upper limit on the cost of *all* staff not directly involved in manufacture, by assigning a figure which leaves an adequate profit when all other unavoidable costs have been taken into consideration. Individual cost limits can be assigned to areas, such as the manufacturing groups, but this may be undesirable as the limit figure does not imply a spending target. The salary and wage bill should be controlled by a regular review of job functions and their roles within the operational structures of the manufacturing groups. This is an area in which sound management is essential. Inefficient operating structures, in which all the jobs *appear* to be necessary, tendencies towards "empire building," and company politics all work directly counter to efficiency.

Forecasting is essential to manpower requirements planning, to enable both the total number and the distribution of people throughout a company to be adjusted without disruption to the company's activities and with a minimum of trauma for the people concerned. The term "natural wastage" has been used to describe the gradual reduction of a work force by the normal processes of people leaving for other jobs, for retirement, and for other reasons. Reduction by natural wastage is only feasible in circumstances where changes are both predictable and gradual, it cannot give the fast response demanded by dramatic and unpredicted economic change. Also, it cannot be expected to result in a desirable distribution of people or skills. Retraining and some hiring of new personnel are usually necessary to close the gaps. There is also the problem of more able people being more mobile, which can result in the reduced personnel being predominantly those of mediocre ability, unless positive steps are taken to avoid this situation.

The emphasis has been placed on the reduction of personnel levels because this is a clear trend in the rubber industry. In the majority of areas the introduction of automatic systems is reducing the manual work content of products. The total number of people employed in the industry can only be increased by an upturn in the demand for rubber products.

11.3.3. Recording Work

Production activities can be recorded or monitored by three distinctly separate techniques:

1. Manual recording of work by operators on worksheets.
2. Manual entry of work into a computer terminal.
3. Automatic monitoring of work by manufacturing equipment.

In each the objective is to provide a feedback of information to enable management and production control staff to do their jobs. One of the prime uses of the feedback of information is to monitor production performance, both against the production plan and to determine where performance can be improved. For both of these, the frequency of reporting and the lag between work being done and being reported, are important.

Manual recording on worksheets is widely used.[2] Until recently, it has been the only viable method of reporting work and for the small company, where communication and supervision problems are minimal, it is usually the most cost-effective method. At the end of the shift or working day the operator will record on the worksheet his or her name, the work-station code, the product or job code(s), and the quantity completed. There may also be a requirement to report nonproductive time.

In a medium-sized or large company, where all communication networks are larger than those of the small company and have to be well planned and structured, the collection and collation of worksheets can be a prolonged business, lengthening the feedback time. This means that the production managers and controllers have a poorer "real-time" picture of production performance than in a small company. In fact, the only way to obtain a measure of current production performance in a company using worksheets for recording is by personal observation and communication. This works well at the section or group level, but is totally inadequate when the consequences of events in one section impinge on the activities in a number of others, requiring a knowledge of current activities throughout the manufacturing areas in order to formulate corrective action. The concept of response time, used in process-control engineering (Section 7.5.2), is useful here. If the response time of a feedback and control system is long in comparison with the rate at which the events to be controlled can change, then instability will result. Circumstances causing a deviation from the production plan can change before corrective action takes effect, resulting in its effects being disruptive rather than corrective. Recording of work is always necessary for accounting purposes, but the response time of the recording method clearly imposes limits on its use as a management tool. Worksheets have to be collected, collated, and the information fed into a computer for analysis. This is a major task which usually cannot be performed more than once a day, resulting in the analyses transmitted to managers and production controllers being a minimum of six hours out of date (the final work by the night shift) and a maximum of 30 hours.

The delays in the collection, collation, and analysis of work-in-progress information can be reduced substantially by using a distributed terminal computer system with recording points at a number of locations through the manufacturing areas. With the delay between reporting and analysis being practically eliminated, the response time for making analyses available to production managers is solely dependent on the frequency of reporting, provided they have access to a computer terminal with a visual display unit.

Computer terminals for the manual entry of work by operators usually serve a number of work stations. Although an increased frequency of reporting is desirable, it must be planned to avoid disruption of operations. Adding time to the natural rest breaks for the purpose of reporting will result in production information being updated at intervals of approximately two hours. In addition, changeovers from one job to another can be reported when they occur, as can breakdowns, shortages, or any other reasons for a work station stopping production. This type of system is also very useful for reporting movements of work in progress which, by implication, quantifies the quantity, type, and locations of work in progress. Many courses of corrective action depend on this information.

A good example of the types of terminals which can be operated effectively in a production environment is provided by the IBM 5230 group of data collection terminals. These include a simple time entry station and two data entry stations which are designed specifically for work reporting, one being for wall mounting and the other for table mounting. The operator is identified to the computer by inserting a personal badge into the terminal and information is entered via a simple keyboard.

Increasing the frequency of reporting obviously provides a more up-to-date picture of the manufacturing operations. It also provides a more discerning one, enabling the variations in performance which occur during the working day or shift to be detected. An unavoidable characteristic of periodic work reporting is the averaging of performance during each period, resulting in greater sensitivity to short-term variations being gained when the periods between reporting are reduced. The detailed information thus gained is extremely useful when improvements in production performance are being sought.

With automatic monitoring of work, the frequency of reporting is generally high enough that the resulting record is essentially continuous, rather than periodic. Its main areas of application are high output, largely automatic processes, which include some mixing and extrusion installations and many molding operations, particularly injection molding. These processes benefit from sophisticated control methods and process monitoring for the quality of performance, as described in Section 7.4. Including the monitoring of quantity completes the package and ensures that the maximum benefit is being gained from the investment in monitoring equipment.

Simple automatic monitoring of machine functions is widely used. A typical example is the activation of a microswitch by the opening of a molding press, to increment a remote counter in the supervisor's office, or to mark a time-base-driven chart, giving a permanent record of output. This type of technique suffers from two main disadvantages. First, it cannot distinguish between a "dry" or nonproductive cycle and a productive one, leaving it open to misinterpretation and abuse. Second, analysis of the record produced involves time-consuming manual operations, even when the information can be fed into a computer.

Computer monitoring systems can be designed and programmed to "recognize" when a process is operating productively by utilizing the signals from a number of sensors, avoiding the possibilities of either misinterpretation or abuse. However, the main strength of a computer monitoring system, be it based on manual data entry, automatic monitoring, or both, is the elimination of the clerical burden from the data collation and analysis. This provides the opportunity for detailed production analyses which are not possible without a computer system.

11.3.4. Analysis and Use of Work-in-Progress Information

When the record of work in progress has been collated in the computer data storage facility there are numerous ways in which it can be extracted, analyzed, and presented to aid production management. In fact, it is quite easy to negate the benefits of a computer system totally by overwhelming managers with a tide of ill-considered and largely unnecessary information. The primary objective in setting up a production information system should be to provide each user of the system with only that information required by the user's job. By following this ideal, the time spent by managers in extracting information they need from that which they do not need is minimized.

It is much easier to provide an effective information and analysis service using a distributed terminal computer system than with a centralized batch-operated system. In the latter case production information is generally distributed daily in hard-copy form to managers. The flexibility of such a system is poor. Interim reports and alternative forms of analyses needed to tackle specific problems are often difficult to obtain on demand. As a result, the periodic delivery of information from the computer group brings with it the problems of a lengthy response time which were identified with worksheets in Section 11.3.3. These can be largely overcome with a distributed terminal system, for which analyses can be obtained on demand, without the expense of producing and delivering a hard copy. However, this rapid response is only beneficial if the information in the computer is being updated continuously, pointing to the use of data entry terminals and automatic production monitoring.

The production control group are prime users of the production information feedback, to record production achievement against expected performance as a basis for adjustments to scheduling and loading plans. In Sections 11.1.5 and 11.1.6 computer-graphics-aided methods of scheduling and loading were described (Figures 11.1 and 11.2). These will be taken up here, to complete the control loop by examining the way in which deviations of actual performance from planned performance are handled.

The movement of the update marker and the adjustment of bar length for a job which will either be completed ahead of schedule or one which will overrun can be implemented directly from the computer, as can the adjustment of all the activities affected by the deviation. These adjustments can result in other jobs being pushed into the overrun or late-delivery categories and cause shortages of work at some work stations. Notification of the jobs thus affected on the text visual display unit adjacent to the graphics terminal is then necessary, since the parts of the loading plan affected may not be included in the current graphics display.

With the current production situation displayed on the graphics and text terminals, the production controller, in cooperation with the production managers concerned, can undertake remedial action. This may simply involve the authorization of overtime working or, in the case of substantial deviations, a reworking of parts of the loading plan. If the loading plan does not contain sufficient flexibility for it to be reworked without disrupting the production program and other product schedules, then the defaulting product goes into the category of arrears, to be dealt with later.

In contrast to the routine updating of the production achievement against the program, via the graphics terminal or, in an alternative system by text tables, most other production analyses are best provided on demand. This can be achieved readily using the "menu" approach. When there is any degree of confidentiality in the information contained in computer files, users are generally issued with individual codes, by which they identify themselves to the computer and gain access to the information they need for their jobs. Logging in and typing the command MENU produces a title list of production analyses appropriate to the user's job, from which a selection can be made.

In general, production analyses are required to emphasize those items which demand and justify attention and action. Overall analyses are only required for certain functions, such as the production planning and control activities previously described, and for accounting purposes. In most other cases it is necessary to adopt a "management-by-exception" philosophy for the in-house design of production analysis programs and for the selection of commercially available program packages. This can often be achieved by simply preparing listings in order of "importance," by whatever criterion is used to define "importance" in the context of a particular analysis. In most cases the criterion is a cost. Analyses prepared in this way generally conform to the so-called 80–20 rule. [7] This states that the first 20% of items in a list arranged in order of importance will account for approximately 80% of total costs. Depending on the type of analysis, this cost could be the overall manufacturing cost, standard cost variance, or scrap product value, each indicating where management effort is best directed.

11.3.5. Functional Analyses and Directions

The term "functional analyses" is used here to differentiate them from those analyses which are concerned with costs. Their main purpose is to supplement the production plan and aid its implementation. A number of examples will be dealt with in this section, but it is important to note that the choice of analyses will depend strongly on the size and nature of the manufacturing facility, and that analyses not covered here may well be required.

Starting with the introduction of jobs into manufacturing, the production manager needs to know when current jobs will be completed on each work station affected. This can be predicted from standard cost or work-study information; but a more accurate method is to forecast the time to completion from the output rate in the current production run, using the techniques described in Section 11.2.2 for inventory planning. In both cases the time to completion is projected forward from the current information on work completed. This type of analysis can be used to update the scheduling and work-station loading records for the production controllers; but when jobs are listed in order of time to completion it constitutes a plan of work for the production manager or supervisor.

When the time to completion indicates that action is required to organize the changeover from one job to another, the supervisor will need to check the remaining stocks of work in progress (WIP) specific to the current job and the availability of WIP specific to the new job. At this stage a listing of the manufacturing routes for both the current and incoming jobs can be helpful, together with the codes of the separate items of WIP associated with them. This identifies to the supervisor which stocks should be checked and, if a physical check is desirable, their location. Alternatively, the stock-level information should be available from WIP records. The WIP remaining in stock will then determine the output to the end of the current job, which can either overshoot or undershoot the production target. Most contracts make some allowance for this situation, but the decision to overrun on completed products or to store or scrap the surplus WIP will need to be taken by the supervisor.

With the changeover time well defined and the availability of WIP for the incoming job confirmed, the resources for achieving the changeover may need to be identified and coordinated, to be available for the duration of the changeover. This is not necessary for operations such as mixing and extrusion, where the changeover is rapid and routine; but in the case of molding and injection molding in particular, the change can involve a number of people and specialized items of equipment. If the people and equipment have substantial work loads (as they should have) and cannot be released from other tasks at short notice, it may be necessary to use simple Gantt scheduling and loading charts to ensure an efficient utilization of their time and to minimize the duration of the changeover. Listings which detail the equipment, procedures, and times for taking out the ending job and installing the starting one are very useful for this purpose, as well as defining responsibilities and ensuring that all the information for the tasks involved is available. This can also be extended to machine settings and specific startup procedures, which are necessary for many changeovers.

Changeovers can bring changes of manufacturing routes, output rates, and the manual work content of work stations. It is often necessary to change the distribution of operators among the work stations, both to maintain operator utilization and efficiency and to ensure a fair and uniform distribution of work loads. This is primarily the responsibility of the production manager although plant layout, with respect to the grouping of work stations and the way in which the production planners allocate jobs to work stations during scheduling and loading, can have a profound effect on the operator productivity which can be achieved. Computer planning aids for this task are extremely valuable, enabling the consequences of different permutations of operators and work stations to be examined. The calculations involved use standard cost or work-study information and are relatively simple; but the form of the VDU display is important for visualizing the relative positions of the groups of work stations. In fact, presenting a schematic of the plant layout is probably the best approach, with each work station being numbered. These can then be associated with job codes, which enables them to be selected for inclusion into the work load of an operator. Presentation of the percentage utilization of the available working time of an operator and the distribution of the nonutilized time in the workload adds to the value of this planning aid.

Even during "normal" production, when there are no changes occurring in the range of products being currently manufactured, the actual output capability of a manufacturing area can vary considerably from the planned capacity, for a number of reasons, each of which need to be swiftly brought to the attention of the production manager. If swift remedial action is not possible, as may be the case with a breakdown or a shortage of WIP, compromise plans have to be formulated, which usually involve concentrating on high-priority jobs at the expense of those having low priority. To aid decisions, a listing of the high-priority jobs and alternative routings for them can be helpful, followed by an analysis of the delays which will be caused on the jobs disrupted, both by the problem and the plan of remedial action. Again, this enables alternative plans to be evaluated before irreversible action is taken.

In addition, jobs falling behind the output schedule detailed on the production plan, for whatever reason, need to be brought to the attention of the production manager concerned. Although this information can be gleaned from the production planning chart or display, a listing of the jobs concerned, in order of the time or quantity they are behind schedule, is more effective and less time consuming. A course of action, such as authorizing overtime working, can then be formulated.

11.3.6. Aging of Arrears and Customer Liaison

The term "arrears" is used to identify orders which have not been completed in time to meet the delivery date quoted to the customer. Arrears occur as a result of natural fluctuations in manufacturing capability, arising from causes such as breakdowns, high levels of absenteeism, and delays on delivery of essential supplies, and are to be expected, together with the occasional early completion of an order, in most companies. In fact, it can be argued that a total absence of arrears indicates serious overmanning and a normally low level of utilization of both personnel and processes.

Having established that arrears *will* occur, procedures have to be established to deal with them and to minimize customer irritation. In Section 11.1.6 on the subject of computer-aided scheduling and loading, two status categories were identified which could result in arrears:

1. Not conforming to schedule but, with the adjustment of "spare" time and rescheduling of activities, able to meet the delivery date.
2. Unable to meet the delivery date using the resources allocated in the schedule.

Although these categories are intended primarily as aids to work-station loading, they serve equally well as predictors of arrears. If any jobs move into them, as a result of delays in production, they become potential arrears. It may be possible to remove jobs from the potential arrears listing by allocating more resources to them or by reorganization of the work-station loading plan. However, in some cases it will have to be accepted that the resources needed are not available or that adjusting the loading plan would disrupt other jobs and cause them to enter the arrears category. Two important points emerge here:

1. It is often possible to predict that a job will be in arrears well in advance of the quoted delivery date.

2. In cases where the cause of arrears is insufficient manufacturing capacity (as opposed to shortage of supplies), it is often possible to select the jobs which will fall into arrears.

Forecasts of arrears enable customers to be informed of the problem in advance of the delivery date, possibly allowing time for implementation of contingency plans to minimize the inconvenience. This ensures that the customer is treated as well as is possible in the circumstances, and minimizes the damage to the company's reputation. A new delivery date can then be renegotiated, if the products are still needed.

When sufficient planning flexibility exists for selection of the jobs to go into arrears, a number of courses of action are possible. The primary guide is the priority allocated to jobs by the sales office. It is usually preferable to let low-priority jobs go into arrears instead of ones with high priority. However, this does depend on the effectiveness of customer liaison and the judgment of those concerned with making the decisions. It may be possible to identify and select customers whose immediate needs can be satisfied by delivery of a part order, to bring the demand for finished products within the manufacturing capability of the company.

When a job, or part of a job, cannot be completed by the due date it will be entered onto the arrears listing, for the attention of the production controller. Those jobs not in current production will require rescheduling, while the disruption caused by those in current production running late will determine if they are allowed to continue to completion. For the former group the priority for rescheduling is usually determined by the default time on the delivery date, hence the term "aging of arrears," although the commencement of jobs waiting on supplies will obviously depend on them becoming available.

The jobs on the arrears listing will be scheduled into production in preference to new orders, and the delivery dates of new orders must take into account the influence of the manufacturing time for the arrears on the forward loading of work stations, to enable the deficit to be worked off. Once arrears have been dealt with, the production management task, which extends from the acceptance of an order to the delivery of the products, is complete.

REFERENCES

1. Starr, M. L., *Production Management: Systems and Synthesis*, 2nd ed., Prentice-Hall, Englewood Cliffs, New Jersey (1972).
2. Moore, F. G., and R. Jablonski, *Production Control*, 3rd ed., McGraw-Hill, New York (1969).

3. Laios, L., and R. Gibson, "A Planning Computer Display for Job-Shop Scheduling," Husat Memo No. 120, Human Sciences Department, Loughborough University, U.K. (1976).
4. Tersine, R. J., and J. H. Campbell, *Modern Materials Management*, North-Holland, New York (1977).
5. Starr, M. K., and D. W. Miller, *Inventory Control: Theory and Practice*, Prentice-Hall, Englewood Cliffs, New Jersey (1962).
6. Coyle, R. G., *Mathematics for Business Decisions*, Nelson, London (1971).
7. Gedye, G. R., *Works Management and Productivity*, Heinemann, London (1979).

Index

ABC analysis 426
Acceptable quality level (AQL) 258, 262
Acceptance criteria 259, 261
Adaptive control systems 246–250
Added value 426
Additives 100, 277, 283
Administration 295
Administrative structure 330, 331, 334
Air traps 172
Analog controller 244
Analog-to-digital converter (ADC) 233
Analog transducers 218, 219
Analysis of variance (ANOVA) 204
Angle of fall 278
Angle of repose 278
Angle of shear 278
Angular displacement measurement 228
Angular slip velocity 24
Angular velocity measurement 229
Angular velocity transducers 229
Appropriation budgets 392
Arrears 440–442
Attributes sampling plan 259
Autoclaves 173
Automatic monitoring 436, 437
Automatic systems 434
Automation 270
Average revenue curve 360
Average total cost curve 361
Average variable cost curve 360

Backrinding 157, 171
Backscatter sensors 230
Baker Perkins MPC/V 105
Balanced manufacture 288
Band formation
  back-roll 115
  front-roll 115, 116

Band formation index 115
Barrier screw design 104
Batch manufacture 270
Big bags 279
Blisters 172
Brainstorming 326
Breakdown maintenance 388
Breaker plate 79
Break-even chart 362
Budget forecasts 377–382
Budget variance 394, 395
Budgetary control 391–395
Buffer stocks 415, 416
Buffer stores 275
Bulk density 277
Bulk liquids 279, 425
Bulk powders 425
Bulk supplies 424
Business center concept 331, 334, 355, 379, 381
Business plans 377–382

Calender configurations and operations 119–121
Calender control 127–129
Calender feeding methods 125–126
Calender nip 123, 124, 126, 127
Calender operation characteristics 126–129
Calender performance factors 126
Calender roll deflection and methods 121–124
Calender roll temperatures 126
Calendering 111, 118, 284
  material utilization 366
  short-run 118
Calenders
  bull-gear 124
  four-roll 119

Calenders (*Cont.*)
  temperature-control systems   118
  three-roll   119
  Z configuration   121
Capillary flow equation   79
Capillary rheometers   25–29, 161
Capital expenditure   384
Car-parking space   296
Cash budget   382–384
Cash-flow management   348, 353–355
Cash-flow working sheet   384
Cavity transfer mixer (CTM)   76, 78
Chronos Richardson loss-in-weight unit   283
Clamping force   157
Closed-loop control systems   241, 246, 250, 251
Color graphics   406, 407, 411
Compactors   104
Company beneficiaries   316–318
Company hierarchy   330
Company organization   316–336
Company performance   356, 367–369
Company philosophy   315–316
Company policy   417
Company structure   329–335
Compounding ingredients   431
Compressed-air services   233
Compression molding   146, 148, 156
Compression molds   155, 165
Compression presses   150–152
Computer-aided standard costing   372
Computer analysis   3
Computer calculations   206–207
Computer control systems   128, 250, 253–254
  interfacing and actuation devices   252–253
Computer files   401
Computer graphics   405, 407
Computer hierarchy   236, 254
Computer information gathering   11
Computer methods   2, 3, 170, 321, 365, 372–373, 380
Computer monitoring systems   436–437
Computer planning aids   440
Computer process control   286
Computer programs   138, 170, 196, 215, 409, 411–412, 420, 427
Computer systems   251, 369, 373, 379, 400, 427–431, 435–437
Computer technology   254
Conduction   132–137
Consignment stocking   417, 418

Constant-speed ac motors   70
Constant strain-amplitude waveform   36
Constant-strain cycle principle   36
Consumable process supplies   273, 274
Containers   279
Continuous mixing   100–109
  in batch systems   107–108
  blended feedstock system   103
  cost analysis   107
  determining operating conditions   105
  feeding from batch-blending
    machines   102–103
  feedstock form and preparation   100–104
  single-screw mixers   104–105
  technical and economic advantages of   107
Contour correction curves   123
Contour plot   210
Control charts   238, 240
Control system layout   193
Convection boundary conditions   137–140
Convective heat transfer   137–140
Convective heat-transfer coefficient   31
Correlation coefficient   420
Cost analysis   356–377
  continuous mixing   107
Cost-benefit analysis   307–309, 381
Cost centers   363–365, 367, 373
Cost changes   376–377
Cost classification   356–359
Cost control   365, 391
Cost identification   356–377
Cost recording   356–359
Costs
  avoidable   357–358
  controllable   358
  differential   359
  direct   356, 357
  and expenses   359
  fixed   308, 361
  incremental   359–362
  indirect   356, 357
  inventory   416
  irrelevant   358
  maintenance   388
  marginal   359–362
  opportunity   358, 416
  optimization   388
  ordering, transport, and handling   413
  out-of-pocket   357
  overhead   364
  per standard rework addition to mixer   366

Costs (*Cont.*)
  quality 389–391
  relevant 358
  scrap and rework 365–367
  standard, *See* Standard costs
  sunk 357–358
  total 416, 438
  unavoidable 357–358
  variable 309, 362
Couette flow 22, 23
Critical-path methods 310
Cross-link insertion rate 39, 40
Cross linking 32–35, 143, 146, 170, 172
Cryogenic trimming 167, 168
Cure curve 143–145
Cure levels 144–146
Cure modeling 143–146
Cure rate 145, 146
Cure simulation 35, 140
Cure state prediction 140–146
Cure time 38–40, 58, 147, 150
Cure trace 36–39, 143, 144
Curemeters 36–40
Curing operation 169
Customer liaison 441–442
Customer request 347
CV process 177

Dancer arm 99
Daniels Hydramould 152
Data acquisition systems 233–237
Data analysis and presentation 237–240
Data bases 401, 411
Data collection 436
Data entry 436
Decision frameworks 4–5
Decision levels 9
Decision making 5, 40, 206, 356, 357
  management 323
  resources used in 323
Defect identification 191–192
Defect levels 264
Defective work 365
Deflashing 167
Delamination 171
Delivery checking 424
Delivery records 425
Delivery validation 424
Delta-Mooney measurement 22
Development contracts 348
Development costs 348

Diagnostic facilities 250
Diagnostic routines 251
Differential scanning calorimetry (DSC) 30, 188
Differential thermal analysis (DTA) 30, 188
Digital controllers 244, 245
Digital transducers 218, 219, 233
Dimensional stability 177
Dimensions specification 191
Dimensions measurement 229–232
Direct current motors 71
Direct digital control (DDC) 246, 247
Direct labor 356
Direct labor allowances 370
Direct material 356
Discounted cash flow 384
Disk encoder 228
Dispersive mixing 48, 49, 59, 64, 100
Displacement measurement 226
Displacement transducer 228–230
Distortion 171
Distributive mixing 48, 49, 100
Downstroking presses 150
Draw ratio 94
Drum vulcanizers 174
Dust-control methods 281

Economic order quantity 412, 414, 415, 418
Economics of manufacturing operations 353–395
Efficiency variance 395
80–20 rule 438
Ejector pins 165
Elastic effects 29
Elastic memory 89
Elastic recovery 89
Electrical resistance heating 152
Emissivity 224
Empirical modeling 203–209
Energy economy 387–389
Energy measurement 232–233
Entrance effects 159–160, 162
Equilibrium of forces for flow through capillary 26
Equipment purchase 384–387
Equipment replacement 269, 289, 349–351, 384–387
Evolutionary operation methods 195
Expenditure 379
Expenses 359
Experiment design 213–215

Exponential mixing   46, 47
Exponential smoothing constant   423
Exponentially weighted moving average   423
Extensional flow   19, 21, 82, 83
Extensional power-law constant   21
Extensional tests   19
Extensional viscosity   81
Extrudate cross-section dimensions   94, 100
Extrudate temperature   99
Extruder barrel   72
Extruder die design characteristics   97
Extruder die land length   89–90
Extruder die preform and land function   90
Extruder die specification   91
Extruder die swell   89
Extruder dies for composite extrudate
        shapes   94
Extruder drives   70–71
Extruder heads   72–73, 79
Extruder line   3
Extruder operation characteristics   94–100
Extruder operation curves   96–97
Extruder speed control   71
Extruders   69–100
    bearings   71
    cold-feed   69, 70, 75–79, 98, 124, 125
        comparison with hot-feed   74
        vented   77–79
    conical   104
    elements of construction   70–73
    feed rate   98
    feed zone   72
    gearboxes   71
    hot-feed   69, 70, 74–79
        comparison with cold-feed   74
        dump or batch   75
        general-purpose   74–75
    modular construction   70
    pin   76
    screw speed   99
    single-screw   95
    temperature control   73
    transmissions   71
Extrusion die design for complex
        extrudates   88–93
Extrusion die swell   87
Extrusion dies   83
    design elements   80–81
    land region flow   85
    lead-in sections   81–84
    one-dimensional flow   81

Extrusion dies (*Cont.*)
    parallel sections   84–86
    slit   85, 86
    tapered   81–84
Extrusion process   284
    material utilization   366
Extrusion trials   196

F distribution   205–206
F ratio   205–207
Factorial experiment design (FED)   169, 195–
        197, 201
Failure behavior   19
Failure warnings   251
Farrel Bridge Banbury machines—F
        Series   45
Farrel Bridge Banbury rotors   52
Farrel Bridge MVX   105, 108
Farrel continuous mixer   107–108
Fault diagnosis   251
Fiber optics   224, 225
FIFO (first in, first out)   427
Fill factor   60–61, 66, 101
Fillers   277, 279
Financial requirements   379
Financial resources   353
Finishing processes   168
Finite difference method   135–138
Finite element method   135
Fixed budgets   393
Flash removal   167
Flashless molding   158
Flexible budgets   393
Flexible manufacturing cell (FMC)   286–288
Flexitime   336
Flow behavior   17, 18
Flow control   124
Flow instabilities in mixing   64–65
Flow monitoring   253
Flow path   161, 162, 164
Flow process charts   288–290, 293, 296
Flow properties   15
    measurement of   21–29
    of raw elastomers and rubber mixes   16–21
Flow rate   17, 28, 83, 84, 161
Flow through capillary, equilibrium of forces
        for   26
Flow velocity profiles   26
Fluidization principles   175
Fluidized beds   174–176
Folding flows   46

Forecasting
  for inventory planning   419–420
  for manpower requirements planning   434
  for production demand   419
Forecasting models   325
Fractional conversion   140
Francis Shaw Intermix Internal Mixers   44
Francis Shaw Intermix rotors   52
Friction ratio   112, 116
Fume extraction   177
Fume venting   176
Functional analysis   439–441
Functional tests   192

Gantt chart   402, 405–407, 440
Geometric-shape factor   91, 93
Göttfert elastograph   36, 37
Granulation   100
Green strength   19, 20, 65
Group technology   270

Handling systems   281
Heat flow   29
Heat-history profile   34
Heat transfer
  conductive   132–137, 152
  convective   137–140
  modes of   131–132
  in vulcanization   134
Heat-transfer coefficient   142, 173
Heat-transfer properties   15, 29–31
Heating curves   142
Hookian spring   20
Hot air ovens   171–172
Hydraulic circuit   154
Hydraulic drive units   150
Hydraulic power unit   152
Hydraulic ram   150, 153
Hydraulic system   155
Hydrodynamic stability   114

Iddon high-intensity mixing scroll   105
Identification codes   401, 431
Idle capacity variance   394
Index of aisle space   306
Index of automatic machine loading   305
Index of direct materials handling   304
Index of distributive mixing   47
Index of floor-area loading density   306
Index of gravity utilization   304
Index of indirect materials handling   304

Index of production line flexibility   305
Index of storage space   306
Index of storage volume utilization   307
Index of work-station flexibility   306
Induction heating   152
Inductive displacement transducer   227
Inflation effects   369, 386, 387
Information feedback   11, 13, 256, 423–424,
  429, 434, 437
Information flows   13
Information requirements   11
Information services   369, 437
Information systems   11
Infrared sensors   223
Injection molding   146, 168
  stages of   169
Inserts   272
Inspection   191
  goods inwards   255, 256, 273
  off-line   254, 255
  by variables   259
Inspection specification   191
Instruction elements   327
Intensive impeller blenders   101
Interactive optimal experiment design   215
Internal mixers
  design elements   48–56
  drive systems   55–56
  gearing arrangements   56
  outline specifications   44
  ram and door configurations   49–52
  ram pressure   61
  rotor and chamber designs   48–49
  rotor shaft seals   53
  steam heating   54
  temperature control   53–54
  variable-speed drive   62
  water-tempering systems   54
Inventory classes   426
Inventory holdings   426, 427
Inventory management   424–427
  computer methods   427–431
Inventory planning   419–424
  forecasting for   419–420
Inventory transaction register   430
Isometric plot   210
Isothermal flow   161
Isovels (contours of constant flow
  velocity)   88

Jobbing companies   337, 398, 402

Labor-in-progress account 374
Labor variance 374, 375
Lack-of-fit F ratio 207
Lack-of-fit mean square (LFMS) 206
Laminar flow 16, 25
Laminar mixing 46, 75
Last-period-demand method 422, 423
Lead times 415
Leakage flow 95
Legislation requirements 270, 272, 295
LIFO (last in, first out) 427
Limit switch 226
Limit values 185
Linear potentiometer 227
Linear variable differential transducer
    (LVDT) 227, 228
Liquid assets 355
Loading 400, 404–412
Loading chart 405, 408–411, 440
Logical methods 324
Logical tree 325
Lot tolerance fraction defective (LTFD) 257
Lumped parameter method 161, 162

Machine-setting specifications 189–190
Management-by-exception 11, 438
Management structure 330
Management system 318–321
    four stages of 320–329
    functional structure of 331
    for one manager 320
    for two levels of management 322
Manpower reuqirements planning 432–434
    forecasting for 434
Manual control 241
Manual operation 240, 284–285, 400
Manual operations specification 190–191
Manufacturing capacity contraction 351
Manufacturing capacity development 349–
    351
Manufacturing capacity expansion 350–351
Manufacturing organization 347
Manufacturing overhead 356, 362–365, 375
Manufacturing performance index 365
Manufacturing specification 185–187
Manufacturing systems 1–13
    basic levels of 1
    fundamental stages 1
Manufacturing versatility 339
Market identification 336–338
Market research 336–339

Market selection 9
Master specification 192, 370, 401, 402, 411
Material allowances 370
Material angles 278
Material behavior and testing 15–41
Material characterization 187–188
Material handling 164–165
Material price variance 373
Material specifications 187–188
Material utilization 366–367
Material utilization variance 374
Maximum standard deviation (MSD) 261
Maxwell model 20, 21, 25, 114
Measures of activity 394
Mechanical handling 277
Medical services 296
Melt fracture 21, 64, 80, 87
'Menu' approach 438
Metering systems 283
Microcomputer application 431
Microprocessor systems 106, 153, 184, 253,
    386
Microwave power absorption ranges 178
Microwave vulcanization 176–177
Mill operational index 115
Milling 111 (see also Two-roll milling)
    continuous 116
    discontinuous 116
    nonsymmetrical 112, 113
    symmetrical 113
Mixed-compound store 275
Mixing 43–68, 280–284 (see also Continuous
    mixing)
    automatic operation 283
    batch discharge criteria 56–57
    circulating water and batch temperature
        effects 61–63
    constant energy 62
    constant-speed 59
    cooling efficiency 67
    fill factor 60–61, 66
    first-batch effect 61
    flow instabilities in 64–65
    in-batch uniformity 58
    integrated systems 105–107
    laboratory simulation 65–67
    material addition criteria 56–57
    material input sequence 57–60
    mechanisms of 44–48
    multistage 64
    operations associated with 280

Mixing (*Cont.*)
  practical variables   56–64
  rotor speed   60
  scaling rule for rotor speed   67
  short-run   62
  tire and conveyor-belt-type compounds   59
  upside-down   59
  wastage   365–366
  water-tempering systems   62
Mixing sequences   63–64
Model fitting   213
Model reference adaptive control   248
Mold cavity   156, 157, 160
Mold cavity plate   157
Mold changing   286
Mold cleaning   167
Mold construction   163
Mold design   160, 166, 170
Mold dimensions   156
Mold filling temperature   170
Mold fouling   167
Mold lubricants   166
Mold opening jigs   165
Mold removal   165
Mold split line position   156
Mold stripping   164–165
Mold surface treatments   166
Mold temperature   159
Molding faults   171–172
Molding process   131, 146–172, 284–288
  operating conditions   168–170
Molding variables   168–169
Molds
  cold-runner   163
  compression   155, 165
  fixed tool   147, 156
  injection   156–164
  loose tool   147, 156
  multicavity   156
  opening   165
  positive- or plunger-type   147
  semipositive   147
  single cavity   157
Monitoring
  automatic   436, 437
  of company performance   367–369
  current production   184
  manufacturing operations   186
  output quality   8, 183
  process   217–240
  process integrity   251

Monitoring (*Cont.*)
  production performance   434
Monsanto oscillating disk curemeter   32, 36
Mooney *ML* (1 + 4) test   22
Mooney viscometer   5, 6, 22, 24, 32
Mooney viscosity   65, 209–211
Moving average   422–423
Multivariable regression analysis
    (MVRA)   170, 196, 204

Network analysis methods   310
New products   339–351, 398
Newtonian dashpot   20
Newtonian fluid   27
Newtonian viscosity   17
Nip-force interaction   121
Non-Newtonian flow   91
Non-Newtonian flow profile   85
Nonstreamline flow   82
Nonsystematic methods   324, 326
Normal distribution   216
NRM Plastiscrew   105

Office allocation   295
Onion skin   171
Open flash molding   146
Open-loop control system   241
Operating characteristic (OC) curve   257
Operating costs   373
Operating point   95–96, 99
Operations research methods   325
Optimal control methods   248
Optimizing models   325
Optimum conditions   194
Optimum productivity   37, 194
Orange peeling   171
Order quantities   418
Ordering requirements   429
Ordering strategy   417
Out-of-stock situation   415
Overhead costs   375, 394
Overhead variance   375
Overmanning   433
Overtime   438
Oxidation   177

Patrol inspection   262–263
Performance fluctuations   13
Perpetual inventory system   426, 427, 431
Personnel facilities   294–296
Personnel requirements   432

Personnel turnover   433
PID controller   242, 244
Piezoelectric effect   225
Planning aids   405
Planning boards   405
Planning group   381
Planning, programming, and budgeting
      systems (PPBS)   378, 379
Plant commissioning and startup   312–313
Plant installation
   implemenation   311
   planning   309–311
Plant layout
   functions of   267–270
   guiding principles for   270–271
   quantitative checklists for   299–304
   scale models   298
   unsatisfactory   269
Plant layout efficiency indices   304–307
Plant layout evaluation   300–303
Plant layout investigations   271
Plant layout planning   288–296
Plant layout synthesis and evaluation   296–
      309
Plant maintenance   387–389
Plant purchase and replacement   384–387
Plasticization time   170
Plasticizing unit   153
Platen heating   152
Plug-like velocity profile   86
Poisseuille equation   27
Polynomial equations   196, 202, 204, 207
Porosity   172
Position monitoring   226
Powder blenders   101
Powder characteristics   278
Powder flowability   278
Powder hopper discharge   278
Powder feed devices   99
Power-law flow characteristic   21
Power-law index   17
Power-law relationship   17, 18
Power measurement   232–233
Precompaction operation   104
Predictive models   325
Preform shape effect   90
Preplasticization machinery   124, 125
Present-value method   384, 386
Pressure drop   161, 163
   in channel of arbitrary cross section   91
   due to extensional flow   82, 83

Pressure drop (Cont.)
   over extruder die   91
   due to screen packs   79–80
   due to shear flow   84
   and volumetric output   96
Pressure measurement   225–226
Pressure vessels   173
Preventive maintenance   388–390
Price analysis   361
Price discounts   415
Price negotiations   347
Pricing policy   419
Probability analysis   423
Probability distribution   423
Process-capability studies   192–217
Process consumables   276
Process control   183–265, 365, 367
   general considerations   241–242
   interaction with quality control   183–184
Process development tests   6
Process integrity monitoring control
      systems   251
Process monitoring   217–240
   elements of   218
   general considerations   217–220
Process performance
   empirical modeling and statistical
      analysis   203–209
   prediction, monitoring, and control   7–8
   representation and optimization   209–213
Process precision   215–217
Process supplies   273, 276
Process variability   8, 215
Processability factors   15
Processes and operations planning   290
Processing trials   194–203
Product design   340–341, 347, 348
Product mix optimization   399
Product performance   341
Product performance specification   184, 192,
      340
Product quality audit specifications   191
Product range maintenance   338–339
Product specification   185, 191, 341
Product viability assessment   340
Production analysis   438
Production control   397–398
   computer files for   401
   principles and objectives of   399–401
Production demand, forecasting for   419
Production groups   331

Production information   436, 437
Production management   397–443
Production management control diagram   10
Production organization and management   8–13
Production performance improvements   436
Production performance monitoring   434
Production plan implementation   431–442
Production planning   397–412
Productionizing process   347
Productivity   8, 349–351, 433
Profit margin   338, 367–369
Profitability   384
Profitability criteria   340
Profitability index   386
Programmable controllers   245
Proportional band   242
Prototype production   341
Proximity sensor   226
Pseudoplastic material   27
Pulse-generating transducers   233
Purchasing information and analysis   429
Purchasing strategy   412–419
Pure-error mean square (PEMS)   206

Quality audit specification   186
Quality-control   6, 8, 183–265, 331, 333, 365, 391, 425, 428–430
  general considerations   254–256
  interaction with process control   183–184
  strategy development   254
  systems approach   255
Quality costs   389–391
Quality factors   7
Quality improvement benefits   390
Quality levels   390
Quality parameter   259
Quantitative checklists for plant layout   299–304
Quantity discounts   414
Quantity factors   7
Quarantine store   425
Quarantine system   274, 275, 425
Quartz crystal   226

Radiation thickness sensors   230
Ram injection molding machines   153
Random inspection   264
Random numbers   263
Rate of return on capital expenditure   385
Recording work   434–437

Redundancy factor   200, 201
Reference viscosity   18, 21
Regression analysis   420, 422
Regression equations   208
Regression F ratio   207
Regression mean square (RGMS)   204
Relaxation time   21
Relay switches   153
Repairs   365
Residual mean square (RSMS)   205
Response equation   202, 205, 208
Response time   435
Return on direct capital employed   367
Return on investment   386
Return on total capital employed   368
Reverse crown roll contour   121–122
Rework cost   366
Rework material   365
Ribbon blenders   102
Risk burden   348
Robots   166, 286
Roll bending   122, 123, 128
Roll crossing   123, 124, 128
Roll float   121
Roll neck bending stresses   124
Roll-separating force   124
Rolling nips, streamlines, and velocity profiles   112–113
Rotary displacement transducers   228
Rotary potentiometer   228
Rotatable experiment designs   198, 204, 208
Rotational rheometers   21–25
Routing identification   432

Safety measures   250–251, 284, 294
Safety standards   270
Salt baths   176
Sampling inspection   256–262
Sampling plan   256–262
Scale models   298
Sched U Graph system   405
Scheduling 400,   402–404, 411–412
Scorch criterion   35
Scorch testing   32
Scorch time   39
Sequence chart   402–404, 407, 409
Sequence control   249–250
Servohydraulic actuators   252
Shading diagrams   210–212, 215
Shear flow   16, 18, 19, 23, 84
Shear head   180–181

Shear modulus   21, 25, 36, 38, 39
Shear rate   17, 18, 22, 24, 25, 27, 85
Shear strain   17
Shear stress   16, 17, 25, 27–29
Sheet cooling and batch-off equipment   125–
    126
Sheet-thickness monitoring methods   229–232
Shift systems   335
Simulation   380
Slip velocity   25, 28, 29
Specific heat   30
Specification limits   259, 261
Specifications   184–192, 367
    writing of   184–187
Staff training   351
Stagnation point   112, 113
Standard costs   369–377, 393, 401, 402, 411
    achievement against   373–376
    establishing   370–373
    estimated   370, 371
    expected   369
    first production   371
    formulation of   372
    general considerations   369–370
    initial   371
    production   371
Standard deviation   186, 215–217, 259, 261,
    422
Statistical analysis   198, 203–209, 420
Statistical methods   422, 423
Steam heating   152
Steam pans   173
Steam raising   233
Steam-tube vulcanization   177–180
Stepper motors   252
Stockblender   117
Stocktaking   427, 430
Stepped cure   173
Storage
    areas   291–292
    conditions   276
    facilities   274
    life limit   430
    locations   428
    methods   271–280
    organization   275
    points   277
    requirements   271
    space   274, 416, 428
    time   275
    unit items   272

Stores organization   425
Strain cycle   38
Strain gauges   225
Strain rate   20
Strategic decision making   11
Stress relaxation   25, 114
Subcomponents   273, 274
    specifications   188–189
    tests   189
Subjective inspection   263–264
Subsidiary companies   334
Surface-active materials   167
Surface finishes   168
Surveys   325, 326
Swelling ratio   87
Switching operations   252
Synergism   326
Synthesis process   4
Systems concept   3–5

Tachogenerators   229
Tactical decision making   9
Temperature
    control systems   118–119, 242, 243
    dependence of apparent viscosity   17–18
    distribution   136, 138
    gradients   29, 131, 222
    measurement   220–225
Temperature-time profile   33, 141
Tensile stress   85, 87
Tensile tests   18–19
Testing methods and results   5–6
Thermal
    conductivity   29–30
    diffusivity   30–31
    expansion   157
    expansion coefficient   156
    properties   15, 29–31
Thermistors   223
Thermocouples   223
Thermoelectric sensors   220–222
Thixotropic effects   22
Time-series analysis   420, 421, 444
Time-temperature-fractional conversion (TTC)
    charts   140–142
Time-temperature-percent scorch (TTS)
    chart   33–35
Tire industry   285
TMS Rheometer   23, 39, 127, 188
Tolerance bands   215–217
Tolerances   185, 191

Torque and angular velocity   24
Total average monetary inventory
         (TAMI)   417, 418
Total variance   376
Tote bins   279
Toxic fumes   173
Training requirements   351
Training-within industry (TWI) system   191
Transducers   218, 221, 225, 232, 233
Transfer
   cylinder   165
   molding   147, 148
   molds   157–159
   ports   157
Transfermix machine   107–108
Transferred-heat-history profile   35
Transit routes
   materials products and equipment   292–294
   personnel   294
Transport methods   271–280
Transport requirements   271
Transport unit items   272
Troutonian behavior   19
Two-roll milling   111
   applications and operations   116–118
   banding and bagging   113–115
   factors influencing mill capacities   111–112
   front-back roll transitions in band
         formation   115–116
   material behavior prediction   113
   operating characteristics of   111–118
   semiautomatic operation   118
   as shaping operation   118

Unvulcanized rubber tests   5–6
Update time bar   410
Upstroking presses   150

Vacuum chambers   152
Validation   262–263
Value engineering   372
Vanzetti system   225
Vaqua blast system   167
Variable speed ac motors   71

Variables sampling plan   259
Variance analysis   373
Velocity measurement   228–229
Velocity profiles   112
Velocity transducers   229
Viscoelastic behavior   19–21, 113–114
Viscoelastic recovery   25
Viscometers   21–25
Viscosity effects   15–16
Viscosity–temperature curve   162
Viscosity–temperature relation   17–18
Viscous dissipation   28
Viscous dissipation temperature rise   24
Viscous flow   16–20, 169
   measurements   36
Vulcanization   131
   batch   172–174
   continuous   174–181
   heat transfer in   134
   material utilization   367
   microwave   176–177
   steam tube   177–180
Vulcanization characteristics   15, 32–40

Wall slip   86–87
   measurements   24
Wallace isothermal curemeter   36
Wallace-Shawbury Curometer   37
Waste material   365
Weigh-hopper filling   282
Weighing accuracy   281
Weighing systems   282
Weight measurement   229–232
Wire-wound resistance thermometers   222,
   223
Work-in-progress (WIP)   274–276, 410, 439
Work-in-progress information   435, 437–438
Work-in-progress inventory   426
Working hours   335–336, 374, 433
Working patterns   335–336
Works order   398
Works order file   401

Young's modulus   21